Plant Resistance to Insects in Major Field Crops

Sarwan Kumar • Michael Furlong
Editors

Plant Resistance to Insects in Major Field Crops

Springer

Editors
Sarwan Kumar
Department of Plant Breeding and Genetics
Punjab Agricultural University
Ludhiana, Punjab, India

Michael Furlong
School of Biological Sciences
The University of Queensland
Brisbane, QLD, Australia

ISBN 978-981-99-7519-8 ISBN 978-981-99-7520-4 (eBook)
https://doi.org/10.1007/978-981-99-7520-4

This Springer imprint is published by the registered company Springer Nature Singapore Pte Ltd.
The registered company address is: 152 Beach Road, #21-01/04 Gateway East, Singapore 189721, Singapore

Paper in this product is recyclable.

Preface

Insects have long been a significant challenge for farmers and agricultural systems worldwide. As the global population continues to expand, the demand for food production has risen dramatically. However, the success of major field crops is threatened by the ever-present and evolving threat of insect pests. This book, '*Plant Resistance to Insects in Major Field Crops*', delves into the crucial topic of plant-insect interactions and the development of resistance mechanisms in crops.

The concept of plant resistance to insects is not new; farmers have been observing and selecting pest-resistant plants for centuries. However, with advances in scientific research and technology, our understanding of these complex interactions has deepened. The integration of classical breeding techniques with molecular and genetic tools has enabled us to unravel the intricacies of plant defense mechanisms, laying the foundation for the development of sustainable pest management strategies.

This book aims to provide a comprehensive overview of the current knowledge and advancements in the field of plant resistance to insects in major field crops. It brings together a diverse group of researchers, scientists, and experts who have contributed their expertise and insights to the different chapters of this volume. Their collective efforts have resulted in a rich compilation of information, covering various aspects of plant resistance, including plant-insect interactions, resistance mechanisms, host-plant resistance breeding, and practical implementation of insect-resistant crops.

The chapters in this book explore a wide range of major field crops, such as maize, rice, wheat, oilseeds and pulses, cotton, and more, highlighting the specific challenges posed by insect pests in each crop and the strategies employed to develop resistant varieties. Additionally, the book explores the roles of molecular biology, genomics, biotechnology, and high-throughput phenotyping in enhancing our understanding of plant resistance mechanisms, offering exciting prospects for future research and development. High-throughput phenotyping has revolutionized the field of plant research by enabling scientists to rapidly and accurately assess the resistance traits of a vast array of crops. By leveraging cutting-edge technologies and advanced imaging systems, this method offers unprecedented speed, precision, and scalability in evaluating the intricate interactions between plants and insect pests.

We recognize that this book is just a snapshot of the vast and ever-evolving field of plant resistance to insects. It is our hope that the knowledge and insights shared within these pages will inspire further research, facilitate collaborations, and aid in the development of sustainable and effective pest management strategies. We also aim to provide valuable information to students, researchers, entomologists, plant breeders, agronomists, and policymakers who are invested in addressing the challenges posed by insect pests in major field crops.

We extend our sincere gratitude to all the authors who have contributed their expertise and knowledge to this book. Their dedication and commitment to advancing the field of plant resistance to insects have made this volume possible.

Finally, we hope that '*Plant Resistance to Insects in Major Field Crops*' serves as a valuable resource for anyone interested in understanding and addressing the complex interactions between plants and insects. Together, let us strive to develop resilient and sustainable agricultural systems that can withstand the challenges of insect pests while ensuring global food security.

Ludhiana, Punjab, India — Sarwan Kumar
Brisbane, QLD, Australia — Michael Furlong

Contents

Editors and Contributors

About the Editors

Sarwan Kumar is Senior Entomologist at Punjab Agricultural University (PAU), Ludhiana, India, and holds a PhD in entomology (2005) from CCS Haryana Agricultural University, Hisar, India. He has more than 17 years of experience as an entomologist at PAU, Ludhiana. His research interests include host plant resistance, insect-plant interactions, plant defense, and pest management in oilseed crops with special reference to oilseed *Brassica*. Before being appointed as Assistant Professor of Entomology at PAU in 2006, he worked as Senior Research Fellow (2005–2006) in the Department of Entomology, PAU, and worked on dissemination of pest management strategy in cotton. Dr. Sarwan Kumar has nearly 84 publications, including 45 research articles in journals of national and international repute and 26 conference/symposia presentations. He has authored eight book chapters in edited books published by leading publishers and three scientific reviews. He has guided the research work of more than 15 PG students in entomology and plant pathology. He is an active member of 'All India Coordinated Research Project on Rapeseed-Mustard' and is coordinating the entomology program of the project. He is the recipient of Australian Government's Prestigious Endeavour Research Fellowship (Australia Awards) and Fellow of Indian Society of Oilseeds Research. Besides this, he is editor of Journal of Insect Science (India) and Journal of Oilseed Brassica and member of the Society of Chemical Industry, UK, and Entomological Society of Iran.

Michael Furlong is Professor of Insect Ecology and Biological Control at the University of Queensland, Australia. He completed his PhD at Imperial College, London, in 1993 and took up Postdoctoral Research Fellow positions at Rothamsted Research, UK (1993–1997); the University of Maine, ME, USA (1997–1999); and the University of Queensland (1999–2005) where he was appointed to a teaching and research lectureship in 2005. Professor Furlong leads large teams of researchers at the UQ and in international collaborations, and his work in the South Pacific has led directly to improved pest control and documented reductions in hazardous pesticide use, providing clear benefits for the environment and society. He studies how pest species interact with crop plants and how ecosystem services can be sustained and enhanced when they are stressed by agricultural intensification and climate change. His research and outreach in the Pacific engage national governments and local communities to develop novel extension strategies to provide vocational training that directly impacts resource-poor rural communities.

Contributors

Liyaqat Ayoub Division of Entomology, Sher-e-Kashmir University of Agricultural Sciences and Technology of Kashmir, Kashmir, India

Lukesh Bansal Department of Entomology, Punjab Agricultural University, Ludhiana, India

Ousmane Boukar International Institute of Tropical Agriculture, Kano, Nigeria

Alamuru Krishna Chaitanya Centre for Crop Health, University of Southern Queensland, Toowoomba, QLD, Australia

Rajesh Chowdary Acharya N.G. Ranga Agricultural University, Guntur, Andhra Pradesh, India

Lucy Egan CSIRO Agriculture and Food, Narrabri, NSW, Australia

Christian Fatokun International Institute of Tropical Agriculture, Ibadan, Nigeria

Michael Furlong School of Biological Sciences, University of Queensland, St. Lucia, Brisbane, QLD, Australia

Sajjan Grover Department of Entomology, University of Nebraska-Lincoln, Lincoln, NE, USA

Audil Gull Division of Genetics and Plant Breeding, Sher-e-Kashmir University of Agricultural Sciences and Technology of Kashmir, Kashmir, India

Sheikh Salma Irshad Department of Entomology, Punjab Agricultural University, Ludhiana, India

Ashok Karauppannasamy ICAR-Indian Institute of Horticultural Research, Bengaluru, India

Chikkappa G. Karjagi Unit Office, ICAR-Indian Institute of Maize Research, Pusa Campus, New Delhi, India

Afsaneh Delpasand Khabbazi Department of Biology, Faculty of Science, Karadeniz Technical University, Trabzon, Turkey

Saber Delpasand Khabbazi Department of Agriculture and Food, Institute of Hemp Research, Yozgat Bozok University, Yozgat, Turkey

Suneetha Kota ICAR-Indian Institute of Rice Research, Hyderabad, India

Sunil Kumaraswamy ICAR-Indian Institute of Horticultural Research, Bengaluru, India

Sarwan Kumar Department of Plant Breeding and Genetics, Punjab Agricultural University, Ludhiana, India

Pritha Kundu Department of Entomology, University of Nebraska-Lincoln, Lincoln, NE, USA

Vartika Mathur Animal Plant Interactions Lab, Department of Zoology, Sri Venkateswara College, New Delhi, India
Fellow of Institute of Eminence, School of Climate Change and Sustainability, University of Delhi, New Delhi, India

Venugopal Mendu Department of Plant Sciences and Plant Pathology, Montana State University, Bozeman, MT, USA

Saba Baba Mohammed International Institute of Tropical Agriculture, Kano, Nigeria

Muthugounder Mohan ICAR-National Bureau of Agricultural Insect Resources, Bengaluru, India

Devasantosh Mohanty Department of Biological Sciences, University of North Texas, Denton, TX, USA

Saifullah Omar Nasif Global Centre for Environmental Remediation, University of Newcastle, Callaghan, NSW, Australia

S. Aneeqa Noor Animal Plant Interactions Lab, Department of Zoology, Sri Venkateswara College, New Delhi, India

Patrick Obia Ongom International Institute of Tropical Agriculture, Kano, Nigeria

A. P. Padmakumari ICAR-Indian Institute of Rice Research, Hyderabad, India

Neha Panwar The Graduate School, Colorado State University, Fort Collins, CO, USA

Heena Puri Department of Entomology, University of Nebraska-Lincoln, Lincoln, NE, USA

Gadi V. P. Reddy USDA-ARS Southern Insect Management Research Unit, Southeast Area, Stoneville, MS, USA

J. C. Sekhar Winter Nursery Centre, ICAR-Indian Institute of Maize Research, Hyderabad, India

Anamika Sharma Centre for Biological Control, College of Agriculture and Food Sciences, Florida A&M University, Tallahassee, FL, USA

Mohmmad Aasif Sheikh Division of Entomology, Sher-e-Kashmir University of Agricultural Sciences and Technology of Kashmir, Kashmir, India

Sanket Shinde Department of Entomology, University of Nebraska-Lincoln, Lincoln, NE, USA

Abu Bakar Siddique Tasmanian Institute of Agriculture, University of Tasmania, Prospect, Tasmania, Australia

Pooja Gokhale Sinha Department of Botany, Sri Venkateswara College, Benito Juarez Marg, Dhaula Kuan, New Delhi, India

P. Lakshmi Soujanya Winter Nursery Centre, ICAR-Indian Institute of Maize Research, Hyderabad, India

Warwick Stiller CSIRO Agriculture and Food, Narrabri, NSW, Australia

S. B. Suby Unit Office, ICAR-Indian Institute of Maize Research, Pusa Campus, New Delhi, India

R. M. Sundaram ICAR-Indian Institute of Rice Research, Hyderabad, India

Manuele Tamò International Institute of Tropical Agriculture, Cotonou, Benin

Sathya Thirumurugan Department of Entomology, Punjab Agricultural University, Ludhiana, India

Abou Togola International Institute of Tropical Agriculture, Kano, Nigeria

Fazil Fayaz Wani Division of Pathology, Sher-e-Kashmir University of Agricultural Sciences and Technology of Kashmir, Kashmir, India

Iain Wilson CSIRO Agriculture and Food, Canberra, ACT, Australia

Cennet Yaman Department of Field Crops, Faculty of Agriculture, Yozgat Bozok University, Yozgat, Turkey

Munazah Yaqoob Division of Entomology, Sher-e-Kashmir University of Agricultural Sciences and Technology of Kashmir, Kashmir, India

K. R. Yathish Winter Nursery Centre, ICAR-Indian Institute of Maize Research, Hyderabad, India

Sobiya Zahoor Division of Entomology, Sher-e-Kashmir University of Agricultural Sciences and Technology of Kashmir, Kashmir, India

Qian-Hao Zhu CSIRO Agriculture and Food, Canberra, ACT, Australia

Introduction

1

Neha Panwar, Lukesh Bansal, Michael Furlong, and Sarwan Kumar

Abstract

Growing world population and changing climate pose a significant threat to food security. A number of biotic and abiotic constraints further aggravate this challenge by reducing the crop productivity. Out of the various biotic stresses faced by plants, insects are the major constraints which are largely managed by synthetic chemical pesticides. This approach comes along with several associated risks like environmental degradation, pesticide residues in food chains, non-target effects, health hazards to the persons applying pesticides, and many more. To address these concerns, there is a growing need to shift towards more sustainable management practices which are safer to the environment as well as human health. Host plant resistance has the potential to provide sustainable solution to pest problems which offers economical, effective, easy, and environment friendly option to manage pests. Plants utilize a plethora of biophysical and biochemical defenses that can be harnessed to develop insect resistant cultivars. Conventional breeding techniques take longer time to develop resistant cultivars but they can be easily integrated with modern breeding techniques to develop resistant cultivars in a shorter period of time. Cultivars with even moderate level of resistance can be integrated into IPM system which can help reduce pesticide load.

N. Panwar
The Graduate School, Colorado State University, Fort Collins, CO, USA

L. Bansal
Department of Entomology, Punjab Agricultural University, Ludhiana, India

M. Furlong
School of Biological Sciences, The University of Queensland, Brisbane, QLD, Australia

S. Kumar (✉)
Department of Plant Breeding and Genetics, Punjab Agricultural University, Ludhiana, India
e-mail: sarwanent@pau.edu

S. Kumar, M. Furlong (eds.), *Plant Resistance to Insects in Major Field Crops*,
https://doi.org/10.1007/978-981-99-7520-4_1

Keywords

Breeding · Plant defense · Tritrophic interactions · Resistance

The global population is rapidly growing resulting in increased demand for food. This demand can be met either by increasing cropped area or by increasing productivity per unit area. Since, there is a limited scope for horizontal area expansion; increased requirements can be met by improving the productivity of agriculturally important crops. This can be challenging due to various biotic and abiotic constraints that limit production. It is believed that about 20–40% crop losses occur due to diseases and pests, with insects being a major contributor to this loss (FAO, Food and Agriculture Organization of the United Nations 2009). Insect pests are a major cause of agricultural losses worldwide caused directly by crop damage and indirectly by the diseases that some species transmit. Farmers often use synthetic insecticides to manage pests, prolonged use of these chemicals can lead to many problems such as insecticide resistance, pest resurgence, detrimental effects on non-target beneficial species, residual activity, human health effects, and environmental pollution. To achieve sustainable food production, it is imperative that innovative techniques that complement or replace the use of broad-spectrum chemical pesticides are adopted. Host plant resistance has the potential to contribute to sustainable food production as well as reduce the use of insecticides (Wiseman and Webster 1999). The earliest examples of utilizing a crop resistant variety date back to 1780s when the "Underhill" variety of wheat (Havens 1792), resistant to Hessian fly, *Mayetiola destructor* (Say) and the "Winter Majetin" variety of Apple (Lindley 1831) resistant to the wooly apple aphid, *Eriosoma lanigerum* (Hausmann), were developed. Application of Mendelian rules of heredity further played a crucial role in the systematic development of resistant breeding programs (Panda and Khush 1995). Successful examples of the development of many insect and disease resistant varieties during green revolution like IR-36 variety of rice further sparked interest of breeders and growers in this field (Smith 2005).

Host plant resistance can offer sustainable solutions to pest management problems with many advantages. Research on host plant resistance has led to the development of arthropod resistant crop varieties resulting in significant improvements in food production, alleviation of hunger, and improved nutrition (Khush 1995). In the past few decades, genetic engineering has further expanded the horizons of plant resistance science. Introgression of genes from crop wild relatives (CWRs) and gene transfer from organisms of non-plant origin like the bacterium *Bacillus thuringiensis* have revolutionized breeding for resistance. Novel techniques like genome editing using CRISPR further hold great potential in achieving the sustainability goal of host plant resistance. In this chapter, we discuss how host plant resistance can be efficiently utilized against insect pests.

1.1 The Types of Host Plant Resistance

Painter (1951) defined host plant resistance as "the relative amount of heritable qualities possessed by a plant which influence the ultimate degree of damage done by an insect." It is very important to note that resistance should always be measured on a relative scale by comparing the amount of resistance in a given variety to that of susceptible control plants that experience more damage under similar experimental settings. Host plant resistance can be categorized into three types: non-preference, antibiosis, and tolerance.

Kogan and Ortman (1978) introduced the term "antixenosis" in place of non-preference as it refers to a plant's ability to resist being a host for insect herbivores, which compels the pests to choose a different host plant. This term describes how the presence of specific plant characteristics, such as chemical or physical factors, can alter insect behavior. Antibiosis refers to the negative effects on the insect lifecycle that occur when an insect feeds on a resistant host plant (Painter 1951). It can be caused by both chemical and physical plant defenses, and severity can range from mild to fatal. The effects of antibiosis include mortality in early life stages, extended periods of development, decreased adult lifespan and fertility, and/or abnormal behavior. Tolerance refers to a plant's inherent genetic capacity to continue growing and reproducing while supporting herbivore population roughly equivalent to that which harms a susceptible plant. Unlike antibiosis, it does not exert any selection pressure on the insect population, reducing the probability of selecting for resistant insect biotypes. To determine if tolerance exists, damage levels, plant growth, and yield can be compared between insect infested and uninfested hosts. However, developing crops/hosts that are tolerant to insects requires that tolerance mechanisms are identified and characterized and that genetic basis that underpins tolerance is understood (Peterson et al. 2018). Additionally, it is very difficult to educate farmers to understand that damage caused to crops by pests doesn't always result in economic losses. Apart from this, tolerant varieties may serve as a reservoir for insect populations that may damage the other non-tolerant crops growing in the vicinity. However, modern precision breeding methods, like gene editing and marker-assisted selection, offer the possibility of achieving greater resistance levels while producing high-quality and high-yielding crops. It is crucial to thoroughly investigate these mechanisms because of the advantages that resistant plants can offer in Integrated Pest Management (IPM) systems. Ideally, cultivars which demonstrate a combination of resistance mechanisms should be developed.

1.2 Bases of Host Plant Resistance

Plants and insects have been engaged in a co-evolutionary arms race for millions of years. As plants have evolved new defenses to protect themselves from herbivores, insects have evolved new adaptations to overcome those defenses. These plant defenses can be either constitutive or induced. Constitutive defenses are expressed in plants even in the absence of herbivores, whereas, induced defenses are generally activated by interactions with herbivores. Plant defenses can also be classified as direct, including physical defenses such as spines, thorns, and tough leaves, and chemical defenses such as toxic compounds, or indirect such as the production of volatile compounds that attract natural enemies of the herbivorous insects. These defenses can make it more difficult for herbivorous insects to feed on the plants, or can make the insects less fit, reducing their ability to reproduce and survive.

1.2.1 Biophysical Bases

A plant's suitability as a host for insect pests is in part governed by a number of physical plant characteristics (in addition to biochemical characters) including leaf waxiness, trichomes, plant shape and size, toughness, etc. Trichomes are short hair-like structures present on plant parts which help plants to defend themselves against different insect pests. Different trichome characteristics like shape, length, density of trichomes, etc. play a crucial role in determining the extent of protection offered against different insect pests. These trichomes generally make the plant unsuitable/ less suitable for oviposition and/or feeding and, therefore, provide protection to plants against herbivory (Dalin et al. 2008). Trichomes are of two types, simple or glandular, depending on whether they secrete defensive fluids or not. Simple trichomes serve as a physical barrier, whereas, glandular trichomes secrete repellant or toxic chemicals to ward off the pests (Handley et al. 2005).

Epicuticular waxes can act as biophysical barriers to herbivores by making it difficult for them to attach to and feed on plant surfaces (Jeffree 1986). They serve as a barrier for insect feeding by eliciting negative tactile and chemical reactions in insects' mechano- and chemoreceptors.

Plant toughness or rigidity can also prove useful in defending plants against various insects. Because of the presence of substances like lignin, cellulose, callose, etc., it becomes extremely difficult for herbivores to pierce plant tissue for oviposition or feeding. For instance, the solid-stemmed variety of wheat and rind hardness of sugarcane imparts resistance against stem fly and internode borer, respectively (Platt and Farstad 1946; Martin et al. 1975).

1.2.2 Biochemical Bases

Plants generate a wide range of secondary metabolites, which are believed to number over 200,000 molecules belonging to different classes such as alkaloids, cyanogenic alkaloids, glucosinolates, phenylpropanoids, terpenes, and aromatic polyketides (Mithofer and Boland 2012). These chemicals can be involved either in direct defenses that are toxic or repellant to herbivores or indirect defense that attract the natural enemies of phytophagous insects. However, some insects have evolved to detoxify or tolerate these secondary metabolites and utilize them as cues for oviposition and/or feeding (Schoonhoven et al. 2005). These secondary metabolites can be either phytoanticipins such as tannins (constitutively present) or phytoalexins such as isothiocyanates (induced after herbivory) in nature (Morant et al. 2008).

Plants may also release volatile organic compounds (VOCs) which play a key role in regulating plant–insect interactions (Blande et al. 2007). Herbivore-induced plant volatiles (HIPVs) are one such example of VOCs which have been implicated in indirect defenses. Upon herbivory, plants may produce volatile compounds which can be utilized by natural enemies to detect the likely presence of herbivore host/prey (Furlong et al. 2018). These chemical signals have been referred to as 'Call or cry for help' (Dicke and Baldwin 2010), and they are integral to indirect defenses that attract parasitoids and predators to herbivore damaged plants.

1.3 Breeding for Insect Resistance

1.3.1 Conventional Breeding

A source of resistance is the first requirement in any breeding programme aimed at development of insect resistant cultivar. For the development of an effective standardized screening protocol to detect resistance, thorough knowledge about pest infestations, seasonal history, as well as pest hot-spot regions is required. The screening for insect resistance is a very long process and since pest infestation may show temporal variations, it becomes very challenging to identify reliable and durable source(s) of resistance.

After the identification of the insect resistance trait, it is transferred to the cultivated background. When the trait of interest in plants is present in plants of the same species or closely related species, traditional breeding methods can be used. A number of breeding methods like selection, pedigree method, hybridization, induced mutagenesis, etc. can be utilized for transferring the desired resistance trait into the cultivated background. However, these methods take a long time to develop resistant varieties, though, marker-assisted selection can aid in accelerating breeding programme (Collard and Mackill 2008). Kumar et al. (2011) attempted to introgress resistance to turnip aphid, *Lipaphis erysimi* (Kaltenbach) from wild *Brassica fruticulosa* to *B. juncea*. Rigorous phenotyping over the years led to identification of resistant introgression lines though with some linkage drag with some undesirable traits. More recent research has attempted to remove this drag through the use of γ-irradiation (Agrawal et al. 2021).

Earlier efforts by plant breeders largely focused on the development of high-yielding cultivars with little attention paid to insect or disease resistance. This led to the loss of many defense-related genes in many of the high-yielding varieties compared to their crop wild relatives (Chen et al. 2015). Most cultivated crops produce fewer plant secondary metabolites (PSMs) and are less toxic to herbivores than their wild relatives (Olsen and Wendel 2013). For example, domestication of tomato has led to the loss of two genes, zFPP and ShZIS, which are known to code for enzymes involved in production of sesquiterpene 7-epizingiberene which has a role in plant defense (Olsen and Wendel 2013). When these genes were reintroduced into the cultivated tomato, the plants became more resistant to various insect pests.

1.3.2 Modern Breeding

For crops that have limited genetic variation in the desired traits, genetic technologies such as genetic transformation and genome editing can expedite the development of cultivars with insect resistance. Researchers studying plant resistance are now utilizing genotyping by sequencing technologies to quickly and accurately discover and map resistance genes. Also, this technique can efficiently track resistance genes during the breeding and selection process (Campbell et al. 2018). While it is important to keep uncovering new sources of resistance, it is equally crucial to maintain resistant genotypes that have already been developed. For example, maize breeding lines developed in the 1970s with polygenic resistance to the fall armyworm are currently being used to address the pest's recent invasion in Sub-Saharan Africa (Feldmann et al. 2019).

Over the last few decades, there has been increased focus on the use of some defense-related genes in the development of transgenic plants resistant to insects. Genes coding for plant lectins are one such example. Plant lectins are a diverse class of proteins which can be utilized to induce defense responses in plants against insect pests. These lectins serve as feeding deterrents for herbivores by binding to the carbohydrate molecules on the insect gut leading to cessation of feeding, diarrhea, and ultimately death (Esch and Schafrath 2017). Rauf et al. (2019) reported that transgenic tobacco plant upregulating Hvt-lectin caused complete mortality of *Spodoptera litura* and *Helicoverpa armigera*.

Protease inhibitors (PIs) are another class of proteins present naturally in plants that defend them against herbivory by inhibiting the activity of digestive proteases in the insect gut. These PIs bind to the active site of digestive proteases leading to inhibition of protein digestion. This is manifested in reduced utilization of protein by insects leading to adverse effect on its development and increased susceptibility to other stress factors such as pathogens and predators. Plant protease inhibitors such as cowpea trypsin inhibitor (CpTI) and soybean trypsin inhibitor (STI) have been widely explored for their involvement in host plant defense to insect herbivores. CpTI and STI both inhibit the action of trypsin-like enzymes in the guts of insect herbivores, resulting in decreased feeding and development (Zhao et al. 2019).

Bacillus thuringiensis (Bt) and *Bacillus sphaericus* have been the most successful group of organisms identified for use in genetic transformation of crops for pest control on a commercial scale (Schuler et al. 1998). Generally, these plants express Bt toxins at levels sufficient to cause high mortality of target pests in the field. Though Bt genes have been found to be successful and hence commercialized, there is a fear of insects developing resistance to Bt gene products (Estruch et al. 1997; Tabashnik et al. 2023). To overcome this difficulty, gene pyramiding (more than one gene) is advocated. Just as it is important to rotate crops in the farm for maintaining a dynamic ecosystem, gene rotation is necessary in biotechnology-based agriculture.

Ribosome-Inactivating Proteins (RIPs) are plant proteins that inhibit protein synthesis in eukaryotic cells by catalyzing the removal of adenine from ribosomal RNA, causing depurination and prevent the production of functional ribosomes (Stirpe 2013). They can be utilized by plants against different insect pests because of their insecticidal properties. When an insect is exposed to RIPs, the RIPs penetrate the insect's cells and limit protein synthesis, resulting in their death. Moreover, RIPs can activate a variety of defense-related genes and pathways in the host plant, resulting in the synthesis of additional defensive chemicals such as phytohormones and plant secondary metabolites. For example, transgenic tobacco plants which upregulated the RIP production genes reported both sublethal and aphicidal effects against *Myzus persicae* (Hamshou et al. 2016).

RNA interference (RNAi) allows for the development of transgenic plants that are more flexible than protein toxins in terms of their selectivity and mode of action. To achieve this, essential gene-specific double-stranded RNA (dsRNA) from an insect pest is taken up into cells, where Dicer enzymes break it down into small interfering RNA (siRNA) molecules. These molecules then direct the RNA-induced silencing complex (RISC), which contains the Argonaute protein, to degrade complementary mRNAs, and sometimes prevent the target mRNA from being translated (Scott et al. 2013). The delivery of dsRNA is typically achieved through plant genetic manipulation (Price and Gatehouse 2008). However, topical application of dsRNA through sprays or drenches can control lepidopteran and hemipteran pests (Li et al. 2015). RNAi can be used for developing insect resistant plants, for example, Pitino et al. (2011) reported that RNAi resulted in reduced feeding of green peach aphids due to silencing of Rack-1 and COO2 gene in transgenic *Arabidopsis* and tobacco plants.

Genome editing technologies like CRISPR (clustered regularly interspersed short palindromic repeats) can also be used to incorporate resistance characteristics into crops (Georges and Ray 2017). The CRISPR/Cas9 bacterial defense against viruses is the most extensively used CRISPR system in eukaryotes, including plants. The Cas9 protein is essentially a nuclease enzyme that cuts DNA at a specific place indicated by a synthesized guide RNA. There are several CRISPR techniques available, some of which can induce site-specific modifications in DNA, such as indels (which frequently result in frameshift mutations), replacement or insertion of particular sequences, and gene expression suppression through the use of a deactivated Cas9. CRISPR is the most preferred method for creating new varieties in response to resistant insect strains that overcame plant resistance. This is due to the fact that there is only a difference of a single or few nucleotides between susceptible and resistant alleles of plant resistance genes. With the help of CRISPR, the susceptible allele can be edited to become a resistance allele, reducing the requirement for traditional approaches which involve extensive crossing and backcrossing.

1.3.3 Potential of Plant Resistance

Resistant cultivars provide an economic advantage to growers as they offer genetically built-in control against pests for only the cost of the seed and no special application technique is required to utilize them. Even when a moderately resistant variety is combined with pesticide treatment, the costs of insecticidal control can be decreased considerably. Moreover, such reduced pesticidal applications provide added ecological benefits such as lower pesticide residues, cleaner water supplies, and less harm to beneficial natural enemies (Smith 2005).

Host plant resistance can serve as a foundation to and be integral component of IPM. Insects that feed on resistant plants generally experience reduced vigor, which can make them more vulnerable to insecticides and ultimately reduce the amount of insecticide required for successful pest control (Stout 2014). Due to reduced insecticidal usage, the economic threshold (ET) for resistant varieties is comparatively higher than that for susceptible varieties of the same crop (van den Berg et al. 1997). This threshold is frequently reached later in the crop growth season, resulting in lower insect populations on these plants throughout the season. Tolerant plants that are resistant to insects work very well in integrated pest management (IPM) programs because they raise the economic threshold without exerting any kind of selection pressure on the pest populations (Peterson et al. 2018).

Resistant varieties can also be integrated with biological control in IPM programs. Their integration can be either additive or synergistic. It is generally believed that resistant varieties may have a negative impact on natural enemies as the pest population is lowered, however, this is largely offset by decreased chemical pesticide usage, which ultimately increases predator and parasitoid populations. Further, moderately resistant varieties which can tolerate pest damage and maintain pest populations offer sufficient food and shelter for beneficial natural enemies (Stout and Davis 2009). Rand et al. (2020) demonstrated that even minimal parasitism (22%) along with solid-stemmed resistant wheat cultivars can effectively manage wheat sawfly.

Insects can cause indirect damage to plants by serving as a vector for diseases, especially viruses. It is very uncommon for plants to be resistant to both a virus and the vector that transmits it, but vector resistance can greatly decrease the incidence and spread of the virus. In the U.S. High Plains, wheat varieties that contain gene (s) for resistance to the wheat curl mite *Aceria tosichella*, the vector for wheat streak mosaic virus (WSMV), have continuously reduced WSMV epidemics for more than 40 years, as about 30% of wheat varieties have this resistance (Carver et al. 2016).

1.3.4 Challenges for Host Plant Resistance

It takes a lot of money, time (years), and expertise to develop a single resistant variety by traditional methods. Alone, such resistant varieties are not able to compete with high-yielding varieties for grower demand. Sometimes, it is very difficult to make growers understand that primary objective of host plant resistance is ecological sustainability which can be achieved by utilizing it as a component of IPM-a method to reduce insecticidal inputs. Modern techniques like market-assisted selection, genetic engineering, genome editing, etc. have significantly reduced the time required to develop a resistant cultivar. Further, the growing demand for sustainable and organic solutions has led to increased interest in this eco-friendly approach.

Both conventional and transgenic insect resistant cultivars may lose their efficacy when insects evolve to feed on these insect resistant cultivars, leading to the development of new insect biotypes. For example, varieties with a single resistance gene to Hessian fly, *M. destructor* and brown planthopper, *Nilaparvata lugens* (Stål) are vulnerable to failure 6–8 years after release (Johnson et al. 2017; Horgan 2018). On the other hand, efficacy of polygenic resistance is long-lasting in most crops. For example, polygenic resistance in raspberry plants has suppressed virulence of raspberry aphid, *Amphorophora idaei* Borner for about 30 years (Jones et al. 2000). Another strategy to prevent or delay virulence in insects can be the use of a combination of conventional and transgenic polygenic resistance. This approach has been used in soybean against lepidopteran insects. Benning ME, a soybean near-isogenic line, has two significant quantitative trait loci, one from linkage group M and one from linkage group E, which provide field resistance against soybean looper, *Chrysodeixis includens* (Walker) and Southern armyworm, *Spodoptera eridania* (Cramer). Moreover, resistance in Benning ME plants is much higher when the cry1Ac transgene is present (Ortega et al. 2016).

1.4 Conclusion

Repeated attempts to control insect pests have shown that there can never be a permanent solution because high selection pressure on insects results in the development of new ways to overcome control measures, whether they are insecticide based or based on host plant resistance (Thrall et al. 2011). A highly or completely resistant variety is neither practical nor desirable. Instead efforts should be directed to how we can utilize moderately resistant varieties which are components of IPM approaches. However, we can still improve the durability of resistance traits in plants by implementing effective resistance management strategies. Polygenic resistance, refugia crops, and gene pyramiding can prove useful in this regard (Carriere et al. 2015). These resistance management strategies can be used along with the newer technologies like genome editing to develop crops that can be modified in real time to combat with evolving insect pests.

References

Agrawal N, Gupta M, Atri C, Akhatar J, Kumar S, Heslop-Harrison PJS, Banga SS (2021) Anchoring alien chromosome segment substitutions bearing gene(s) for resistance to mustard aphid in *Brassica juncea–B. fruticulosa* introgression lines and their possible disruption through gamma irradiation. Theor Appl Genet 134(10):3209–3224. https://doi.org/10.1007/s00122-021-03886-z

Blande J, Pickett J, Poppy G (2007) A comparison of semiochemically mediated interactions involving specialist and generalist brassica feeding aphids and the braconid parasitoid *Diaeretiella rapae*. J Chem Ecol 33:767–779. https://doi.org/10.1007/s10886-007-9264-7
Campbell ZC, Acosta-Gamboa LM, Nepal N, Lorence A (2018) Engineering plants for tomorrow: how high-throughput phenotyping is contributing to the development of better crops. Phytochem Rev 17:1329–1343. https://doi.org/10.1007/s11101-018-9585-x
Carriere Y, Crickmore N, Tabashnik BE (2015) Optimizing pyramided transgenic Bt crops for sustainable pest management. Nat Biotechnol 33:161–168. https://doi.org/10.1038/nbt.3099
Carver BF, Smith CM, Chuang WP, Hunger RM, Edwards JT, Yan L, Bowden RL (2016) Registration of OK05312, a high-yielding hard winter wheat donor for wheat curl mite resistance. J Plant Regist 10:75–79. https://doi.org/10.3198/jpr2015.04.0026crg
Chen YH, Gols R, Benrey B (2015) Crop domestication and its impact on naturally selected trophic interactions. Annu Rev Entomol 60:35–58. https://doi.org/10.1146/annurev-ento-010814-020601
Collard BC, Mackill DJ (2008) Marker-assisted selection: an approach for precision plant breeding in the twenty-first century. Philos Trans R Soc Lond Ser B Biol Sci 363:557–572. https://doi.org/10.1098/rstb.2007.2170
Dalin P, Agren J, Bjorkman C, Huttunen P, Karkkainen K (2008) Leaf trichome formation and plant resistance to herbivory. In: Schaller A (ed) Induced plant resistance to herbivory. Springer, Dordrecht
Dicke M, Baldwin IT (2010) The evolutionary context for herbivore induced plant volatiles: beyond the 'cry for help'. Trends Plant Sci 15(3):167–175
Esch L, Schafrath U (2017) An update on Jacalin-like lectins and their role in plant defense. Int J Mol Sci 18:1592. https://doi.org/10.3390/ijms18071592
Estruch JJ, Carozzi NB, Desai N, Duck NB, Warren GW, Koziel MG (1997) Transgenic plants: an emerging approach to pest control. Nat Biotechnol 15:137–141. https://doi.org/10.1038/nbt0297-137
FAO (Food and Agriculture Organization of the United Nations) (2009) Global agriculture towards 2050. High level expert forum. FAO, Rome
Feldmann F, Rieckmann U, Winter S (2019) The spread of the fall armyworm *Spodoptera frugiperda* in Africa-what should be done next? J Plant Dis Prot 126:97–101. https://doi.org/10.1007/s41348-019-00204-0
Furlong MJ, Ang Gurion CK, Rehan S, Zalucki MP (2018) Bringing ecology back: how can the chemistry of indirect plant defenses against herbivore be manipulated to improve pest management? Front Plant Sci 9:1436. https://doi.org/10.3389/fpls.2018.01436
Georges F, Ray H (2017) Genome editing of crops: a renewed opportunity for food security. GM Crops Food 8:1–12. https://doi.org/10.1080/21645698.2016.1270489
Hamshou M, Shang C, Smagghe G, van Damme EJM (2016) Ribosome-inactivating proteins from apple have strong aphicidal activity in artificial diet and in planta. Crop Prot 87:19–24. https://doi.org/10.1016/j.cropro.2016.04.013
Handley R, Ekbom B, Agren J (2005) Variation in trichome density and resistance against a specialist insect herbivore in natural populations of *Arabidopsis thaliana*. Ecol Entomol 30: 284–292. https://doi.org/10.1111/j.0307-6946.2005.00699.x
Havens JN (1792) Observations on the Hessian fly. Trans New York Soc Agron Pt 1:89–107
Horgan FG (2018) Integrating gene deployment and crop management for improved rice resistance to Asian planthoppers. Crop Prot 110:21–33. https://doi.org/10.1016/j.cropro.2018.03.013
Jeffree CE (1986) The cuticle, epicuticular waxes and trichomes of plants, with reference to their structure, functions and evolution. In: Juniper BE, Southwood TRE (eds) Insects and the plant surface. Edward Arnold, London, pp 23–64
Johnson AJ, Moniem HEMA, Flanders KL, Buntin GD, Reay-Jones FPF, Reisig D, Stuart JJ, Subramanyam S, Shukle RH, Schemerhorn BJ (2017) A novel, economical way to assess virulence in field populations of Hessian fly (Diptera: Cecidomyiidae) utilizing wheat resistance gene H13 as a model. J Econ Entomol 110:1863–1868. https://doi.org/10.1093/jee/tox129

Jones AT, McGavin WJ, Birch ANE (2000) Effectiveness of resistance genes to the large raspberry aphid, *Amphorophora idaei* Borner, in different raspberry [*Rubus idaeus* L.] genotypes and under different environmental conditions. Ann Appl Biol 136:107–113. https://doi.org/10.1111/j.1744-7348.2000.tb00014.x

Khush GS (1995) Modern varieties—their real contribution to food supply and equity. GeoJournal 35:275–284

Kogan M, Ortman EE (1978) Antixenosis—a new term proposed to replace Painter's "nonpreference" modality of resistance. Bull Entomol Soc Am 24:175–176. https://doi.org/10.1093/besa/24.2.175

Kumar S, Atri C, Sangha MK, Banga SS (2011) Screening of wild crucifers for resistance to mustard aphid, *Lipaphis erysimi* (Kaltenbach) and attempt at introgression of resistance gene (s) from *Brassica fruticulosa* to *Brassica juncea*. Euphytica 179:461–470. https://doi.org/10.1007/s10681-011-0351-z

Li H, Guan R, Guo H, Miao X (2015) New insights into RNAi approach for plant defence against piercing-sucking and stem-borer insect pests. Plant Cell Environ 38:2277–2285. https://doi.org/10.1111/pce.12546

Lindley G (1831) A guide to the orchard and kitchen garden. Longman, p 597

Martin FA, Richard CA, Hensley SD (1975) Host resistance to *Diatraea saccharalis* (F.): Relationship of sugarcane internode hardness to larval damage. Environ Entomol 4:687–688. https://doi.org/10.1093/ee/4.5.687

Mithofer A, Boland W (2012) Plant defense against herbivores: chemical aspects. Annu Rev Plant Biol 63:431–450. https://doi.org/10.1146/annurev-arplant-042110-103854

Morant AV, Jorgensen K, Jorgensen C, Paquette SM, Perez RS, Moller BL, Bak S (2008) β-Glucosidases as detonators of plant chemical defense. Phytochemistry 69:1795–1813. https://doi.org/10.1016/j.phytochem.2008.03.006

Olsen KM, Wendel JF (2013) A bountiful harvest: genomic insights into crop domestication phenotypes. Annu Rev Plant Biol 64:47–70. https://doi.org/10.1146/annurev-arplant-050312-120048

Ortega MA, All JN, Boerma HR, Parrott WA (2016) Pyramids of QTLs enhance host-plant resistance and Bt-mediated resistance to leaf-chewing insects in soybean. Theor Appl Genet 129:703–715. https://doi.org/10.1007/s00122-015-2658-y

Painter RH (1951) Insect resistance in crop plants. University Press of Kansas, Lawrence, p 520

Panda N, Khush GS (1995) Host plant resistance to insects. CABI, Wallingford, p 431

Peterson RKD, Higley LG, Pedigo LP (2018) Whatever happened to IPM? Am Entomol 64:146–150. https://doi.org/10.1093/ae/tmy049

Pitino M, Coleman AD, Maffei ME, Ridout CJ, Hogenhout SA (2011) Silencing of aphid genes by dsRNA feeding from plants. PLoS One 6(10):e25709. https://doi.org/10.1371/journal.pone.0025709

Platt AW, Farstad CM (1946) The reaction of wheat varieties to wheat stem sawfly attack. Sci Agric 26:231–247

Price DR, Gatehouse JA (2008) RNAi-mediated crop protection against insects. Trends Biotechnol 26:393–400. https://doi.org/10.1016/j.tibtech.2008.04.004

Rand TA, Richmond CE, Dougherty ET (2020) Modeling the combined impacts of host plant resistance and biological control on the population dynamics of a major pest of wheat. Pest Manag Sci 76:2818–2828. https://doi.org/10.1002/ps.5830

Rauf I, Javaid S, Naqvi RZ, Mustafa T, Amin I, Mukhtar Z, Jander G, Mansoor S (2019) *In-planta* expression of insecticidal proteins provides protection against lepidopteran insects. Sci Rep 9:6745. https://doi.org/10.1038/s41598-019-41833-7

Schoonhoven LM, van Loon JJA, Dicke M (2005) Insect-plant biology. Oxford University Press, Oxford

Schuler TH, Poppy GM, Kerry BR, Denholm I (1998) Insect-resistant transgenic plants. Trends Biotechnol 16:168–175. https://doi.org/10.1016/S0167-7799(97)01171-2

Scott JG, Michel K, Bartholomay LC, Siegfried BD, Hunter WB, Smagghe G, Zhu KY, Douglas AE (2013) Towards the elements of successful insect RNAi. J Insect Physiol 59:1212–1221. https://doi.org/10.1016/j.jinsphys.2013.08.014

Smith CM (2005) Plant resistance to arthropods: molecular and conventional approaches. Springer, Dordrecht, 423p

Stirpe F (2013) Ribosome-inactivating proteins: from toxins to useful proteins. Toxicon 67:12–16. https://doi.org/10.1016/j.toxicon.2013.02.005

Stout MJ (2014) Host-plant resistance in pest management. In: Abrol DP (ed) Integrated pest management: current concepts and ecological perspectives. Elsevier, pp 1–21

Stout M, Davis J (2009) Keys to the increased use of host plant resistance in integrated pest management. In: Peshin R, Dhawan AK (eds) Integrated pest management: innovation-development process. Springer, Dordrecht, pp 163–181

Tabashnik BE, Fabrick JA, Carrière Y (2023) Global patterns of insect resistance to transgenic Bt crops: the first 25 years. J Econ Entomol 116:297–309. https://doi.org/10.1093/jee/toac183. Erratum in: J Econ Entomol. 2023 Jan 20: PMID: 36610076

Thrall PH, Oakeshott JG, Fitt G, Southerton S, Burdon JJ, Sheppard A, Russell RJ, Zalucki M, Heino M, Denison RF (2011) Evolution in agriculture: the application of evolutionary approaches to the management of biotic interactions in agro-ecosystems. Evol Appl 4:200–215. https://doi.org/10.1111/j.1752-4571.2010.00179.x

van den Berg J, van Rensburg GDJ, van der Westhuizen MC (1997) Economic threshold levels for *Chilo partellus* [Lepidoptera: Pyralidae] control on resistant and susceptible sorghum plants. Bull Entomol Res 87:89–93

Wiseman BR, Webster JA (1999) Economic, environmental, and social benefits of resistance in field crops. Entomological Society of America, Lanham, p 189

Zhao H, Ullah H, McNeill MR, Du G, Hao K, Tu X, Zhang Z (2019) Inhibitory effects of plant trypsin inhibitors Msti-94 and Msti-16 on *Therioaphis trifolii* (Monell) (Homoptera: Aphididae) in alfalfa. Insects 10:154. https://doi.org/10.3390/insects10060154

Unraveling the Coevolutionary Arms Race: Insights into the Dynamic Interplay of Plants, Insects and Associated Organisms

2

Vartika Mathur, Pooja Gokhale Sinha, and S. Aneeqa Noor

Abstract

Insect–plant interactions are complex and dynamic relationships that have evolved over millions of years. Plants have developed various adaptations to deter insect herbivores, including physical, chemical, and induced defences. In response, insects have evolved detoxification mechanisms, behavioural adaptations, and physiological adaptations to overcome these defences. This coevolutionary arms race has shaped the interactions between plants and insects, leading to a diverse array of strategies and counter-strategies. Additionally, other associated organisms such as endosymbionts and rhizosphere microbes have been shown to play a critical role in these interactions. Endosymbionts can alter the nutritional quality of plant tissue and confer resistance to environmental stressors, while rhizosphere microbes can influence plant growth and nutrient uptake. Understanding the coevolutionary arms race and the role of associated organisms in insect–plant interactions has important implications for plant protection and management. By leveraging these relationships, we can develop sustainable and eco-friendly approaches to crop protection and pest management.

V. Mathur (✉)
Animal Plant Interactions Lab, Department of Zoology, Sri Venkateswara College, New Delhi, India

Fellow of Institute of Eminence, School of Climate Change and Sustainability, University of Delhi, New Delhi, India
e-mail: vmathur@svc.ac.in

P. G. Sinha
Department of Botany, Sri Venkateswara College, New Delhi, India

S. A. Noor
Animal Plant Interactions Lab, Department of Zoology, Sri Venkateswara College, New Delhi, India

S. Kumar, M. Furlong (eds.), *Plant Resistance to Insects in Major Field Crops*,
https://doi.org/10.1007/978-981-99-7520-4_2

Keywords

Insect–plant interactions · Plant defence · Induced defence · Endosymbionts · Rhizosphere · Crop protection

2.1 Introduction

2.1.1 Overview

The term 'Coevolution' was introduced by Ehrlich and Raven (1964) to describe the close interaction between two or more species in an ecosystem, resulting in genetic adaptations in one or both species due to natural selection imposed by the other interacting species. This process is a product of evolution and can have reciprocal impacts (Janzen 1980; Berenbaum and Zangerl 1998; Woolhouse et al. 2002). The interaction between any groups of interacting species, including mutualists, pathogens, competitors, and predators and prey, can lead to coevolution. The nature and intensity of their interaction determine the selection pressures exerted by each species on the other. Previous research has demonstrated the significance of the relationship between two species in determining the selection pressures on each other (Hochberg et al. 2000; Thompson and Cunningham 2002).

Interaction between plants and insects is a constantly changing system. These interactions have a long history of coevolution, with each group engaged in an ongoing arms race, thus influencing the other's evolutionary trajectory (Labandeira 1998). The theory of coevolution developed by Ehrlich and Raven (1964) forms the foundation for the current understanding of the dynamic interplay between plants and insects. The relationship between insects and plants has been dynamically occurring for over 400 million years, resulting in a complex system of interactions that includes herbivory, pollination, and other mutualistic relationships (Labandeira 2013). Insects perform essential functions for plants, such as defence and pollination, while plants provide critical resources such as shelter, oviposition sites, and food for insect growth and reproduction. Nonetheless, herbivores can pose a significant threat to plants and exert strong selection pressure to evolve strategies to tolerate or resist them (Panda and Khush 1995).

The coevolution of insect–plant interactions has resulted in a complex system of adaptations and counter-adaptations, with each group influencing the other's evolutionary trajectory. This ongoing process has led to the diversity and complexity of the natural world we see today, with insects and plants playing critical roles in ecosystem function and biodiversity.

2.1.2 Coevolution in Shaping Various Insect–Plant Interactions

One of the most prominent examples of coevolution between insects and plants is the evolution of morphological adaptations and specialized feeding strategies. Many insects have evolved to feed on specific plants. For example, some insects have evolved specialized mouthparts to pierce and suck plant tissue, while others have developed chewing mouthparts to consume leaves (Krenn 2019). However, excessive herbivory on leaves, stems, flowers, and other plant parts can lead to reduced growth and reproduction, and in some cases, can even kill the plant. Accordingly, plants have evolved specific mechanisms to deter or tolerate herbivory. They have evolved various structures such as thorns, spines, tough leaves, and other physical barriers to deter herbivores (War et al. 2012). Some plants also produce/increase toxic chemicals that are harmful to herbivores, in response to which insects have developed mechanisms to detoxify these chemicals (Gatehouse 2002). This arms race between plants and insects has resulted in evolution of a vast array of chemical compounds and biochemical pathways (Schoonhoven et al. 2005; Howe and Jander 2008). Plants also recruit other organisms such as predators or parasitoids to attack the herbivores.

The coevolution of insect–plant interactions has also resulted in various mutualistic relationships (Bronstein and Huxman 1997). Many plants rely on insects for pollination, and in return, provide them with a source of nectar or pollen. This mutualistic relationship has led to the coevolution of floral morphology and insect behaviour. Plants have evolved to form floral structures, shapes and colours (Willmer 2011), and chemical signals (Raguso 2008) to attract specific pollinators. Accordingly, insect pollinators perceive and interpret these signals by using their visual and olfactory sense organs (Chittka and Raine 2006). These insects feed on the nectar produced by plants (Heil 2011) and aid in pollination by developing specialized adaptations for collecting and transferring pollen, such as long tongues or hairy bodies (Johnson and Steiner 2000).

Coevolutionary mutualistic relationships between insects and plants can extend beyond just pollination. In some cases, insects live in close association with the plant and providing the plant with protection from predators in exchange for nutrients (Bronstein et al. 2006). This type of relationship has led to the evolution of specialized structures in plants to accommodate the insects.

One such structure is called a gall, which is a growth on the plant caused by the interaction between the plant and an insect, mite, or other arthropod. Gall-forming insects, such as wasps and flies, lay their eggs in plant tissues, which induces the plant to form a protective structure called a gall around the developing insect. The gall provides a safe environment for the insect to develop and feed, and in exchange, the insect secretes chemicals that alter the physiology of the plant to make it more suitable for its needs (Stone and Schönrogge 2003; Jansen-González et al. 2012). Another specialized structure is the domatia, which are small cavities or pouches that are formed in the leaves or stems of some plants, and are inhabited by mutualistic arthropods such as mites, spiders or insects, which provide plant with protection (Agrawal and Karban 1997). Similarly, many ant species depend on plants for food and housing in exchange for protection against predators (Heil and McKey 2003; Nelson et al. 2019).

Furthermore, the coevolution of insect–plant interactions has also driven the evolution of mimicry. Some insects have evolved to mimic the physical appearance or chemical signals of other insects to avoid being detected by predators or to attract prey. In turn, some plants have evolved to mimic the physical appearance or scent of other plants to deceive herbivores or attract specific pollinators (Schaefer and Ruxton 2009).

These types of plant–insect relationships have likely evolved through a process of coevolution, where the plants and insects have adapted to each other's presence over time. As with pollination, the specific details of these relationships can vary widely depending on the species involved. However, in this review, we will focus on the coevolutionary dynamics of herbivore insect–plant relationship.

2.2 Coevolutionary Arms Race during Insect Herbivory

The concept of coevolutionary arms race describes the evolutionary adaptations that occur between two species, where each species is under selective pressure to evolve in response to the adaptations of the other. In the context of plant–insect herbivore interactions, this refers to the adaptations that occur between plants and the insect herbivores that feed on them (Ehrlich and Raven 1964).

Insect herbivory has been one of the major driving forces in the evolution of plant defence mechanisms. Plants have developed a variety of physical and chemical defences to deter herbivores, including thorns, spines, tough fibrous leaves and toxic secondary metabolites, such as alkaloids and terpenoids (Hanley et al. 2007; Wöll et al. 2013). Moreover, depending on the intensity and kind of an insect damage, plants modify the synthesis and distribution of these defence chemicals. Insects, in turn, have evolved mechanisms to overcome these defences, such as specialized mouthparts for feeding, specific digestive enzymes, detoxification mechanisms, and behavioural adaptations (War et al. 2018). This dynamic relationship between insects and plants has driven the coevolution of traits in both groups, resulting in a diverse array of interactions (Sharma et al. 2021).

The interaction between *Asclepias syriaca* (milkweed) and *Danaus plexippus* (monarch butterfly) is a typical example involving coevolution through toxic cardenolides. Milkweed plants produce toxic cardenolides that deter most insect herbivores, while monarch butterflies have evolved to store these toxins in their bodies, making them unpalatable to predators. As a response, milkweed plants have evolved to produce more complex mixtures of cardenolides, making it harder for monarch butterflies to sequester them (Agrawal et al. 2012).

Thus, many herbivores have evolved the ability to break down and detoxify plant toxins. In response, plants have evolved more complex and diverse chemical defences to counteract these adaptations. In some cases, plants have even evolved mutualistic relationships with other organisms, such as parasitoids, ants or fungi, which provide additional protection against herbivores (Ali and Agrawal 2012; Mathur et al. 2013a).

2.3 Plant Adaptations to Insect Herbivory

The coevolution of insects and plants results in plants acquiring adaptations that improve their own chances of surviving and reproducing in an environment with multiple kinds of herbivory. Plant defence against insect herbivory can be broadly classified into two genetic strategies, namely resistance and tolerance mechanisms. These strategies enable plants to protect themselves from damage caused by herbivorous insects through the production of chemical or physical barriers, as well as by regenerating lost or damaged tissues (Agrawal 2000; Stowe et al. 2001).

Resistance mechanisms involve the production of compounds that deter or harm herbivores, such as toxic chemicals, physical barriers or structural defences. These compounds may be constitutive, meaning they are present in the plant at all times, or induced, meaning they are produced in response to herbivory (Karban and Myers 1989). Plants have evolved various physical features, such as resins, wax, silica and lignins, which serve as direct defence mechanisms against herbivorous insects. These morphological traits, together with secondary metabolites, constitute a crucial aspect of the plant–insect interaction (Hanley et al. 2007; Belete 2018).

Tolerance mechanisms involve the ability of plants to recover from herbivory by regenerating damaged tissues, reallocating resources to compensate for lost tissues, or by increasing photosynthesis to produce more energy. Tolerant plants can often withstand higher levels of herbivory without experiencing significant reductions in growth or reproduction (Strauss and Agrawal 1999). For example, some plants may increase the growth rate of undamaged leaves after herbivory to compensate for the loss of damaged leaves (Garcia and Eubanks 2019).

Plants may use a combination of resistance and tolerance mechanisms to defend themselves against insect herbivores. The specific mechanisms used by a plant depend on factors such as the type of herbivore, the intensity of herbivory, and the availability of resources for growth and repair.

In addition, plants employ indirect defence systems such as the production of extrafloral nectaries and volatile chemicals to attract predators and other enemies of herbivores to assist in their defence (Heil and Karban 2010). Over time, the diversity and complexity of plant secondary metabolites have increased, placing increased adaptive pressure on herbivores.

2.3.1 Morphological Features

Plants possess a cuticle covered by epicuticular waxes which form films and crystals (Koch et al. 2004). These waxes provide protection against desiccation and pathogens, as well as increasing the slipperiness of the cuticle, discouraging non-specialized insects from populating leaf surfaces (Muller et al. 2007). The biosynthesis and composition of these waxes vary during plant development, and their physical–chemical properties respond to changes in temperature and season (Howe and Schaller 2008). Even changes in the wax composition due to egg deposition were found to increase fatty acid tetratriacontanoic acid (C34) and a decrease tetracosanoic acid (C24), thereby attracting the egg parasitoid (Blenn et al. 2012).

Moreover, plants have various structures such as thorns and spines that protect them mainly from mammals, and hairs called trichomes that protect against insects (Karban and Myers 1989). When trichomes are removed, herbivorous insects feed and grow more easily. Insect feeding has been observed to cause an increase in trichome density (Fordyce and Agrawal 2001; Mathur et al. 2011).

Scientific evidence suggests that the toughness of leaves can prevent the penetration by insects with piercing-sucking mouthparts, while increasing the wear and tear on the mandibles of herbivores with biting-chewing mouthparts (Raupp et al. 2008). While younger leaves may contain higher levels of chemical defences, mature leaves can be strengthened with various macromolecules such as lignin, cellulose, suberin, and callose, as well as small organic molecules like phenolics and even inorganic silica particles (Schoonhoven et al. 2005). When roots are consumed by insect herbivores, they exhibit significant regrowth in both density and quantity. Furthermore, genotypes with long and fine roots are less susceptible to herbivory than those with short and thick roots (Belete 2018).

Many plant species have laticifers and resin ducts in their vascular tissues that store latex and resins under internal pressure. When the channels are broken, the substances are secreted and can entrap or intoxicate the herbivore (Pickard 2008). Laticifers are found in over 10% of angiosperms and are most common in tropical regions (Agrawal and Konno 2009). Among more than 50 plant families that are known to produce latex, Asclepias, commonly referred to as milkweed, stands out as one of the most extensively studied plant groups (Dussourd and Hoyle 2000; Buttery and Boatman 1976). Specialist herbivores that feed on latex-producing plants can block the flow of latex by cutting veins or trenches in the leaves (Carroll and Hoffman 1980). Conifers produce oleoresins which are mixture of terpenoids and phenolics stored in high pressurized intercellular spaces called ducts (Phillips and Croteau 1999). According to scientific studies, when herbivores cause damage to a tree, the flow of resin can remove stem-boring bark beetles from the borehole (Mumm and Hilker 2006). The resin acids contain highly volatile monoterpenes and sesquiterpenes, which can deter insects while the wound is healing. However, specialist insects have evolved to overcome this defence mechanism by cutting across resin ducts, and some even use the resin as an olfactory cue to locate and select their preferred host plant (Raffa et al. 2016).

2.3.2 Chemical Defences

Plants synthesize a variety of chemical compounds that are categorized into primary and secondary metabolites based on their biological roles. Primary metabolites are essential for basic life processes such as growth, development, and reproduction. Conversely, secondary metabolites, also known as bioactive compounds, serve several ecological functions such as defence against herbivores and microbial pathogens, attraction of pollinators and seed-dispersing animals, and facilitation of competitive and symbiotic interactions with other plants and microbes (Jain et al. 2019; Salam et al. 2023). These chemical compounds have evolved to specifically target the unique biological systems of herbivores, such as their nervous, digestive, and endocrine organs, and can be produced either constitutively or upon induction (Senthil-Nathan 2013; Karban and Myers 1989). Additionally, they contribute significantly to the sensory properties of plants, including their odours, tastes, and colours. Generally, bioactive specialized compounds act as repellents for generalist insects while attracting specialist insects. Toxic compounds are more likely to intoxicate generalist herbivores, while specialists will need to allocate resources to detoxify them, thereby slowing down their growth and development (Kessler and Baldwin 2002; Macel 2011).

More than 100,000 plant secondary compounds, such as phenolics, terpenoids, alkaloids, cyanogenic glucosides, and glucosinolates have been identified (Hadacek 2002; Howe and Schaller 2008), with each plant species produces only a small, but unique combination of these compounds. While many secondary metabolites are found in multiple plant families, some compounds are specific to certain plant families or genera. A good example of this is the Solanaceae family, which contains alkaloids such as nicotine and solanine that are not commonly found in other plant families (Fiesel et al. 2022). Similarly, the Brassicaceae family contains glucosinolates such as sinigrin and brassicanapin that are not typically found in other plant families (Nguyen et al. 2020). The presence of unique secondary metabolites in different plant families can be attributed to their evolution and adaptation to specific environmental pressures and ecological niches. For instance, certain plant families may have evolved unique secondary metabolites to defend themselves against specific herbivores or pathogens that are prevalent in their native habitats.

2.3.2.1 Alkaloids

Alkaloids are bioactive natural products that are widely distributed and can be found in over 15,000 different plants. They are primarily found in Leguminosae, Liliaceae, Solanaceae, and Amaryllidaceae and have evolved as a defence against insect herbivory (Wink 2020). They are characterized by their alkaline properties and nitrogen-containing heterocyclic rings, which are synthesized from amino acids in the roots and then accumulated above ground. The heterocyclic ring structure includes pyridines, pyrroles, indoles, pyrrolidines, isoquinolines, and piperidines. Some alkaloids, such as caffeine and solanidine, are alkaline but not derived from amino acids, while others like mescaline are alkaline and derived from amino acids but do not contain nitrogen in a heterocycle (Thawabteh et al. 2019; Phukan et al. 2023). Plants containing demissine and solanine alkaloids have been found to be resistant to certain herbivores, although some can detoxify them. Pyrrolizidine alkaloids (PAs) occur naturally in many plants as non-toxic N-oxides but become toxic when reduced in the digestive tracts of certain insect herbivores. Some herbivores, such as *Utetheisa ornatrix*, can detoxify PAs and use them as a defence against their own predators (Wink 2019; Bezzerides et al. 2004).

2.3.2.2 Cyanogenic Glucosides

Cyanogenic glucosides (CNglcs) are present in various plant species, including angiosperms, monocotyledons, dicotyledons, gymnosperms, and pteridophytes, with more than 2600 species from over 550 genera and 150 families (Yulvianti and Zidorn 2021). These amino acid derived glucosides come from tyrosine, valine, isoleucine, and phenylalanine and are stored in vacuoles (Gleadow and Møller 2014). Damage of plant tissue due to insect herbivory causes exposure to β-glucosidases leading to hydrolysis and the formation of toxic hydrogen cyanide (HCN) and a ketone or aldehyde (Boter and Diaz 2023). HCN is a potent toxin that can disrupt cellular respiration by inhibiting cytochrome c oxidase in the mitochondrial respiratory pathway. This can result in severe health complications and even death in high enough concentrations (Manoj et al. 2020). In addition to their toxic properties, CNglcs can also serve as nitrogen storage compounds or osmoprotectants, and their presence may increase a plant's resistance against certain herbivores while acting as phagostimulants or oviposition cues for others. However, the production of CNglcs comes at a cost, as it can be energetically expensive and may hinder plant growth and development. Additionally, the release of HCN during CNglc production can harm the plant itself by inhibiting the production of phytoalexins, which are essential for defending against microbial pathogens (Vetter 2000).

2.3.2.3 Glucosinolates

Glucosinolates (GLS) are compounds containing sulphur and nitrogen that are found in Brassicaceae and Capparales. They are derived from amino acids, and there are over 200 different known structures (Ishida et al. 2014; Rosa-Téllez 2023). They are categorized into four groups based on their amino acid precursor of the side chain: aliphatic glucosinolates (50%) derived from methionine, indole glucosinolates (10%) synthesized from tryptophan, aromatic glucosinolates (10%) derived from phenylalanine or tyrosine, and structures synthesized from several different amino acids (30%) or with unknown biosynthetic origin (Hopkins et al. 2009). These compounds are more abundant in roots than in shoots, and different tissues have different dominant glucosinolates (Van Dam et al. 2009). When they are hydrolysed by myrosinases upon tissue disruption, they break down into toxic breakdown products such as isothiocyanates, nitriles, and thiocyanates, which are responsible for the flavours of several vegetable foods (Eisenschmidt-Bönn et al. 2019; Wieczorek et al. 2018). These breakdown products act as both herbivore toxins and feeding repellents (Jeschke et al. 2016). Different insects respond to glucosinolates differently (Mathur et al. 2013b). The metabolic diversity in toxin production by individual plants can also provide defence against herbivores with different feeding strategies or resistance mechanisms (Speed et al. 2015; Wittstock et al. 2003).

2.3.2.4 Phenolics

Phenolics of several classes are synthesized by a wide variety of plants. This group comprises approximately 10,000 distinct chemical entities such as tannins, vanillin, ferulic acid, and caffeic acid (Alamgir and Alamgir 2018; de la Rosa et al. 2019). They can be synthesized using either the malonic acid pathway or the Shikimic acid pathway (Saltveit 2017). Phenols serve numerous functions for plants, such as herbivore defence, pollinator attraction, mechanical support, systemic acquired resistance (SAR), and allelopathy (Divekar et al. 2022). There are multiple examples of phenolic compounds acting as defence mechanisms against insects. They can act as feeding deterrents by reducing the palatability and digestibility of plant tissues or directly inhibit insect growth and development or even cause mortality. Additionally, some phenolics attract natural enemies of insect herbivores, such as parasitoids or predators, which can help to reduce insect populations (Rehman et al. 2012).

2.3.2.5 Terpenoids

Terpenoids are a class of organic compounds synthesized in plants from either acetyl-CoA or glycolytic intermediates. They are classified based on the number of isoprene units they contain, ranging from monoterpenes with ten carbons to polyterpenes with more than 40 carbons (Ninkuu et al. 2021). Terpenes play various roles in plants, including defence against herbivores and pathogens, attraction of pollinators, response to abiotic stress, synthesis of plant hormones, and release of volatile organic compounds (Abbas et al. 2017). These functions contribute to the adaptation and survival of plants in different environments (Aharoni et al. 2005). Essential oils, a blend of volatile monoterpenes and sesquiterpenes with well-known insecticidal properties, are found in many plants. Leafcutter ants are repelled by the terpenoid limonene, which citrus trees generate (Fürstenberg-Hägg et al. 2013). Coniferous plants generate monoterpenes, which are poisonous to many insects, including bark beetles. While certain terpenoid amide derivatives can function as insect juvenile hormone analogs, phytoecdysones, which are steroids found in common fern, prevent insect moulting by imitating moulting hormones (Canals et al. 2005). Additionally, when combined, several terpenoids have synergistic effects that have a higher impact on insect fatality rates.

Many secondary metabolites are constitutively present in a certain amount irrespective of presence of herbivorous insects or other stressors. However, upon herbivore attack, plants may recognize elicitor molecules, which are then transduced into the cell to activate specific genes for the biosynthesis of diverse defensive chemicals with unique chemical activities. This process is mediated by complex cellular transduction mechanisms that regulate the genetic expression of biochemical pathways (Bonaventure et al. 2011; Maffei 2010; Maffei et al. 2012; Mithoefer and Boland 2012). These routes can diversify since they are not required for growth and development (Hartmann 2007). The diversity of metabolic products in plants can be attributed to several molecular processes such as gene and genome duplications, accumulation of point mutations, and multi-loci control. These mechanisms result in variations in gene expression, enzyme activity, and substrate specificity, leading to the production of diverse secondary metabolites (Kroymann 2011; Weng et al. 2012; Moore et al. 2014).

2.4 Insect Adaptations to Plant Defences

Insect herbivores have evolved various adaptations to cope with the complex challenges posed by their plant hosts, including the timing of plant growth and reproduction, the specific nutrient composition, and the chemical and physical defences of the plant. As a consequence of this coevolution, herbivorous insects have developed specialized feeding behaviours and are often restricted to a narrow range of host plants that are related phylogenetically and/or share similar biochemical compositions. This adaptation enables herbivorous insects to efficiently feed on their host plants, while minimizing the risks of consuming toxic or unsuitable plant material (Simon et al. 2015; War et al. 2018).

In response to selective pressure from plants that produce direct defence compounds, specialist herbivores have evolved various defence mechanisms to mitigate the toxicity of these compounds. Unlike generalist herbivores, they have developed the ability to detoxify the hazardous chemicals through enzymatic inactivation or sequestration. This adaptation allows them to tolerate the ingestion of plant material that would otherwise be toxic to them (Nishida 2002; Peng et al. 2007; Petschenka and Agrawal 2016). Additionally, specialized herbivores can use certain compounds, such as phagostimulants, to locate suitable host plants (del Campo et al. 2001; Picaud et al. 2003). Some herbivores can even store these protective compounds and use them for defence against predators and parasites or to attract mates (Cogni et al. 2012). This pattern of specialization highlights the strong evolutionary relationship between host plants and specialized herbivores and has been observed in numerous studies (Becerra 2007; Bandeili and Müller 2010; Richards et al. 2010; Agrawal et al. 2012).

2.4.1 Behavioural Adaptations

Insects have the ability to avoid consuming poisonous plants by utilizing either genetically predetermined or learned avoidance mechanisms that are triggered through visual, olfactory, or tactile exposure (Chapman 2003). Female insects possess genetically programmed oviposition behaviour that can prevent them from laying eggs on unsuitable plants (Fox et al. 2004). However, in some cases, larvae may need to migrate to locate a suitable host plant (Cotter and Edwards 2006). Furthermore, insects may consume non-toxic parts of plants or seek out environments that are free from toxins, as well as utilizing plants that are not currently producing toxins (Nealis and Nault 2005). Insects can exhibit diverse responses to toxins depending on their surrounding environmental conditions. For instance, solitary phase locusts are deterred by a bitter-tasting plant alkaloid, whereas gregarious-phase locusts are attracted to it. This difference in feeding behaviour is reflective of various anti-predator strategies, such as changes in colouring and grouping tendencies (Glendinning et al. 2002; Despland 2021; Després et al. 2007).

Insects utilize chemosensation as a means of detecting and avoiding potentially harmful secondary compounds present in their environment. Gustatory and olfactory receptor neurons allow insects to detect these poisons through their senses of taste, smell, and touch. Gustatory receptors are categorized into sweet, bitter, umami, salt, and carbon dioxide receptors, and bitter receptors are involved in sensing secondary compounds that inhibit insect intake by activating downstream signaling pathways. This aversive mechanism may be inherited or acquired (Zunjarrao et al. 2020). Some insects limit their diet to organs of non-toxic plants or eat only on plants with low levels of toxins to avoid hazardous substances (Fox et al. 2004). However, insects also avoid bitter substances even though they are not always harmful, which limits their host plant options and increases the cost of avoidance (Pelden and Meesawat 2019). To circumvent plant defences, insects have developed various strategies, such as recognizing previously induced reactions, feeding on diverse plant tissues, weaving silk over spines, making trenches across leaves before eating, and puncturing leaf veins to block the passage of poisons (Perkins et al. 2013).

2.4.2 Manipulation of Plant Chemical Defences

While avoiding plant allelochemicals is generally the most effective way for insects to minimize their exposure to harmful compounds, they have evolved additional mechanisms to cope with occasional consumption of toxic substances.

Many insects can affect a plant's chemical defences by making tunnels in leaves to release pressure in secretory canals and remove poisonous exudates near their feeding spot (Helmus and Dussourd 2005; Becerra 2003). Insects have developed the ability to inhibit the plant's defence mechanisms by secreting elicitors that can lower the expression of regulatory genes that are crucial for initiating the plant's defences. This allows the insects to better exploit the plant and continue feeding on it without triggering a defensive response (Bede et al. 2006; Divekar et al. 2022). Gall-inducing insects, such as sawflies, have developed the ability to reduce the levels of toxic phenolic chemicals found in the galls where their larvae grow. This adaptation is thought to facilitate frequent switching between host plants, which can increase the survival and reproductive success of these insects (Nyman and Julkunen-Tiitto 2000). By reducing the levels of toxic chemicals in their feeding sites, gall-inducing insects can better tolerate a wider range of plant species and exploit new host plants as they become available, ultimately enhancing their ecological flexibility and evolutionary potential.

2.4.3 Sequestration of Plant Defence Compounds by Insects

Sequestration is a common defence strategy employed by insects that can have significant impacts on their interactions with plants. This strategy involves the selective absorption, transport, endogenous metabolism, and excretion of plant chemicals into different organs of the insect. Through this process, insects can accumulate and store large quantities of plant compounds, often in specialized tissues, for use in a variety of contexts, such as chemical defence against predators or as a source of nutrients. The sequestration of plant chemicals can also lead to coevolutionary dynamics between insects and plants, as insects may evolve mechanisms to overcome or exploit plant defences, while plants may evolve counter-strategies to defend against insect herbivory (Pentzold et al. 2014; Nishida 2002). Insects are known to sequester a diverse array of compounds from their host plants, including aromatic compounds, nitrogen-containing secondary compounds such as alkaloids, cyanogenic glycosides, glucosinolates, and isoprenoids such as cardiac glycosides, cucurbitacins and iridoid glycosides (Opitz and Müller 2009). These sequestered compounds can serve various functions, including as defensive agents against predators or parasites, as feeding stimulants or repellents, and as precursors for the biosynthesis of sex pheromones or other signaling molecules (Dobler 2001). The specific suite of sequestered compounds can vary widely among insect taxa, reflecting differences in the evolutionary history, feeding behaviour, and ecological pressures faced by each group.

Studies have revealed numerous mechanisms by which insects can acquire and store plant chemicals, including a shift from de novo synthesis of detoxifying compounds to sequestration of host phytochemicals. Sequestration is widespread in several insect groups because it is thought to have a lower metabolic cost than de novo production (Beran and Petschenka 2022). In spite of this, there has been limited experimental research on it since the evolutionary background and natural history of the study system aren't well known.

Insects that sequester defence compounds from plants have evolved diverse mechanisms to accumulate these compounds in various parts of their bodies, such as the haemolymph, defence glands, or integument. Insects absorb plant defence chemicals from the gut lumen into the hemocoel through the peritrophic matrix and gut epithelium. The type of absorption mechanism depends on the chemical properties of the substances, with absorption occurring either through passive diffusion or carrier-mediated transport. The permeability of the epithelial layer also plays a role in absorption and may be modulated by efflux pumps that restrict the uptake of certain substances (Dobler et al. 2015; Sorensen and Dearing 2006). This selective accumulation is achieved through processes such as selective absorption through the gut, transport within the body, endogenous metabolism, and excretion via the malpighian tubules (Petschenka and Agrawal 2016). The cardiac glycoside digitoxin is one example. It was discovered to passively diffuse over the midgut of the milkweed insect *Oncopeltus fasciatus*, but not across the midguts of the desert locust *Schistocerca gregaria* and the American cockroach, *Periplaneta americana* (Scudder and Meredith 1982). This suggests that mechanisms that either aid or hinder the absorption of plant defence compounds across the gut are a key factor in the development of sequestration. It was demonstrated that the cytochrome P450 monooxygenases, a different group of membrane proteins, are implicated in nicotine transportation from the gut into the haemolymph of the tobacco hornworm, *Manduca sexta* (Kumar et al. 2014).

In addition, insects may modify the amounts and composition of sequestered defence chemicals based on the chemical composition of their food plant. For example, some insects can selectively sequester certain types of glucosinolates or iridoid glycosides depending on the plant species they consume (Baden et al. 2013; Beran et al. 2014; Yang et al. 2020). Nevertheless, because the processes of sequestration have only been studied in a small number of insect species, the significance of each of these aspects is not entirely known.

2.4.4 Detoxification

Insects have developed sophisticated detoxification mechanisms to biotransform and excrete secondary metabolites, which have the potential to cause toxicological harm (Erb and Robert 2016). This enables insects to minimize the deleterious effects of these compounds and maintain their ability to feed on plants that contain secondary metabolites. Some insects have specialized gut enzymes or gut microorganisms that detoxify ingested plant chemicals, allowing them to consume otherwise toxic plant material (van den Bosch and Welte 2017).

Insects possess a variety of enzymes that aid in detoxifying plant toxins through oxidation, reduction, hydrolysis, or conjugation of molecules (Birnbaum and Abbot 2018). The three primary types of enzymes involved in detoxification are cytochrome P450 monooxygenases (also known as CYPs or P450s), glutathione S-transferases (GSTs), and carboxylesterases (COEs) (Feyereisen 2005). Berenbaum and Zangerl (1998) demonstrated that the ability of insects to metabolize furanocoumarins, a class of compounds produced by plants, was found to be associated with the varying levels of furanocoumarin production among different plant species. This suggests a tight genetic linkage between the genes responsible for the insect's capacity to utilize host plants and those involved in metabolizing these chemical compounds. The cytochrome P450-dependent monooxygenase gene is a notable example of a gene implicated in both plant–insect interactions and insect physiology. Insects, such as *Papilio polyxenes*, have developed adaptations to thrive on host plants that contain toxins by undergoing diversification of P450 enzymes, which play a key role in the detoxification process (Scott and Wen 2001).

2.4.5 Further Processing of Plant Toxins

A multitude of insects have developed the ability to eliminate or excrete detrimental plant compounds, and in some instances, these compounds are sequestered and utilized for diverse purposes, such as synthesizing pheromones, defending against infections or predators, and generating pigments for adult coloration (Beran and Petschenka 2022; Robinson et al. 2023). Selective transport and storage mechanisms that keep the poison from interfering with the insect's physiological functions are necessary for sequestration (Kuhn et al. 2004). Molecular studies on leaf beetles have shown that a minor modification in ancestral defence mechanism has facilitated the utilization of plant components for predator defence. This adaptation has emerged convergently in multiple leaf beetle lineages and is metabolically advantageous due to its low energy cost (Kuhn et al. 2004; Termonia et al. 2001).

2.5 Involvement of Higher Trophic Level

Insects that feed on plants release volatile organic compounds (VOCs) and extrafloral nectar as a signal to be found and attacked by parasitoids and other natural enemies of the insect (Heil 2008, 2015; Shah et al. 2015). Insect-caused plant damage results in the induction of volatile and nectar and extrafloral nectar secretion to entice the herbivore's natural enemies, such as ants, wasps, and parasitoids (Heil 2011, 2015; Mathur et al. 2013a, b). These natural enemies may exert selective pressure on insect herbivores, resulting in the evolution of various traits, such as behavioural changes or morphological adaptations that enable herbivores to avoid predation (Dicke and Baldwin 2010). These changes, in turn, can influence herbivore feeding behaviour and plant defence strategies, potentially leading to coevolutionary interactions between herbivores and their host plants (Vale et al. 2018). In contrast, when predators consume herbivores that are not well adapted to the plant's chemical defences, it can lead to the selection for herbivores that can more efficiently detoxify or sequester plant compounds. This can result in an arms race between herbivores and plants, where plants evolve more potent chemical defences, and herbivores continue to adapt their detoxification and sequestration mechanisms to overcome them. The pressure from predators can, therefore, indirectly impact the coevolutionary dynamics between herbivores and their host plants, driving the evolution of novel traits that can enhance herbivore fitness and promote coevolutionary interactions between plants and insects.

2.6 Role of Symbiotic Microorganisms

Insects and plants have a mutually beneficial relationship with their symbiotic microorganisms. This relationship creates a complex web of interactions that have both ecological and evolutionary implications. These microbes are critical components of the intricate relationships among plants, insects, and their environments. The interaction between insects and plants involves an indirect interaction with the microorganisms that inhabit each other. In other words, insects and plants have a shared relationship with the microorganisms living within them, which plays a critical role in shaping their interactions and the broader ecological and evolutionary consequences (Sharma et al. 2021; Sugio et al. 2015). The microorganisms that inhabit both plants and insects are highly diverse and can be found in various parts of their respective hosts. Endophytes, microorganisms that live within the tissues of plants, can be found in the roots, stem, leaves, seeds, and fruits of plants, while insects harbour the symbiotic microorganisms throughout their bodies (Compant et al. 2021; Provorov and Onishchuk 2018).

Endophytes are known to play a crucial role in insect–plant interactions by influencing the behaviour, fitness, and survival of their insect hosts (Shikano et al. 2017; Grunseich et al. 2019). Endophytes can produce various secondary metabolites that have insecticidal properties, which can help to deter herbivorous insects from feeding on the plant. The interdependent relationship between plants and endophytes has been demonstrated to promote plant growth and improve their physiological condition, as well as providing a vital defence mechanism against harsh environmental conditions (Malinowski et al. 2000; Peschiutta et al. 2018; Mathur and Ulanova 2022). They contribute to not only growth-promoting phytohormones such as indole acetic acid and gibberellic acids, but also various metabolic compounds such as alkaloids, terpenoids, and flavonoids (Mukherjee et al. 2022; Sharma et al. 2023; Tan and Zou 2001). Endophytes can also influence the nutritional quality of the plant for insect herbivores, by altering the levels of carbohydrates, proteins, and other nutrients in the plant tissue. This can affect the performance and development of herbivorous insects that feed on the plant. Furthermore, endophytes can also confer resistance to environmental stressors such as drought, salinity, and temperature fluctuations, which can improve the overall health and survival altering the levels of carbohydrates, proteins, and other nutrients in the plant tissue and its associated insect community (White et al. 2019). Root-associated microorganisms also contribute to the increased emission of volatiles. The rhizospheric microbiome is susceptible to changes caused by herbivory, which, in turn, can affect the relationship between the plant microbiome and the herbivore (Venturi and Keel 2016).

Insects harbour microorganisms both internally and externally that may be divided into two categories: primary and secondary symbionts. Primary symbionts are crucial for the survival and reproduction of insects and are typically obligatory and vertically transferred from mother to offspring (Bright and Bulgheresi 2010; Szklarzewicz and Michalik 2017). Secondary symbionts, on the other hand, are not essential for host survival and can reside in various insect tissues. They provide a range of benefits to their hosts, such as protection against environmental stresses, natural enemies, and plant toxins (Brownlie and Johnson 2009; van den Bosch and Welte 2017). Secondary symbionts reproduce only in association with their hosts, but they can undergo horizontal transfer. Many studies have highlighted the importance of secondary endosymbionts in insect–plant interactions, including recognition, defence, digestion, and fitness of the insect (Frago et al. 2012; Paniagua Voirol et al. 2018).

Microbial symbionts play a crucial role in insect–plant interactions in several ways. They are known to aid in the recognition of host plants by altering the sensory abilities of insects, such as their ability to detect plant volatiles. Moreover, they can provide protection against natural enemies of insects, such as predators, parasites, and pathogens, which can increase the survival and fitness of the insect. They also help insects to digest plant material and extract nutrients from it, which can be difficult to break down without the aid of microorganisms. Additionally, they confer resistance against plant toxins, allowing the insect to consume a wider range of plant species. These symbiotic microorganisms can also contribute to the overall health and fitness of the insect, which can affect its reproductive success and ability to survive in its environment (Frago et al. 2012; Frago and Zytynska 2023; Hansen and Moran 2014). Overall, microbial symbionts have a significant impact on the interactions between insects and plants, and understanding these relationships can provide insights into the ecology and evolution of both groups of organisms.

2.7 Role of Rhizosphere in Shaping Insect–Plant Interactions

Rhizosphere is a complex and dynamic system that plays a crucial role in shaping insect–plant interactions. One way in which the rhizosphere influences these interactions is by providing a habitat for beneficial microorganisms, such as bacteria and fungi, that can promote plant growth and health. These microorganisms can produce plant growth-promoting substances, such as phytohormones and enzymes, that stimulate plant growth and increase the nutritional quality of the plant for herbivorous insects (Berendsen et al. 2012). As a result, plants growing in healthy and diverse rhizospheres can be more attractive to herbivorous insects, which can lead to increased herbivory and pest pressure. However, these rhizosphere microbes can modulate insect herbivory in different manners and are powerful drivers of insect–plant coevolution (Rasmann et al. 2017; Van der Putten et al. 2001).

In contrast, studies have reported that by creating volatiles, the microbial populations found in the roots help plants survive a variety of biotic and abiotic stressors (Garbeva and Weisskopf 2020). Depending on the insect's feeding guild and specialization, these microorganisms have varying effects on insect herbivory (Pineda et al. 2010). By boosting sugar and protein production, causing metabolic changes, and lowering root pathogen infestation, these bacteria give the plant a selective advantage during insect attacks (Westman et al. 2019). Studies have demonstrated that in order to survive without their insect prey, soil-dwelling entomopathogenic fungi like *Metarhizium* and *Beauveria* have evolved a symbiotic association with plants (Hu and St. Leger 2002). It is hypothesized that other fungi connected to plants may have provided these microorganisms with the genes for insect disease. Rhizospheric microbial communities, such as PGPR, PSB, and VAM, have been reported to boost the antioxidant activity of plants after insect damage (Song et al. 2014; Kousar et al. 2020; Sharma and Mathur 2020).

Thus, interactions between plants, insects, and the rhizosphere can provide insights into the ecology and evolution of these organisms and can help to develop sustainable strategies for managing insect pests in agricultural and natural ecosystems.

2.8 Conclusion

Over the period of 400 million years, several strategies for dealing with chemical defences of their hosts have been developed by insects that feed on plants. The response of plants to insect herbivores has significant implications not only for the interacting players themselves but also for the interactions between responding plants and other organisms associated with plants. This coevolutionary arms race between plants and insect herbivores can have significant ecological consequences. It can influence the structure and diversity of plant communities, as well as the evolution of insect herbivore species. The interactions between plants, insects, and their biotic and abiotic factors such as parasitoids and predators, endosymbionts, and the rhizosphere associated with them are shaped by coevolutionary processes, which have driven the adaptation of these organisms to each other over time. Plants and insects have co-evolved complex mechanisms of defence and counter-defence that allow them to interact in a dynamic and constantly evolving way. It can also have economic impacts, as some insect herbivores are pests that can cause significant damage to crops.

In summary, the coevolutionary arms race between plants and insect herbivores is an ongoing process of adaptation and counter-adaptation, where each species evolves in response to the adaptations of the other. This process can have significant ecological and economic implications and provides a fascinating example of the complexity of evolutionary interactions between species. Understanding the coevolutionary dynamics between plants, insects, and the rhizosphere can provide valuable insights into the ecology and evolution of these organisms and can help to develop sustainable strategies for managing insect pests in agricultural and natural ecosystems.

References

Abbas F, Ke Y, Yu R, Yue Y, Amanullah S, Jahangir MM, Fan Y (2017) Volatile terpenoids: multiple functions, biosynthesis, modulation and manipulation by genetic engineering. Planta 246:803–816

Agrawal AA (2000) Benefits and costs of induced plant defense for *Lepidium virginicum* (Brassicaceae). Ecology 81:1804–1813

Agrawal AA, Karban R (1997) Domatia mediate plant arthropod mutualism. Nature 387:562–563

Agrawal AA, Konno K (2009) Latex: a model for understanding mechanisms ecology and evolution of plant defense against herbivory. Annu Rev Ecol Evol Syst 40:311–331

Agrawal AA, Petschenka G, Bingham RA, Weber MG, Rasmann S (2012) Toxic cardenolides: chemical ecology and coevolution of specialized plant–herbivore interactions. New Phytol 194: 28–45

Aharoni A, Jongsma MA, Bouwmeester HJ (2005) Volatile science? Metabolic engineering of terpenoids in plants. Trends Plant Sci 10(12):594–602

Alamgir ANM, Alamgir ANM (2018) Secondary metabolites: secondary metabolic products consisting of C and H; C, H, and O; N, S, and P elements; and O/N heterocycles. In: Therapeutic use of medicinal plants and their extracts: volume 2: phytochemistry and bioactive compounds, pp 165–309

Ali JG, Agrawal AA (2012) Specialist versus generalist insect herbivores and plant defense. Trends Plant Sci 17:293–302

Baden CU, Franke S, Dobler S (2013) Host dependent iridoid glycoside sequestration patterns in *Cionus hortulanus*. J Chem Ecol 39:1112–1114

Bandeili B, Müller C (2010) Folivory versus florivory—adaptiveness of flower feeding. Naturwissenschaften 97:79–88

Becerra JX (2003) Synchronous coadaptation in an ancient case of herbivory. Proc Natl Acad Sci 100:12804–12807

Becerra JX (2007) The impact of herbivore–plant coevolution on plant community structure. Proc Natl Acad Sci 104:7483–7488

Bede JC, Musser RO, Felton GW, Korth KL (2006) Caterpillar herbivory and salivary enzymes decrease transcript levels of *Medicago truncatula* genes encoding early enzymes in terpenoid biosynthesis. Plant Mol Biol 60:519–531

Belete T (2018) Defense mechanisms of plants to insect pests: from morphological to biochemical approach. Trends Tech Sci Res 2:555–584

Beran F, Petschenka G (2022) Sequestration of plant defense compounds by insects: from mechanisms to insect–plant coevolution. Annu Rev Entomol 67:163–180

Beran F, Pauchet Y, Kunert G, Reichelt M, Wielsch N, Vogel H, Heckel DG (2014) *Phyllotreta striolata* flea beetles use host plant defense compounds to create their own glucosinolate-myrosinase system. Proc Natl Acad Sci 111:7349–7354

Berenbaum MR, Zangerl AR (1998) Chemical phenotype matching between a plant and its insect herbivore. Proc Natl Acad Sci 95:13743–13748

Berendsen RL, Pieterse CM, Bakker PA (2012) The rhizosphere microbiome and plant health. Trends Plant Sci 17(8):478–486

Bezzerides A, Yong TH, Bezzerides J, Husseini J, Ladau J, Eisner M, Eisner T (2004) Plant-derived pyrrolizidine alkaloid protects eggs of a moth (Utetheisa ornatrix) against a parasitoid wasp (Trichogramma ostriniae). Proc Natl Acad Sci 101(24):9029–9032

Birnbaum SSL, Abbot P (2018) Insect adaptations toward plant toxins in milkweed–herbivores systems—a review. Entomol Exp Appl 166:357–366

Blenn B, Bandoly M, Küffner A, Otte T, Geiselhardt S, Fatouros NE, Hilker M (2012) Insect egg deposition induces indirect defense and epicuticular wax changes in *Arabidopsis thaliana*. J Chem Ecol 38:882–892

Bonaventure G, VanDoorn A, Baldwin IT (2011) Herbivore-associated elicitors: FAC signaling and metabolism. Trends Plant Sci 16(6):294–299

Boter M, Diaz I (2023) Cyanogenesis, a plant defence strategy against herbivores. Int J Mol Sci 24 (8):6982

Bright M, Bulgheresi S (2010) A complex journey: transmission of microbial symbionts. Nat Rev Microbiol 8(3):218–230

Bronstein JL, Huxman TE (1997) The evolution of plant–insect mutualisms. New Phytol 136:373–384

Bronstein JL, Alarcón R, Geber M (2006) The evolution of plant–insect mutualisms. New Phytol 172:412–428

Brownlie JC, Johnson KN (2009) Symbiont-mediated protection in insect hosts. Trends Microbiol 17(8):348–354

Buttery BR, Boatman SG (1976) Water deficits and flow of latex in soil water measurement plant responses and breeding for drought resistance. Academic Press, pp 233–289

Canals D, Irurre-Santilari J, Casas J (2005) The first cytochrome P450 in ferns: evidence for its involvement in phytoecdysteroid biosynthesis in Polypodium vulgare. FEBS J 272(18):4817–4825

Carroll CR, Hoffman CA (1980) Chemical feeding deterrent mobilized in response to insect herbivory and counter adaptation by *Epilachna tredecimnotata*. Science 209:414–416

Chapman RF (2003) Contact chemoreception in feeding by phytophagous insects. Annu Rev Entomol 48:455–484

Chittka L, Raine NE (2006) Recognition of flowers by pollinators. Curr Opin Plant Biol 9:428–435

Cogni R, Trigo JR, Futuyma DJ (2012) A free lunch? No cost for acquiring defensive plant pyrrolizidine alkaloids in a specialist arctiid moth (*Utetheisa ornatrix*). Mol Ecol 21:6152–6162

Compant S, Cambon MC, Vacher C, Mitter B, Samad A, Sessitsch A (2021) The plant endosphere world—bacterial life within plants. Environ Microbiol 23(4):1812–1829

Cotter SC, Edwards OR (2006) Quantitative genetics of preference and performance on chickpeas in the noctuid moth, *Helicoverpa armigera*. Heredity 96:396–402

de la Rosa LA, Moreno-Escamilla JO, Rodrigo-García J, Alvarez-Parrilla E (2019) Phenolic compounds. In: Postharvest physiology and biochemistry of fruits and vegetables. Woodhead Publishing, pp 253–271

del Campo ML, Miles CI, Schroeder FC, Mueller C, Booker R, Renwick JA (2001) Host recognition by the tobacco hornworm is mediated by a host plant compound. Nature 411:186–189

Despland E (2021) Selection forces driving herding of herbivorous insect larvae. Front Ecol Evol 9: 760–806

Després L, David JP, Gallet C (2007) The evolutionary ecology of insect resistance to plant chemicals. Trends Ecol Evol 22:298–307

Dicke M, Baldwin IT (2010) The evolutionary context for herbivore-induced plant volatiles: beyond the 'cry for help'. Trends Plant Sci 15:167–175

Divekar PA, Narayana S, Divekar BA, Kumar R, Gadratagi BG, Ray A, Behera TK (2022) Plant secondary metabolites as defense tools against herbivores for sustainable crop protection. Int J Mol Sci 23:2690

Dobler S (2001) Evolutionary aspects of defense by recycled plant compounds in herbivorous insects. Basic Appl Ecol 2:15–26

Dobler S, Petschenka G, Wagschal V, Flacht L (2015) Convergent adaptive evolution—how insects master the challenge of cardiac glycoside-containing host plants. Entomol Exp Appl 157:30–39

Dussourd DE, Hoyle AM (2000) Poisoned plusiines: toxicity of milkweed latex and cardenolides to some generalist caterpillars. Chemoecology 10:11–16

Ehrlich PR, Raven PH (1964) Butterflies and plants: a study in coevolution. Evolution 18:586–608

Eisenschmidt-Bönn D, Schneegans N, Backenköhler A, Wittstock U, Brandt W (2019) Structural diversification during glucosinolate breakdown: mechanisms of thiocyanate, epithionitrile and simple nitrile formation. Plant J 99(2):329–343

Erb M, Robert CA (2016) Sequestration of plant secondary metabolites by insect herbivores: molecular mechanisms and ecological consequences. Curr Opin Insect Sci 14:8–11

Feyereisen R (2005) Insect P450. In: Comprehensive molecule insect science. Elsevier, Oxford

Fiesel PD, Parks HM, Last RL, Barry CS (2022) Fruity, sticky, stinky, spicy, bitter, addictive, and deadly: evolutionary signatures of metabolic complexity in the Solanaceae. Nat Prod Rep 39 (7):1438–1464

Fordyce JA, Agrawal AA (2001) The role of plant trichomes and caterpillar group size on growth and defence of the pipevine swallowtail *Battus philenor*. J Anim Ecol 70:997–1005

Fox CW, Stillwell RC, Amarillo-s AR, Czesak ME, Messina FJ (2004) Genetic architecture of population differences in oviposition behaviour of the seed beetle *Callosobruchus maculatus*. J Evol Biol 17:1141–1151

Frago E, Zytynska S (2023) Impact of herbivore symbionts on parasitoid foraging behaviour. Curr Opin Insect Sci 57:101027

Frago E, Dicke M, Godfray HCJ (2012) Insect symbionts as hidden players in insect–plant interactions. Trends Ecol Evol 27(12):705–711

Fürstenberg-Hägg J, Zagrobelny M, Bak S (2013) Plant defense against insect herbivores. Int J Mol Sci 14(5):10242–10297

Garbeva P, Weisskopf L (2020) Airborne medicine: bacterial volatiles and their influence on plant health. New Phytol 226(1):32–43

Garcia LC, Eubanks MD (2019) Overcompensation for insect herbivory: a review and meta-analysis of the evidence. Ecology 100:e02585

Gatehouse JA (2002) Plant resistance towards insect herbivores: a dynamic interaction. New Phytol 156:145–169

Gleadow RM, Møller BL (2014) Cyanogenic glycosides: synthesis, physiology, and phenotypic plasticity. Annu Rev Plant Biol 65:155–185

Glendinning JI, Davis A, Ramaswamy S (2002) Contribution of different taste cells and signaling pathways to the discrimination of "bitter" taste stimuli by an insect. J Neurosci 22:7281–7287

Grunseich JM, Thompson MN, Aguirre NM, Helms AM (2019) The role of plant-associated microbes in mediating host-plant selection by insect herbivores. Plants (Basel) 9(1):6

Hadacek F (2002) Secondary metabolites as plant traits: current assessment and future perspectives. Crit Rev Plant Sci 21:273–322

Hanley ME, Lamont BB, Fairbanks MM, Rafferty CM (2007) Plant structural traits and their role in anti-herbivore defence. Perspect Plant Ecol Evol Syst 8:157–178

Hansen AK, Moran NA (2014) The impact of microbial symbionts on host plant utilization by herbivorous insects. Mol Ecol 23(6):1473–1496

Hartmann T (2007) From waste products to ecochemicals: fifty years research of plant secondary metabolism. Phytochemistry 68(22–24):2831–2846

Heil M (2011) Nectar: generation regulation and ecological functions. Trends Plant Sci 16:191–200

Heil M (2015) Extrafloral nectar at the plant-insect interface: a spotlight on chemical ecology, phenotypic plasticity, and food webs. Annu Rev Entomol 60:213–232

Heil M, Karban R (2010) Explaining evolution of plant communication by airborne signals. Trends Ecol Evol 25:137–144

Heil M, McKey D (2003) Protective ant-plant interactions as model systems in ecological and evolutionary research. Annu Rev Ecol Evol Syst 34:425–553

Helmus MR, Dussourd DE (2005) Glues or poisons: which triggers vein cutting by monarch caterpillars? Chemoecology 15:45–49

Hochberg ME, Gomulkiewicz R, Holt RD, Thompson JN (2000) Weak sinks could cradle mutualistic symbioses-strong sources should harbour parasitic symbioses. J Evol Biol 13:213–222

Hopkins RJ, van Dam NM, van Loon JJ (2009) Role of glucosinolates in insect-plant relationships and multitrophic interactions. Annu Rev Entomol 54:57–83

Howe GA, Jander G (2008) Plant immunity to insect herbivores. Annu Rev Plant Biol 59:41–66

Howe GA, Schaller A (2008) Direct defenses in plants and their induction by wounding and insect herbivores. In: Induced plant resistance to herbivory. Springer, Dordrecht, pp 7–29

Hu G, St. Leger RJ (2002) Field studies using a recombinant mycoinsecticide (*Metarhizium anisopliae*) reveal that it is rhizosphere competent. Appl Environ Microbiol 68(12):6383–6387

Kroymann J (2011) Natural diversity and adaptation in plant secondary metabolism. Curr Opin Plant Biol 14(3):246–251

Ishida M, Hara M, Fukino N, Kakizaki T, Morimitsu Y (2014) Glucosinolate metabolism, functionality and breeding for the improvement of Brassicaceae vegetables. Breed Sci 64(1):48–59

Jain C, Khatana S, Vijayvergia R (2019) Bioactivity of secondary metabolites of various plants: a review. Int J Pharm Sci Res 10(2):494–504

Jansen-González S, Teixeira SDP, Pereira RAS (2012) Mutualism from the inside: coordinated development of plant and insect in an active pollinating fig wasp. Arthropod Plant Interact 6: 601–609

Janzen DH (1980) When is it coevolution? Evolution 34:611–612

Jeschke V, Gershenzon J, Vassão DG (2016) Insect detoxification of glucosinolates and their hydrolysis products. In: Advances in botanical research, vol 80. Academic Press, pp 199–245

Johnson SD, Steiner KE (2000) Long-tongued fly pollination and evolution of floral spur length in the *Disadraconis* complex (Orchidaceae). Evolution 54:1713–1722

Karban R, Myers JH (1989) Induced plant responses to herbivory. Annu Rev Ecol Syst 20(1):331–348

Kessler A, Baldwin IT (2002) Plant responses to insect herbivory: the emerging molecular analysis. Annu Rev Plant Biol 53(1):299–328

Koch K, Neinhuis C, Ensikat HJ, Barthlott W (2004) Self-assembly of epicuticular waxes on living plant surfaces imaged by atomic force microscopy (AFM). J Exp Bot 55:711–718

Kousar B, Bano A, Khan N (2020) PGPR modulation of secondary metabolites in tomato infested with *Spodoptera litura*. Agronomy 10(6):778

Krenn HW (2019) Insect mouthparts: form, function, development and performance, vol 5. Springer Nature, Cham

Kuhn J, Pettersson EM, Feld BK, Burse A, Termonia A, Pasteels JM, Boland W (2004) Selective transport systems mediate sequestration of plant glucosides in leaf beetles: a molecular basis for adaptation and evolution. Proc Natl Acad Sci 101:13808–13813

Kumar P, Rathi P, Schöttner M, Baldwin IT, Pandit S (2014) Differences in nicotine metabolism of two *Nicotiana attenuata* herbivores render them differentially susceptible to a common native predator. PLoS One 9(4):e95982

Labandeira CC (1998) Early history of arthropod and vascular plant associations. Annu Rev Earth Plants Sci 26(1):329–377

Labandeira CC (2013) A paleobiologic perspective on plant–insect interactions. Curr Opin Plant Biol 16:414–421
Macel M (2011) Attract and deter: a dual role for pyrrolizidine alkaloids in plant–insect interactions. Phytochem Rev 10:75–82
Maffei ME (2010) Sites of synthesis, biochemistry and functional role of plant volatiles. S Afr J Bot 76(4):612–631
Maffei ME, Arimura GI, Mithöfer A (2012) Natural elicitors, effectors and modulators of plant responses. Nat Prod Rep 29(11):1288–1303
Malinowski DP, Alloush GA, Belesky DP (2000) Leaf endophyte *Neotyphodium coenophialum* modifies mineral uptake in tall fescue. Plant Soil 227(1–2):115–126
Manoj SR, Karthik C, Kadirvelu K, Arulselvi PI, Shanmugasundaram T, Bruno B, Rajkumar M (2020) Understanding the molecular mechanisms for the enhanced phytoremediation of heavy metals through plant growth promoting rhizobacteria: a review. J Environ Manag 254:109779
Mathur V, Ulanova D (2022) Microbial metabolites beneficial to plant hosts across ecosystems. Microb Ecol 86:25–48
Mathur V, Ganta S, Raaijmakers CE, Reddy AS, Vet LE, van Dam NM (2011) Temporal dynamics of herbivore-induced responses in *Brassica juncea* and their effect on generalist and specialist herbivores. Entomol Exp Appl 139:215–225
Mathur V, Tytgat TO, Hordijk CA, Harhangi HR, Jansen JJ, Reddy AS, Van Dam NM (2013a) An ecogenomic analysis of herbivore-induced plant volatiles in *Brassica juncea*. Mol Ecol 22(24):6179–6196
Mathur V, Wagenaar R, Caissard JC, Reddy AS, Vet LE, Cortesero AM, Van Dam NM (2013b) A novel indirect defence in Brassicaceae: structure and function of extrafloral nectaries in *Brassica juncea*. Plant Cell Environ 36:528–541
Mithöfer A, Boland W (2012) Plant defense against herbivores: chemical aspects. Annu Rev Plant Biol 63:431–450
Moore BD, Andrew RL, Külheim C, Foley WJ (2014) Explaining intraspecific diversity in plant secondary metabolites in an ecological context. New Phytol 201(3):733–750
Mukherjee A, Gaurav AK, Singh S, Yadav S, Bhowmick S, Abeysinghe S, Verma JP (2022) The bioactive potential of phytohormones: a review. Biotechnol Rep 35:e00748
Muller B, Bourdais G, Reidy B, Bencivenni C, Massonneau A, Condamine P, Rolland G, Conéjéro G, Rogowsky P, Tardieu F (2007) Association of specific expansins with growth in maize leaves is maintained under environmental, genetic, and developmental sources of variation. Plant Physiol 143:278–290
Mumm R, Hilker M (2006) Direct and indirect chemical defence of pine against folivorous insects. Trends Plant Sci 11:351–358
Nealis VG, Nault JR (2005) Seasonal changes in foliar terpenes indicate suitability of Douglas-fir buds for western spruce budworm. J Chem Ecol 31:683–696
Nelson AS, Acosta NC, Mooney KA (2019) Plant chemical mediation of ant behavior. Curr Opin Insect Sci 32:98–103
Nguyen VT, Stewart J, Lopez M, Ioannou I, Allais F (2020) Glucosinolates: natural occurrence, biosynthesis, accessibility, isolation, structures, and biological activities. Molecules 25 (19):4537
Ninkuu V, Zhang L, Yan J, Fu Z, Yang T, Zeng H (2021) Biochemistry of terpenes and recent advances in plant protection. Int J Mol Sci 22(11):5710
Nishida R (2002) Sequestration of defensive substances from plants by *Lepidoptera*. Annu Rev Entomol 47:57–92
Nyman T, Julkunen-Tiitto R (2000) Manipulation of the phenolic chemistry of willows by gall-inducing sawflies. Proc Natl Acad Sci 97:13184–13187
Opitz SE, Müller C (2009) Plant chemistry and insect sequestration. Chemoecology 19:117–154
Panda N, Khush GS (1995) Host plant resistance to insects. CAB International, Wallingford
Paniagua Voirol LR, Frago E, Kaltenpoth M, Hilker M, Fatouros NE (2018) Bacterial symbionts in Lepidoptera: their diversity, transmission, and impact on the host. Front Microbiol 9:556

Pelden D, Meesawat U (2019) Foliar idioblasts in different-aged leaves of a medicinal plant (*Annona muricata* L.). Songklanakarin J Sci Technol 41:414–420

Peng J, Li Z, Huang Y (2007) Plant species specific defense signal communication differentially regulates glutathione S-transferase activity and gene expression in the *Helicoverpa armigera* (Hubner). J Plant Interact 2:93–99

Pentzold S, Zagrobelny M, Rook F, Bak S (2014) How insects overcome two-component plant chemical defence: plant β-glucosidases as the main target for herbivore adaptation. Biol Rev 89: 531–551

Perkins LE, Cribb BW, Brewer PB, Hanan J, Grant M, de Torres M, Zalucki MP (2013) Generalist insects behave in a jasmonate-dependent manner on their host plants, leaving induced areas quickly and staying longer on distant parts. Proc R Soc B Biol Sci 280:20122646

Peschiutta ML, Scholz FG, Goldstein G, Bucci SJ (2018) Herbivory alters plant carbon assimilation, patterns of biomass allocation and nitrogen use efficiency. Acta Oecol 86:9–16

Petschenka G, Agrawal AA (2016) How herbivores coopt plant defenses: natural selection, specialization, and sequestration. Curr Opin Insect Sci 14:17–24

Phillips MA, Croteau RB (1999) Resin-based defenses in conifers. Trends Plant Sci 4(5):184–190

Phukan MM, Sangma SR, Kalita D, Bora P, Das PP, Manoj K et al (2023) Alkaloids and terpenoids: synthesis, classification, isolation and purification, reactions, and applications. In: Handbook of biomolecules, pp 177–213

Picaud F, Bonnet E, Gloaguen V, Petit D (2003) Decision making for food choice by grasshoppers (Orthoptera: Acrididae): comparison between a specialist species on a shrubby legume and three graminivorous species. Environ Entomol 32:680–688

Pickard WF (2008) Laticifers and secretory ducts: two other tube systems in plants. New Phytol 177:877–888

Pineda A, Zheng SJ, van Loon JJ, Pieterse CM, Dicke M (2010) Helping plants to deal with insects: the role of beneficial soil-borne microbes. Trends Plant Sci 15(9):507–514

Provorov NA, Onishchuk OP (2018) Microbial symbionts of insects: genetic organization, adaptive role, and evolution. Microbiology 87(2):151–163

Raffa KF, Andersson MN, Schlyter F (2016) Host selection by bark beetles: playing the odds in a high-stakes game. Adv Insect Physiol 50:1–74

Raguso RA (2008) Wake up and smell the roses: the ecology and evolution of floral scent. Annu Rev Ecol Evol Syst 39:549–569

Rasmann S, Bennett A, Biere A, Karley A, Guerrieri E (2017) Root symbionts: powerful drivers of plant above-and belowground indirect defenses. Insect Sci 24(6):947–960

Raupp MJ, Ahern R, Onken B, Reardon R, Bealmear S, Doccola J, Webb R (2008) Comparing hemlock woolly adelgid and elongate hemlock scale control in ornamental and forest hemlocks. In: Onken B, Reardon R [compilers] FourthSympos. Hemlock Woolly Adelgid Eastern United States FHTET-2008-01

Rehman F, Khan FA, Badruddin SMA (2012) Role of phenolics in plant defense against insect herbivory. In: Chemistry of phytopotentials: health, energy and environmental perspectives, pp 309–313

Richards LA, Dyer LA, Smilanich AM, Dodson CD (2010) Synergistic effects of amides from two *piper* species on generalist and specialist herbivores. J Chem Ecol 36:1105–1113

Robinson ML, Weber MG, Freedman MG, Jordan E, Ashlock SR, Yonenaga J, Strauss SY (2023) Macroevolution of protective coloration across caterpillars reflects relationships with host plants. Proc R Soc B 290:20222293

Rosa-Téllez S, Casatejada-Anchel R, Alcántara-Enguídanos A, Torres-Moncho A, Dohgri M, Martínez-Serra C et al (2023) Serine metabolic networks in plants

Salam U, Ullah S, Tang ZH, Elateeq AA, Khan Y, Khan J et al (2023) Plant metabolomics: an overview of the role of primary and secondary metabolites against different environmental stress factors. Life 13(3):706

Saltveit ME (2017) Synthesis and metabolism of phenolic compounds. In: Fruit and vegetable phytochemicals: chemistry and human health, 2nd edn, pp 115–124

Senthil-Nathan S (2013) Physiological and biochemical effect of neem and other Meliaceae plants secondary metabolites against Lepidopteran insects. Front Physiol 4:359

Schaefer HM, Ruxton GD (2009) Deception in plants: mimicry or perceptual exploitation? Trends Ecol Evol 24:676–685

Schoonhoven LM, van Loon JJA, Dicke M (2005) Insect–plant biology, 2nd edn. Oxford University Press, Oxford

Scott JG, Wen Z (2001) Cytochromes P450 of insects: the tip of the iceberg. Pest Manag Sci 57: 958–967

Scudder GGE, Meredith J (1982) The permeability of the midgut of three insects to cardiac glycosides. J Insect Physiol 28:689–694

Shah MMR, Zhang S, Liu T (2015) Whitefly, host plant and parasitoid: a review on their interactions. Asian J Appl Sci Eng 4:48–61

Sharma G, Mathur V (2020) Modulation of insect-induced oxidative stress responses by microbial fertilizers in *Brassica juncea*. FEMS Microbiol Ecol 96(4):fiaa040

Sharma G, Malthankar PA, Mathur V (2021) Insect–plant interactions: a multilayered relationship. Ann Entomol Soc Am 114:1–16

Sharma G, Agarwal S, Verma K, Bhardwaj R, Mathur V (2023) Therapeutic compounds from medicinal plant endophytes: molecular and metabolic adaptations. J Appl Microbiol 134: lxad074

Shikano I, Rosa C, Tan CW, Felton GW (2017) Tritrophic interactions: microbe-mediated plant effects on insect herbivores. Annu Rev Phytopathol 55:313–331

Simon JC, d'Alencon E, Guy E, Jacquin-Joly E, Jaquiery J, Nouhaud P et al (2015) Genomics of adaptation to host-plants in herbivorous insects. Brief Funct Genomics 14(6):413–423

Song YY, Ye M, Li C, He X, Zhu-Salzman K, Wang RL, Zeng RS (2014) Hijacking common mycorrhizal networks for herbivore-induced defence signal transfer between tomato plants. Sci Rep 4(1):3915

Sorensen JS, Dearing MD (2006) Efflux transporters as a novel herbivore counter mechanism to plant chemical defenses. J Chem Ecol 32:1181–1196

Speed MP, Fenton A, Jones MG, Ruxton GD, Brockhurst MA (2015) Coevolution can explain defensive secondary metabolite diversity in plants. New Phytol 208(4):1251–1263

Stone GN, Schönrogge K (2003) The adaptive significance of insect gall morphology. Trends Ecol Evol 18:512–522

Stowe KA, Marquis RJ, Hochwender CG, Simms EL (2001) The evolutionary ecology of tolerance to consumer damage. Annu Rev Ecol Syst 31:565–595

Strauss SY, Agrawal AA (1999) The ecology and evolution of plant tolerance to herbivory. Trends Ecol Evol 14:179–185

Sugio A, Dubreuil G, Giron D, Simon JC (2015) Plant–insect interactions under bacterial influence: ecological implications and underlying mechanisms. J Exp Bot 66:467–478

Szklarzewicz T, Michalik A (2017) Transovarial transmission of symbionts in insects. In: Oocytes: maternal information and functions. Springer, Cham, pp 43–67

Tan RX, Zou WX (2001) Endophytes: a rich source of functional metabolites. Nat Prod Rep 18(4):448–459

Termonia A, Hsiao TH, Pasteels JM, Milinkovitch MC (2001) Feeding specialization and host-derived chemical defense in Chrysomeline leaf beetles did not lead to an evolutionary dead end. Proc Natl Acad Sci 98:3909–3914

Thawabteh A, Juma S, Bader M, Karaman D, Scrano L, Bufo SA, Karaman R (2019) The biological activity of natural alkaloids against herbivores, cancerous cells and pathogens. Toxins 11 (11):656

Thompson JN, Cunningham BM (2002) Geographic structure and dynamics of coevolutionary selection. Nature 417:735–738

Vale PF, Siva-Jothy J, Morrill A, Forbes M (2018) The influence of parasites on insect behavior. In: Insect behavior: from mechanisms to ecological and evolutionary consequences. Oxford University Press, Oxford, pp 274–291

Van Dam NM, Tytgat TO, Kirkegaard JA (2009) Root and shoot glucosinolates: a comparison of their diversity, function and interactions in natural and managed ecosystems. Phytochem Rev 8:171–186

van den Bosch TJ, Welte CU (2017) Detoxifying symbionts in agriculturally important pest insects. Microb Biotechnol 10:531–540

Van der Putten WH, Vet LE, Harvey JA, Wäckers FL (2001) Linking above-and belowground multitrophic interactions of plants, herbivores, pathogens, and their antagonists. Trends Ecol Evol 16(10):547–554

Venturi V, Keel C (2016) Signaling in the rhizosphere. Trends Plant Sci 21(3):187–198

Vetter J (2000) Plant cyanogenic glycosides. Toxicon 38(1):11–36

War AR, Paulraj MG, Ahmad T, Buhroo AA, Hussain B, Ignacimuthu S, Sharma HC (2012) Mechanisms of plant defense against insect herbivores. Plant Signal Behav 7:1306–1320

War AR, Taggar GK, Hussain B, Taggar MS, Nair RM, Sharma HC (2018) Plant defence against herbivory and insect adaptations. AoB Plants 10(4):ply037

Weng JK, Philippe RN, Noel JP (2012) The rise of chemodiversity in plants. Science 336 (6089):1667–1670

Westman SM, Kloth KJ, Hanson J, Ohlsson AB, Albrectsen BR (2019) Defence priming in Arabidopsis—a meta-analysis. Sci Rep 9(1):13309

White JF, Kingsley KL, Zhang Q, Verma R, Obi N, Dvinskikh S, Kowalski KP (2019) Endophytic microbes and their potential applications in crop management. Pest Manag Sci 75(10):2558–2565

Wieczorek MN, Walczak M, Skrzypczak-Zielińska M, Jeleń HH (2018) Bitter taste of Brassica vegetables: the role of genetic factors, receptors, isothiocyanates, glucosinolates, and flavor context. Crit Rev Food Sci Nutr 58(18):3130–3140

Willmer P (2011) Pollination and floral ecology. Princeton University Press, Princeton

Wink M (2019) Quinolizidine and pyrrolizidine alkaloid chemical ecology–a mini-review on their similarities and differences. J Chem Ecol 45(2):109–115

Wink M (2020) Evolution of the angiosperms and co-evolution of secondary metabolites, especially of alkaloids. In: Coevolution of secondary metabolites, pp 151–174

Wöll S, Kim SH, Greten HJ, Efferth T (2013) Animal plant warfare and secondary metabolite evolution. Nat Prod Bioprospect 3:1–7

Woolhouse ME, Webster JP, Domingo E, Charlesworth B, Levin BR (2002) Biological and biomedical implications of the co-evolution of pathogens and their hosts. Nat Genet 32:569–577

Wittstock U, Kliebenstein DJ, Lambrix V, Reichelt M, Gershenzon J (2003) Chapter five Glucosinolate hydrolysis and its impact on generalist and specialist insect herbivores. Recent Adv Phytochem 37:101–125

Yang ZL, Kunert G, Sporer T, Körnig J, Beran F (2020) Glucosinolate abundance and composition in Brassicaceae influence sequestration in a specialist flea beetle. J Chem Ecol 46:186–197

Yulvianti M, Zidorn C (2021) Chemical diversity of plant cyanogenic glycosides: an overview of reported natural products. Molecules 26(3):719

Zunjarrao SS, Tellis MB, Joshi SN, Joshi RS (2020) Plant–insect interactions: the saga of molecular coevolution. In: Co-evolution of secondary metabolites. Springer, Cham, pp 19–45

Host Plant Resistance to Insects in Cotton

3

Lucy Egan, Qian-Hao Zhu, Iain Wilson, and Warwick Stiller

Abstract

Across the globe, cotton production is limited by endemic and introduced insect pests. The incorporation of host plant resistance (HPR), defined as the ability of a plant to defend itself against targeted biotic stresses, has been a key breeding target for many cotton breeding programs. The development of resistant cultivars provides a cost effective and environmentally sustainable solution for cotton growers. The key insect pests to cotton production include *Helicoverpa* spp., pink bollworm, *Spodoptera* spp., silverleaf whitefly, various plant and stink bugs, aphids, and thrips. Cotton plants have a range of native and genetically modified traits that infer HPR. However, breeding for HPR in cotton is a slow and complex process, with cultivar development commonly requiring more than 15 years to reach the market. Often, the incorporation of native HPR traits has a negative impact on yield and important agronomic characteristics, so it is not economically viable. Therefore, determining the value of a native HPR trait under multiple scenarios is a critical element for research and development investment. This chapter aims to review the major pests to cotton production and the current status of native HPR in commercial cultivars to each pest, outline the process of breeding for HPR in cotton, and discuss the economic value of HPR traits.

Keywords

Gossypium · Host plant resistance · Breeding · Germplasm · Native traits · Economic value

L. Egan (✉) · W. Stiller
CSIRO Agriculture and Food, Narrabri, NSW, Australia
e-mail: Lucy.Egan@csiro.au

Q.-H. Zhu · I. Wilson
CSIRO Agriculture and Food, Canberra, ACT, Australia

S. Kumar, M. Furlong (eds.), *Plant Resistance to Insects in Major Field Crops*,
https://doi.org/10.1007/978-981-99-7520-4_3

Abbreviations

ASAL	*Allium sativum* leaf agglutinin
Bt	*Bacillus thuringiensis*
DNA	Deoxyribonucleic acid
dsRNA	Double-stranded RNA
GM	Genetically modified
HMGS	Host-mediated silencing of insect genes
HPR	Host plant resistance
miRNA	Micro RNA
mRNA	Messenger RNA
RNA	Ribonucleic acid
RNAi	RNA interference
ROI	Return on investment
SAT	Shoot apical meristem transformation
siRNA	Small interfering RNA
SLW	Silverleaf whitefly
tasiRNA	Trans-acting siRNA
T-DNA	Transfer DNA
TSSM	Two-spotted spider mite
USA	United States of America

3.1 Introduction

Cotton is an important global crop and is grown in over 70 countries for fibre, oil, and protein (Leff et al. 2004). The cotton genus, *Gossypium*, includes >50 species. Out of these, four species are cultivated for cotton production: *Gossypium arboreum* (Desi cotton), *Gossypium herbaceum* (Levant or Arabian cotton), *Gossypium barbadense* (Pima, Egyptian, or Sea Island cotton), and *G. hirsutum* (Upland cotton) (Wendel et al. 2009; Constable et al. 2015). The most predominant species for commercial production is *Gossypium hirsutum,* and primary, secondary, and tertiary gene pools for the *Gossypium* genus are based on crossing compatibility with *G. hirsutum.* Over 90% of global cotton production utilises *G. hirsutum* (Constable et al. 2015).

The *Gossypium* genus consists of diploid (A–G and K genomes, $2n = 2x = 26$) and tetraploid species (AD genomes, $2n = 4x = 52$) (Lubbers and Chee 2009). The genetic resources of cotton are spread across five continents, and Munro (1994) suggests that the difference in genomes is due to the geographical isolation. The A, B, and E genomes are found across Africa and Asia, the C, G, and K genomes are confined to Australia, the D genomes to America, and the F and G genomes are only found in a single species each (Stiller and Wilson 2014).

On average, the annual global yield of cotton cultivars is increasing. The increase is due to a range of factors including improved farming management practices and improved higher performing cultivars (Liu et al. 2013; Voora et al. 2020). However contrary to the global trend, yields in Pakistan are decreasing (Shahzad et al. 2022; USDA 2022). China, India, the United States of America (USA), Pakistan, Uzbekistan, and Brazil are the leading producers of cotton by volume, but the Australian cotton industry has the highest yields of any production region in the world (Constable and Bange 2015).

Under current cotton management systems, a large proportion of the global crop is grown with cultivars expressing genetically modified (GM) traits developed by multinational corporations such as Bayer, BASF, Corteva, and Syngenta. These GM cultivars express insecticidal protein(s) encoded by gene(s) from *Bacillus thuringiensis* (Bt), a gram-positive soil bacterium. The technology has provided a sustainable solution for the control of major Lepidopteran cotton insect pests, including cotton bollworm (*Helicoverpa* spp.) that became increasingly difficult to control during the late 1980s and the early 1990s due to increased insect populations and resistance to insecticides (Wilson et al. 2018). For example, in Australia, between 1993 and 2019, pesticide use to control *Helicoverpa* reduced by 95% due to the uptake of GM cotton cultivars (Constable et al. 2011; Cotton Australia 2023a). The successful deployment of GM cotton cultivars has been aided by the constant monitoring of resistance development of targeted insects and the planting of conventional cotton to allow for the dilution of resistant alleles (Downes et al. 2017). However, GM cotton is not the 'silver bullet' for pest management. Bt cotton does not control all pests, and the uptake of GM cotton cultivars has increased the threat of previously categorised 'secondary pests' to cotton. The rise in secondary pests is largely due to the reduction in broad-spectrum pesticides used to control Lepidopteran pests which provided some control against secondary pests. Trapero et al. (2016) stated that host plant resistance (HPR) traits may help support the resistance management of the Bt cottons, largely due to the Lepidopteran species interaction with secondary pests that could be controlled by HPR traits.

Although the GM trait packages are critical to the cotton industry, the backbone of cotton breeding programs is conventional (non-GM) germplasm enhancement. Advances in breeding techniques and the uptake of molecular tools have facilitated increases in yield and fibre quality characteristics in cotton (Liu et al. 2013). However, despite the advancement of the crop, yield is often limited by pest damage (Egan and Stiller 2022). Cotton has seven major pests that impact production: Lepidopteran spp. (*Helicoverpa*, *Heliothis,* and *Spodoptera*), plant bugs, stink bugs, silverleaf whitefly (SLW), aphids, and thrips (Luttrell et al. 1994; Trapero et al. 2016; Cotton Australia 2023b). These pests are present across cotton production regions globally, and the severity of the pests changes due to differing environmental conditions across geographical areas. Egan and Stiller (2022) stated that when a pest is widespread, causing significant yield loss, and there is no cost effective or durable management solution, HPR (where available) is the most appropriate method of control.

HPR is commonly defined as the plant's ability to defend itself against a biotic stress, namely a pest or disease (Stout 2014). HPR traits can be in the form of native, from the primary, secondary, or tertiary gene pools, or GM traits. The types of plant resistance to pests have been categorised into three major groups: antibiosis, antixenosis (non-preference), and tolerance (Koch et al. 2016). Antibiosis induces resistance through the plant's biology to negatively impact the pest's biology and can result in pest death. Non-preference trait is a host trait that negatively impacts the pest, but the pest can still survive on the host, or makes the host become less attractive to the pest. Tolerance traits reduce the impact/damage to the plant by the pest whilst maintaining pest population levels similar to those seen on susceptible plants. A key example in cotton is the correlation between the number of extrafloral nectaries and the indirect defence of the plant (Llandres et al. 2019). Whilst the three groups of resistance are beneficial for reducing yield loss due to pest damage, there is concern that it could lead to pest populations that develop resistance to one or more mechanisms. Due to the fast life cycle of many pests and the slow nature of plant breeding, there is an increased opportunity for the evolution of resistance in pest populations. However, this is heavily dependent on the nature of the insect resistance mechanism.

This chapter aims to outline the major pests to cotton production and their current HPR status, summarise the process of breeding for HPR in cotton, and discuss the economic challenges when breeding for HPR. The content will take a somewhat narrow view of HPR and focus on the native or non-GM traits that are available to infer resistance but will briefly discuss GM approaches when necessary and how GM could be a guide as to what needs to be targeted via native HPR.

3.2 Major Pests to Cotton Production

The types of sucking and chewing pests that affect cotton differ throughout the world. Several reviews summarise the major pests to cotton and the available sources of resistance and different traits that could be utilised in incorporating HPR (Jenkins and Wilson 1996; Trapero et al. 2016). Specifically, Trapero et al. (2016) summarised the genetic sources of HPR and identified traits in cotton that provide resistance to secondary pests. Therefore, building on those reviews, the major pests and the current status of HPR in commercial cultivars to each pest are summarised below, and Table 3.1 provides a summary of whether the HPR traits were commercially released in a cultivar.

3.2.1 *Helicoverpa* Species

Helicoverpa is a major pest to conventional (non-GM) cotton (Downes et al. 2017). The two *Helicoverpa* species that impact cotton are *H. armigera* and *H. punctigera* and cause similar damage to cotton crops. However, *H. armigera* has developed a high level of resistance to insecticides and is of growing concern to the industry

Table 3.1 The native host plant resistance (HPR) traits available to the major cotton pests and the associated commercially released variety (if applicable)

Insect	Native HPR trait	Deployed into a variety (✓/✗)	Name of variety and who released it
Helicoverpa species	Increased gossypol	✗	
	Biochemical compounds	✗	
	Yellow/orange pollen	✗	
	Frego bract	✓	• **Sicot 3**, CSIRO (Thomson 1984)
	Nectariless and glabrous	✗	
	Red plant colour	✗	
	Extreme hairiness	✗	
	Nectariless	✓	• **Stoneville 731N**, Stoneville Pedigreed Seed Company (USDA 2016) • **Stoneville 825**, Stoneville Pedigreed Seed Company (USDA 2016) • **DP 0935 B2RF**, Monsanto Technology LLC (USDA 2016)
Pink bollworm	Nectariless	✓	• **Stoneville 731N**, Stoneville Pedigreed Seed Company (USDA 2016) • **Stoneville 825**, Stoneville Pedigreed Seed Company (USDA 2016) • **DP 0935 B2RF**, Monsanto Technology LLC (USDA 2016)
	Okra	✗	
	Glabrous	✗	
Spodoptera species	Increased gossypol	✗	
Silverleaf whitefly	Okra leaf shape	✗	
	Reduced leaf trichomes	✗	
	Increased phenolic compounds	✓	• **Ravi**, Cotton Research Institute, Pakistan (Perveen et al. 2001; Yasmin et al. 2008)
	Reduced soluble sugars and proteins	✓	• **Niab-Kiran**, Nuclear Institute for Agriculture and Biology, Pakistan (Rizwan et al. 2021; Nuclear Institute for Agriculture and Biology 2022)
Plant and stink bugs	High gossypol	✗	
	Nectariless	✓	• **Stoneville 731N**, Stoneville Pedigreed Seed Company (USDA 2016) • **Stoneville 825**, Stoneville Pedigreed Seed Company (USDA 2016)

(continued)

Table 3.1 (continued)

Insect	Native HPR trait	Deployed into a variety (✓/×)	Name of variety and who released it
			• **DP 0935 B2RF**, Monsanto Technology LLC (USDA 2016) • **UA212ne**, Arkansas Agricultural Experiment Station (Bourland and Jones 2020)
	Frego bract	×	
	Leaf hair	×	
	Glandless	×	
Aphids	Reduced leaf hair	×	
Thrips	Pubescence	×	
	Glandless	×	

The name of the variety is emboldened in the last column

(Joußen et al. 2012). The species composition in the crop varies with seasonal fluctuations, time of the year, and location. Often the *H. armigera* population overwinter towards the end of the season and emerge in the early stages of the next growing season. *H. punctigera* is often the dominant species in the early stages of the growing season but depending on the season, *H. armigera* can have early infestations and become the dominant species (Cotton Australia 2023b). The *Helicoverpa* adults produce a large number of eggs and are mobile, which under favourable conditions can lead to a rapid population growth. *Helicoverpa* attacks all stages of plant growth, and the chewing damage can lead to secondary plant infections (Riaz et al. 2021). Key damage symptoms include chewing holes on soft leaves, squares, and bolls and often lead to shedding off of fruiting bodies. Recent reports show that hybridisation has occurred between *H. armigera* and *H. zea* (native to Brazil) in South America (Cordeiro et al. 2020; Rios et al. 2022). This hybrid breeding between the native and invasive pest could severely impact the current resistance management strategies in cotton (Cordeiro et al. 2020).

Currently, *Helicoverpa* is well controlled due to the deployment of Bt cotton. Three generations of Bt cotton (Bollgard I, II and III) have been developed and commercially adopted. Bollgard I (BG I) contains a single Bt gene Cry1Ac, whereas BG II and III contain two (Cry1Ac and Cry2Ab) and three (Cry1Ac, Cry2Ab, and Vip3A) Bt genes, respectively, providing a higher level and broader spectrum of protection from Lepidopteran pests than BG I. Transgenic cotton cultivars co-expressing Cry1A and CpTI, a cowpea-derived gene encoding trypsin inhibitor, are also being used in commercial cotton production in China (Cui et al. 2011; Li et al. 2020). However, if resistance management protocols such as a requirement for refuge cotton (cotton crops that are planted to generate numbers of susceptible moths that have not been exposed to the Bt proteins in cotton) are not followed correctly, then there will be a need for increased HPR through either native or GM traits.

Before Bt cotton was deployed, large research efforts were invested into finding native traits that conferred resistance and focused on traits such as high gossypol (Jenkins and Wilson 1996), biochemical compounds (Jenkins and Wilson 1996), yellow or orange pollen (Bailey 1981), nectariless (the removal of the nectaries) (Lukefahr et al. 1971), glabrous leaves (Lukefahr et al. 1971), and varying gland densities (Scheffler et al. 2012). Most of these traits confer similar levels of resistance across the Lepidopteran spp. that affect cotton (Jenkins and Wilson 1996) and are summarised below.

- Gossypol, a yellow pigment that is produced in the glands of the plant, is an important allelochemical that confers resistance. Bottger et al. (1964) were the first to report the effect of gossypol in relation to pest resistance and found that cotton germplasm that lacked gossypol glands suffered increased damage from bollworms. Breeding lines with high levels of gossypol have shown reduced insect damage and slower larval growth, and it is largely attributed to the toxicity of gossypol to Lepidopteran species (Lukefahr and Houghtaling 1969; Parrott 1990).
- Biochemical compounds that provide resistance from plant defence mechanisms are commonly identified in the literature. Bi et al. (1997) showed that the biochemical defence compounds of peroxidase, ascorbate oxidase, lipoxygenase, and diamine oxidase are found in higher levels after foliar and pest damage. Therefore, genetic manipulation to overexpress the genes involved in biosynthesis of those defensive compounds could enhance resistance.
- Pollen that is yellow or orange has been shown to reduce the growth of *Heliothis* larvae (Hanny et al. 1979; Bailey 1981). There are no known breeding efforts to use these in the development of cultivars. However, since most Upland cotton cultivars have cream pollen, it may indicate that other pollen colours have a detrimental effect on yield (Jenkins and Wilson 1996).
- Frego bract (narrow and twisting flower bract) has been reported in the literature to provide resistance through the reduction in bollworm oviposition (Maxwell et al. 1969; Lincoln et al. 1971; Thomson 1984). However, there are conflicting reports to the validity of the resistance, and it has been concluded that frego bract is not preferred by the bollworm when other food sources are available. However, in the absence of other food sources, the bollworm will still overwhelm the crop and overcome the resistance. Furthermore, the frego bract trait increases the susceptibility of the genotypes to the 'plant bug' complex. To the best of our knowledge, there has only been one commercially produced cultivar with frego bract; Sicot 3 (Thomson 1984).
- The traits of nectariless and glabrous (reduced or no leaf trichomes) leaves reduce the oviposition of the *Heliothis* moths (Lukefahr et al. 1965, 1971). Lukefahr et al. (1965) identified that nectariless cotton genotypes had a reduction of up to 60% in egg deposition by *Heliothis* spp. Genotypes that were both nectariless and glabrous had up to an 80% reduction.
- Red plant colour is a minor researched trait, but Hunter et al. (1965) identified that non-preference resistance was inferred in red leaf types compared to green

foliage. The resistance was largely related to the intensity of the red; the darker the red, the less preference by the pest (Jones et al. 1970, 1989).

- Wannamaker (1957) was one of the first to evaluate the effect of leaf hair density in relation to feeding and reproduction of the boll weevil (Jenkins and Wilson 1996). Extreme hairiness significantly reduces damage from the boll weevil and the plant bug complex. However, although the trait provides resistance, it is of limited value due to the mechanical complications of harvesting hairy leaves. Further, increased hairiness is preferred by the bollworms complex. Hairy leaves have also been associated with short, coarse fibres that are less desirable for spinning.

3.2.2 Pink Bollworm

Pink bollworm (*Pectinophora gossypiella*) is a critical pest to the global cotton industry. The pest is native to Asia but is now present in many of the global cotton producing regions. The pest damages squares and bolls resulting in severe yield and fibre penalties. The pest became present in the USA in 1917 but has since been eradicated through the use of Bt cotton and release of sterile males (Tabashnik et al. 2021). In cotton producing areas of India, pink bollworm has become resistant to the first release of Bt cotton, which can be attributed to the lack of adoption of refuge cotton to some extent (Tabashnik and Carrière 2019).

Bolls of nectariless plants have been noted to have significantly less damage from pink bollworm and have provided a low but consistent level of resistance (Niles 1980; Adjei-Maafo et al. 1983). On the other hand, the glabrous trait is not sufficient to provide resistance (Wilson and George 1981). Wilson and Flint (1987) suggested that the okra leaf shape could provide resistance to pink bollworm as it changes the microclimate around the boll which negatively impacts the pest. In the USA, several okra leaf and nectariless cultivars have been commercially released, but were largely targeted for growing areas where pink bollworm is not a major problem.

3.2.3 *Spodoptera* Species

The *Spodoptera* spp. pests are of growing concern to cotton production (Barros et al. 2010). Cluster caterpillar (*Spodoptera litura*) is found frequently in tropical and coastal environments from India to the Pacific (Maes 2014). The common name comes from the caterpillar's high-density clustering of young larvae up to the first three instars. The young larvae feed on the surface of the leaf and scrape the leaves, whilst the older larvae can completely devour the leaves, they often feed on reproductive tissue of cotton. Large infestations of the cluster caterpillar can lead to total defoliation and have a severe impact on production (Bhattacharyya et al. 2007; Maes 2014). A clear example of this is the collapse of the cotton industry in Northern Australia in the 1970s of which *Spodoptera* spp. were a key pest. Currently, control is provided by Bt cotton through Cry1Ac and Cry2Ab.

Lesser armyworm (*Spodoptera exigua*) originated from Southeast Asia and is now found in several cotton producing countries. The pest activity is impacted heavily by climatic conditions and is more prevalent in warmer climates. Although it is considered a minor pest for young cotton crops, the lesser armyworm can be a severe defoliator to fruiting cotton by chewing all of the green tissue of leaf lamina except the veins. Fall armyworm (*Spodoptera frugiperda*) is a tropical insect that is an endemic pest to many cotton producing regions of the world. Damage by fall armyworm can range from minimal to severe, and fruiting bodies are often the most damaged part of the crop. De-topping of cotton plants and total defoliation can occur under high infestation levels (Vyavhare 2017).

HPR research against *Spodoptera* spp. has not been as focused as for *Helicoverpa* spp. However, traits such as increased levels of terpene and aldehydes, e.g. gossypol, can be deployed in management programs for *S. frugiperda* (Jesus et al. 2014). Cotton cultivars containing only Cry1Ac have been shown to be not much more efficient in controlling *S. frugiperda* (Adamczyk Jr et al. 1998) and can increase tolerance of individuals feeding on the cultivar (Adamczyk Jr and Sumerford 2001).

Although the above-mentioned native traits were developed in an attempt to provide resistance to Lepidopteran pests, they were largely unsuccessful as the level of resistance they provided was simply not high enough to prevent significant pest damage to the crop. Given that Bt cotton provides stable and robust resistance to Lepidopteran pests, developing HPR cotton cultivars in the form of GM has by far seen the largest investment and is the most successful deployment of an HPR trait in cotton. In 2019, 18 countries cultivated genetically engineered Bt cotton with a total area of ~25.7 million hectares (ISAAA 2019). The major insecticidal Bt genes that are deployed in commercial cotton cultivars are Cry1Ac, Cry2Ab, and Vip3A. But, in the past decade, several new entomotoxic Cry proteins from Bt have been expressed in cotton individually or in different combinations with other insecticidal proteins to evaluate their efficacy in resistance to different species of insect pests; including non-lepidopteran insects that are not targeted by the Bt genes currently used (Table 3.2). For instance, transgenic cotton plants expressing Cry10Aa are highly resistant to *Anthonomus grandis* (cotton boll weevil), with a mortality rate of 90–100% (Ribeiro et al. 2017). For the Bt proteins with low insecticidal efficacy, amino acid optimisation has been used to change the protein characteristics and to enhance their efficacy. A good example of such optimisation is the creation of Cry51Aa2.834_16, a new variant of Cry51Aa2 (Gowda et al. 2016). The transgenic cotton plants expressing Cry51Aa2.834_16 can reduce populations of both *Lygus hesperus* and *L. lineolaris* more significantly than those expressing Cry51Aa2. Commercially acceptable transgenic events have been identified and will be introgressed into elite commercial cotton cultivars for release. However, the Bt genes have specificities which only allows it to be a good defence against specific classes and is not a silver bullet for pest management.

Table 3.2 Genes from *Bacillus thuringiensis*, plants and insects being evaluated for insect resistance in cotton

Transgene	Recipient cultivar	Target pests	Transformation approach	References
Insecticidal genes				
Single gene				
Vip3A or *tvip3A*	Zhongmiansuo 35	*Spodoptera frugiperda, S. exigua,* and *Helicoverpa armigera*	AMT	Wu and Tian (2019)
ASAL	NC-601	*Amrasca devastans* and *Bemisia tabaci*	SAT	Vajhala et al. (2013)
Cry51Aa2.834_16	DP393	*Lygus hesperus* and *L. lineolaris*	AMT	Gowda et al. (2016)
Cry2AX1	Coker 310	*H. armigera*	AMT	Dhivya et al. (2016)
Cry1Ac	Coker 310FR	*H. armigera*	AMT	Singh et al. (2016)
Cry1Ia12	BRS Cedro	*S. frugiperda* and *A. grandis*	PMT	de Oliveira et al. (2016)
Tma12	Coker 312	*B. tabaci*	AMT	Shukla et al. (2016)
Cry10Aa	BRS 372	*Anthonomus grandis*	AMT	Ribeiro et al. (2017)
S. rolfsii lectin	Sahana	*Aphis gossypii* and *S. litura*	SAT	Vanti et al. (2018)
Cry2ax1	Coker 310	*H. armigera*	AMT	Jadhav et al. (2020)
Cry1AcF or *Cry2Aa*	Pusa 8–6	*H. armigera*	SAT	Karthik et al. (2021)
Cry5Aa	n/a	*H. armigera*	PMT	Zhao et al. (2021)

Cry1Ia12	GS-01	*Pectinophora gossypiella*	SAT	Tariq et al. (2022)
Multiple genes				
Vip3A or Vip3a + Cry1Ab	n/a	*H. zea* and *H. virescens*	AMT	Bommireddy et al. (2011)
Cry1Ac + CpTI	CRI23	*H. armigera*	PMT	Cui et al. (2011)
NaPI + StPin1A	Coker 315	*H. armigera*	AMT	Dunse et al. (2010)
Cry9C + Cry1Ac or Cry9C + Cry2A	Simian 3	*S. litura*	AMT + cross	Li et al. (2014)
Vip3AcAa + Cry1Ac	Coker 312	*S. litura, S. exigua,* and *Agrotis ipsilon*	AMT	Chen et al. (2017)
Cry1Be + Cry*1Fa*	n/a	*S. litura and Ostrinia nubilalis*	AMT	Meade et al. (2017)
Vip3AcAa + Cry1Ac	Coker 312	*H. armigera*	AMT	Chen et al. (2018)
Cry1Ac + GNA	GSN-12	*A. gossypii* and *S. littoralis*	SAT	Khabbazi et al. (2018)
Cry1Ac + Cry2Ab	Coker 312	*Spodoptera litura*	AMT	Siddiqui et al. (2019)
Cry2Ab + Cry1F + Cry1Ac	Narasimha	*H. armigera* and *S. litura*	SAT	Katta et al. (2020)
Vip3Aa + ASAL	CEMB33	*B. tabaci*	SAT	Din et al. (2021)
Cry1Fa + Cry32Aa + AtPME + AnPME	Eagle-2	*H. armigera, P. gossypiella,* and *Earias fabia*	SAT	Razzaq et al. (2021)
Cry3Bb1 + Cry3	Eagle-2	*P. gossypiella* and *H. armigera*	SAT	Zafar et al. (2022)
Vip3Aa	CKC-01 (with *Cry1Ac + Cry2Ab*)	*H. armigera*	SAT	Shad et al. (2021)
Host-mediated gene silencing (HMGS) by RNAi				

(continued)

Table 3.2 (continued)

Transgene	Recipient cultivar	Target pests	Transformation approach	References
Single gene				
dsCYP6AE14	R15	*H. armigera*	AMT	Mao et al. (2011)
dsHMG-CoA reductase	YZ1	*H. armigera*	AMT	Tian et al. (2015)
dsNDUFV2	R15	*H. armigera*	AMT	Wu et al. (2016)
dsCYP392A4	Jimian 14	*Tetranychus cinnabarinus*	AMT	Shen et al. (2017)
dsHaHR3	F15	*H. armigera*	PMT	Han et al. (2017)
JHMT or *JHBP*	W0	*H. armigera*	AMT	Ni et al. (2017)
Ghr-miR166b	HS6	*B. tabaci*	SAT	Wamiq and Khan (2018)
dsAILIM	Jin668	*Apolygus lucorum*	AMT	Liang et al. (2021)
dsAgCYP6CY3-P1	n/a	*A. gossypii*	n/a	Zhang et al. (2023)
Multiple genes				
dsCYP6AE14 + *GhCP1*	R15	*H. armigera*	AMT	Mao et al. (2013)
JHMT or *JHBP* + *Cry1Ac*	W0	*H. armigera*	AMT	Ni et al. (2017)
COPβ + *EcR*	Pusa8–6	*A. biguttula*	SAT	Karthik et al. (2023)

AMT Agrobacteria-mediated transformation followed by tissue culture, *PMT* pollen-tube-mediated transformation, *SAT* shoot apical meristem transformation, *n/a* not available

3.2.4 Silverleaf Whitefly

SLW (*Bemisia tabaci* B-biotype) is a serious pest to cotton due to its resistance to many insecticides and ability to vector serious cotton viruses. SLW feeds on the plant sap and causes wilting, dropping of leaves, and in extreme cases, may result in death of the plant. However, the primary damage is caused by sucking large quantities of plant sap and excretion of honeydew which contaminates the lint and creates problems during processing, and ultimately results in an economic loss to the grower (Johnson et al. 1982; Rao et al. 1989; Hequet and Abidi 2002).

There is currently no known resistance to SLW in current commercial cotton cultivars. Morphological resistance has been identified through an okra leaf shape (deeply invaginated leaf lobes) and reduced leaf trichomes (Thomson et al. 1987; Butter and Vir 1989; Fitt et al. 1992; Miyazaki et al. 2013). Although the resistance has been identified, the deployment of the non-preference trait is limited by the impact on yield. An okra leaf shape has less leaf area and consequently reduced photosynthetic capacity and can result in a yield penalty. Whilst the literature supports the possibility of a yield reduction associated with the okra-glabrous trait, Thomson (1994) showed that high-yielding okra leaf breeding populations could be developed. High-yielding commercial okra leaf cultivars have also been released (Reid 1998). Several studies have identified that resistance can also be conferred by an increase of phenolic compounds, tannins (Perveen et al. 2001), flavonoids, and a reduction in soluble sugars and proteins (Rizwan et al. 2021). Breeding for increases in phenolic compounds has seen some success in the literature in other crops (Balyan et al. 2013; Atak et al. 2021), yet is often correlated with a negative impact on other commercially important traits (Kaushik et al. 2015). Leaf gossypol glands have also been linked with increased resistance to whitefly (Khalil et al. 2017; Rizwan et al. 2021).

3.2.5 Plant and Stink Bugs

A range of various bugs (Suborder Heteroptera) can cause damage to cotton, including *Lygus, Creontiades, Nezara, Campylomma,* and *Dysdercus.*

Lygus (*Lygus* spp.) feed on cotton terminals and fruiting bodies. Several *Lygus* species can occupy a cotton crop which include the western tarnished plant bug (*Lygus hesperus*), tarnished plant bug (*Lygus lineolaris*), and the pale legume bug (*Lygus elisus*). The damage to cotton includes stunted growth, deformed bolls, shedding of squares and bolls, and damaged seed (Layton 2000).

Mirids (*Creontiades* spp.) are a sucking pest to cotton and both the adults and nymphs feed by piercing the plant tissue and releasing a chemical that damages the nearby cells and results in a wide variety of symptoms including leaf damage, wilting, deformity, shedding and deformity of bolls, and reduced yield and fibre quality. Mirids are a challenging pest to control as they are extremely mobile, and the population numbers can fluctuate rapidly (Cotton Australia 2023b).

Three stink bug species primarily affect cotton; the green stink bug (*Acrosternum hilare* Say), the southern green stink bug [*Nezara viridula* (L.)], and the brown stink bug [(*Euschistus servus* (Say)]. Population levels of the three species vary extensively across geographical regions and are widely distributed throughout the world. Due to the reduction in applications of broad-spectrum insecticides to control *Lepidopteran* pests, the stinkbug has increased in prevalence in recent years. They feed on bolls of cotton plants, and the majority of the yield loss comes from damage to the younger bolls (<10 days old). Excessive stink bug damage can cause a reduction of fibre quality, in the form of stained lint and poor colour grades (Kamminga et al. 2012).

There are very few commercial cultivars with targeted incorporated HPR to plant bugs (Jenkins and Wilson 1996). Germplasm has been identified that is resistant to *L. hesperus* which includes species related to *G. hirsutum* (Tingey et al. 1975) and germplasm with high gossypol content (Tingey et al. 1975; Schuster and Frazier 1976). Benedict et al. (1983) confirmed that the absence of gossypol in seeds and plant tissues increased the susceptibility to plant bugs. The nectariless trait has conferred some resistance to plant bugs by causing reduced oviposition and reduced nectar as a food source for the nymphs (Schuster et al. 1976; Bailey et al. 1984; Jenkins and Wilson 1996). Attempts to combine the frego bract trait with the nectariless trait have shown increased resistance to plant bugs (Jones et al. 1989), although the frego bract trait by itself increases susceptibility to plant bugs (Milam et al. 1985). The incorporation of the nectariless trait has shown promise in the past as there has been no major negative impact on agronomic or fibre quality characteristics linked with the trait (Meredith Jr and Bridge 1977; McCarty Jr et al. 1983).

As mentioned previously, the development of *Lygus* resistant cultivars has shown promise through a transgenic approach (Gowda et al. 2016). Transgenic cotton that expressed a variant of a Bt crystal protein (Cry15Aa2) significantly reduced the *Lygus* spp. in field trials, and the germplasm containing the variant has been moved forward to the cultivar development pipeline. The trait has been named 'ThryvOn'™ (Bayer 2023).

3.2.6 Aphids

The cotton aphid (*Aphis gossypii*) causes damage to all plant parts and is known to transmit various viral diseases including the economically devastating cotton blue disease (Aboughanem-Sabanadzovic et al. 2019; Bag et al. 2021; Parkash et al. 2021; Tarazi and Vaslin 2022) and cotton bunchy top disease (Reddall et al. 2004; Ellis et al. 2013, 2016). Similar to SLW, aphids also secrete honeydew which causes lint contamination and challenges for lint processing. If large aphid populations are left uncontrolled for several weeks, significant yield loss can occur. Dunnam and Clark (1939) reported that the number of aphids increased in correlation with the number of trichomes on the lower leaf surface and concluded that a reduced hair pattern on the leaf could be a potential HPR trait.

3.2.7 Thrips

A number of thrips species cause damage to cotton, including *Frankliniella* spp. and *Thrips* spp. They are commonly early seedling pests, and primarily feed on the leaves and early squares. Damage can result in silvering of the leaves and damage to the plant terminal (Cotton Australia 2023b). Under some conditions, heavy infestations may reduce stands, stunt plants, and delay fruiting and maturity. However, under favourable growing conditions, cotton can often recover completely from early thrips damage. In areas where seedling emergence typically occurs under warm conditions, thrips are usually of little concern. Despite being a pest, thrips are a useful control of two-spotted spider mites (TSSM) as they predate on TSSM eggs.

Pubescence (the presence of trichomes) is associated with resistance to thrips at the seedling stage (Jenkins and Wilson 1996). However, the deployment of the trait increases oviposition of bollworms and whitefly and has been associated with trash and mechanical harvesting issues (Byrd et al. 2016), so is not an effective nor practical HPR solution in countries that use mechanical harvesting. Miyazaki et al. (2017) identified that some diploid cottons provide resistance to thrips.

3.3 Breeding for Host Plant Resistance to Pests in Cotton

3.3.1 Breeding HPR Traits by Conventional Approaches

Modern cotton breeding programs aim to exploit novel variation for breeding targets and select stable cultivars that are adapted to the local environment. The overarching themes of most cotton breeding programs are similar—increase yield and fibre quality, increase pest and disease resistance, and enhance adaptation to regional constraints. Primarily, yield is the most prominent breeding target in cotton breeding programs as this is what the grower is paid for. The breeding methodology depends on the breeding target, heritability of the trait, technology available for phenotypic and genotypic assessments, and available germplasm. Pure line development is a key goal and results in stable and uniform cultivars. India and China engage in cotton hybrid development, but the majority of programs in other countries focus on pure line development. The generalised cotton breeding methodology begins with the recombination of desired traits from individual lines by hybridisation (pedigree and backcross breeding methods are the most common) and consequent inbreeding to create variation within a population. Successive rounds of selections are performed to select for the desired phenotype and to fix traits. The breeding lines are then evaluated across multiple environments (or multiple different sites within a target environment) and compared to current commercial cultivars. Breeding lines that outperform (for one or more traits) what is currently available on the market are commercially released to the industry.

Breeding for HPR to a biotic stress is generally only considered if there is no other suitable management strategy due to the time and cost it takes to identify and introgress the resistance into an elite background. Egan and Stiller (2022) outlined the factors that must be present for HPR to be considered as a viable investment decision. In summary, the pest or pathogen must be widespread, causes significant yield loss and there be no cost effective and durable management strategy. For a HPR trait to be pursued by a breeder, there has to be native genetic variation for the trait. Commonly for polygenic traits such as pest resistance, variation in resistance is often continual rather than discrete. If there is no variation present in the germplasm, then it must be attempted to be created through mutagenesis or molecular techniques, e.g. gene editing. In the situation where a breeder is unable to identify HPR in their germplasm, they may look to import/exchange germplasm with international genebanks or institutions. However, licensing and biosecurity requirements can make this process long and complex.

Traits that provide resistance to pests include one or more defence mechanisms that often function in a complex manner and have strong interaction with the environment. Resistance is often defined as chemical or morphological, or genomic regions are identified that control the resistance. Chemical resistance is often observed with compounds such as flavonoids, tannins, and terpenoids (Wink 1988; Hagenbucher et al. 2013), while morphological resistance has been seen in plant architecture and trichome manipulation, such as the okra leaf shape (Miyazaki et al. 2013).

The most common approach to the development of a resistant cultivar can be categorised into three steps: (1) evaluate germplasm to identify sources of resistance, (2) introgress the resistance into an elite background, and (3) deploy the cultivar into current crop management systems. Due to the long timelines and large investment associated with integrating HPR into an elite background, the selection pressure for resistance in the material is high to ensure that only the most resistant material is moved forward for advancement.

In the case of pest resistance, the first step is to thoroughly understand the interaction of the crop and pest. Once this is understood, a wide screen of germplasm (often from diverse genetic backgrounds and includes landraces, international cultivars, and related species) is conducted in a field trial that is purposefully infested with the pest. This is the most efficient way of identifying if any genetic background has resistance to the pest. If a breeder is unsuccessful in identifying resistance to the pest, they may look outside of their available germplasm collection for resistance alleles. Trapero et al. (2016) summarised the genetic sources of HPR and identified traits employed in cotton against pests that are traditionally considered secondary.

Once resistant material is identified, a high-throughput screening method (often in a glasshouse or growth cabinet) is developed to accurately screen for resistance that is reflective of field observations. The methods often involve uniformly infesting the germplasm with the pest and evaluating survival, damage, and other phenotypic scores. High-throughput screening methods allow germplasm to be accurately screened at scale and aid in decreasing the long timelines associated with HPR breeding activities. However, it is critical that the results from the controlled environment screening method are reflective of field resistance. Field resistance is defined as the ability of germplasm to resist a stress compared to control cultivars. However, this can be complex to replicate in a glasshouse environment, and in some case not possible, as often pest behaviour differs in controlled conditions in comparison to the field. Although resistance is often genetically controlled, it is heavily influenced by environmental conditions, including temperature, rainfall, light intensity, crop nutrition, genotype, plant growth stage, and the presence of other pests. The breakdown or lack of identified field resistance in controlled conditions could be attributed to the plant being stressed by factors that are not present in the field. In controlled environment pest screenings, often the intensity of the pest infestation is higher than what is seen under commercial production conditions, and this could cause the resistance to breakdown. Despite the challenges, several screening techniques have been developed to identify resistance to different insects in cotton; thrips (Conzemius et al. 2023), Jassid (Batra and Gupta 1970), whitefly (Sukhija et al. 1986; Butter and Vir 1989), and leafhopper (Mekala 2004).

The ease of the introgression of resistance into an elite background is largely determined by where the source of the resistance originated from, i.e. primary, secondary, or tertiary germplasm, in combination with any ploidy differences of the elite background and the source of resistance, and the genetic complexity of the trait. In cotton, resistance to a pest is commonly polygenic, and often the variation in the level of resistance is continual. The introgression is complicated by the environmental interaction between the resistance traits and the genetic variability of the pest. Due to the polygenic nature of many resistance traits, during the recombination process, resistance may be partially or completely lost, and segregating generations may have to be regenerated to recover the resistance in an elite background.

To aid in the efficiency and speed of HPR breeding activities, molecular tools, such as deoxyribonucleic acid (DNA) markers, can be developed once a source of resistance and a desired phenotype have been identified. One of the approaches that can rapidly locate the genetic region(s) responsible for the resistance is mapping-by-sequencing, which uses two DNA samples pooled separately from the resistant and susceptible individuals of a segregating population. Molecular markers within the genetic region(s) and linked to the resistance are then designed and adopted to guide selection of resistant progeny (Zhu et al. 2017). Egan and Stiller (2022) stated that as pest resistance is often found in wild or unadapted germplasm, the development and uptake of molecular markers are beneficial for breeding efforts—particularly with the introgression into elite backgrounds that contain industry relevant GM trait packages. As GM trait packages with resistance to target pests and herbicides are critical in many cotton producing countries, the introduction of the GM package can only begin once the source of resistance is introgressed into the predominantly cultivated species, *G. hirsutum*, and has stable inheritance.

The current genetic limitations on the use of HPR may be overcome by new genetic engineering techniques. These tools will be helpful to create new diversity in germplasm but will likely add to the already long breeding timeline (Maxwell 1985).

3.3.2 Incorporating HPR Traits in Cotton by Genetic Engineering

The incorporation of genes encoding insecticidal proteins or of ribonucleic acid (RNA) molecules generating small RNAs targeting genes essential for pest survival into crop genomes by genetic engineering provides crops resistance against insect pests. Genetic engineering is thus the fast track for improving HPR of crops, including cotton, to insect pests. But successful implementation of genetic engineering for HPR in cotton requires information on the genes encoding insecticidal proteins and the genes critical for the growth and development of the pests, as well as efficient gene transformation technologies.

Apart from Bt entomotoxic Cry proteins, plant-derived insecticidal proteins have been expressed in cotton to confer resistance to insect pests (Table 3.2). Protease inhibitors (e.g., CpTI) and lectins (e.g., *Allium sativum* leaf agglutinin, ASAL) are the two major types of proteins being used in cotton, because they have been shown to confer resistance to major sap-sucking insects in other plants. A recent study has also investigated the insecticidal efficacy of pectin methyl esterase (PME) in cotton. PME catalyses de-esterification of pectin in cell wall to regulate the response of plants to insect infestation. Bioassay using transgenic cotton expressing two PME genes (from *Arabidopsis thaliana* and *Aspergillus niger*) and two Cry genes (Cry1Fa and Cry32Aa) showed 100%, 95%, and 70% mortality rate for larvae of *H. armigera*, *Earias fabia* (spotted bollworm), and *Pectinophora gossypiella* (pink bollworm), respectively, which was higher than the transgenic cotton expressing only the two Cry genes (Razzaq et al. 2021). Another study found that constitutive expressing Tma12, a chitin-binding domain containing protein from the fern *Tectaria macrodonta*, in cotton can control >99% of a whitefly population. However, it is unclear whether the insecticidal activity of Tma12 is contributed by its chitinase activity (Shukla et al. 2016).

RNA interference (RNAi) is a sequence-specific gene silencing mechanism by which small RNAs repress the activity of target gene post-transcriptionally by cleavage or translational repression of messenger RNA (mRNA). Host-mediated silencing of insect genes (HMGS) by RNAi for control of insect pests is achieved by insect ingestion of host plant tissue containing RNA molecules generating small RNAs that target essential insect genes. The first commercial transgenic double-stranded RNA (dsRNA) product is maize SmartStax® Pro that is engineered to express dsRNA targeting *Diabrotica virgifera* Dvsnf7 and was approved in Canada (2016) and the United States (2017) (Head et al. 2017). The success of using HMGS in pest control depends on the choice of target genes that are critical for the survival of insect pests, the adoption of suitable RNA precursors for generating small RNAs required for RNAi and ensuring high expression of the small RNAs in insect cells.

Many insect genes have been investigated for their suitability to be used as RNAi targets in cotton, such as genes encoding juvenile hormone methyltransferase (Ni et al. 2017) and a subunit of mitochondrial complex I that catalyses NADH dehydrogenation in respiratory chain (Wu et al. 2016). Effectors are released by insects upon feeding on host plants to conquer host defence. Repressing the effector genes abolishes the insect's anti-defence capability against host defence, so insect genes encoding effectors, such as Highly Accumulated Secretory Protein1 and R-like Protein 1 from *H. armigera* (Chen et al. 2019, 2023), are potentially ideal targets of HMGS. Other potential target-insect-genes of HMGS are those acquired by insects from other species via horizontal transfer, because the horizontally transferred genes tend to be insect unique and important for growth and survival of insects, such as the gene encoding squalene synthase in whitefly (Feng et al. 2022).

The most commonly used RNA molecule for delivering RNAi small RNAs in cotton is double-stranded RNA (dsRNA) (Table 3.2). Upon intake transgenic cotton tissues and/or sap by insects, dsRNA is processed by the insect RNA machinery to generate small RNAs to silence the target insect gene. But several other types of RNA precursors have been tested for their efficacy in HMGS by RNAi in cotton. Overexpressing the cotton endogenous microRNA (miRNA) (ghr-miR166b) that is predicted to target mitochondrial ATP synthase of *B. tabaci* (whitefly) in cotton has been shown to reduce the whitefly population and to suppress the spread of viral disease transmitted by whitefly (Wamiq and Khan 2018). A recent study also demonstrated the possibility of using artificial trans-acting siRNAs (tasiRNAs) in simultaneous control of multiple pests and pathogens in cotton. In the study, the Arabidopsis miR173 and the target-insect-genes engineered with a 22-nt upstream sequence recognised by miR173 were integrated into a single construct and overexpressed in cotton, and the endogenous target-insect-genes are repressed by the artificial tasiRNAs generated from the overexpressed target-insect-genes as a result of miR173-mediated cleavage of the engineered target-insect-genes (Karthik et al. 2023).

For HMGS by RNAi to be effective in insect pests, a high quantity of the RNA precursors needs to be delivered into insect cells. One way to achieve that is to mitigate and/or avoid degradation of the RNA precursors by host plant RNAi machinery. Expressing RNA precursors from the chloroplast genome can escape degradation imposed by host plant RNAi machinery that is present in nuclei but absent in chloroplast, leading to accumulation of RNA precursors (Jin et al. 2015; Wu et al. 2022). Alternatively, the small RNA, such as artificial miRNA, targeting insect genes can be incorporated into RNA precursor from insects to avoid host plant RNAi machinery and to deliver the intact RNA precursors into insect cells, ensuring production of high quantity of small RNAs in insect cells to effectively repress target-insect-genes (Bally et al. 2020).

Pest control by genetic engineering requires the transfer DNA (T-DNA) containing the insecticidal gene and/or the RNA precursor generating siRNA (small interfering RNA), miRNA, or tasiRNA targeting insect gene to be stably integrated into the host plant genome. Thereby, generation of transgenic cotton expressing insecticidal gene and/or RNA precursor is the first step to fulfill the goal. Production of transgenic cotton plants involves two major processes: gene transformation and plant regeneration, both are the bottlenecks for genetic engineering in cotton when using *Agrobacterium*-mediated transformation followed by tissue culture, as cotton is a plant species recalcitrant for *Agrobacterium*-mediated gene transformation and plant regeneration, and only few genotypes are transformable. Genotype dependency of cotton transformation hinders direct application of genetic engineering in elite commercial cotton cultivars. As a result, all commercial Bt cotton cultivars were bred by introgression of the Bt gene(s) from transgenic donors into conventional commercial cotton cultivars through a backcross breeding strategy, which usually takes several years to accomplish, compromising rapid adoption of novel gene technologies.

Deploying tissue-culture independent *in planta* transformation methodologies bypasses genotype dependency of cotton transformation. One such methodology is the shoot apical meristem transformation (SAT) that targets T-DNA to the growing shoot apical meristem in vitro by co-cultivation of mature embryos with *Agrobacterium* and allows the generation of transgenic plants ex vitro (Vajhala et al. 2013). This approach has been adopted by the cotton community, particularly in the past a few years, to investigate the functionality of insecticidal genes and the efficiency of HMGS by RNAi in diverse cotton genotypes (Table 3.2), including commercial Bt cotton cultivar with *Cry1Ac* and *Cry2A* (Shad et al. 2021). The major disadvantage of SAT is the low frequency of transgenic events, which can be compensated by using an increased number of mature embryos.

Pollen magnetofection is a novel gene transformation methodology developed recently and has been successfully used to integrate transgenes into cotton genome without tissue culture (Zhao et al. 2017). This approach uses positively charged magnetic nanoparticles as the cargos of recombinant DNA and magnetic field as the force to deliver the recombinant DNA into pollen grains, which are then used in pollination to produce transgenic seeds. Like SAT, this approach suffers from low transformation rate. But with the rapid advancement in the development of diverse nanomaterials, pollen magnetofection or similar technologies is anticipated to be widely adopted by the cotton community for gene transformation because of their relatively low technique demand compared to the approach of *Agrobacterium*-mediated transformation followed by tissue culture.

3.4 How Valuable Is an HPR Trait?

Determining the value of a HPR trait can be complex. Breeding for HPR is a long-term and large investment for a breeding program. There is substantial literature published that highlights the basic research of identifying a trait the confers resistance, but the follow through of the incorporation of these traits into commercial cultivars is minimal. The lack in uptake of the resistance traits is largely governed by the efficacy of the trait together with the practicality of deploying the trait, i.e. the associated deleterious effects and/or negative impact on yield and/or agronomic performance. As previously discussed, the HPR traits that provide resistance can increase susceptibility to other pests or have a severe impact on yield. The impact that a trait has on overall performance is deemed the 'trait trade-off'. Trait trade-offs are common when breeding for novel breeding targets, and the value is ultimately decided by the return on investment (ROI).

Private breeding programs and companies are all largely driven by revenue, so the economic incentive of pursuing a trait is a key factor in their decision-making. The driver of investment decisions in breeding programs and companies is most commonly yield. For pests that do not cause significant damage and for which sustainable control options are in place, the breeding programs are more likely to breed for better local adaption and higher yield. The incentive to develop HPR to a pest has to be large enough that it would cause a loss of revenue for the company, or in a public breeding program, the benefit has to be large enough to justify the time and resources. The seasonal variation of pest/host–plant complexes can complicate the landscape when justifying the ROI for a HPR trait. All breeding programs and companies have defined resources and they must allocate them when and where appropriate. Three scenarios are outlined below to illustrate.

- Scenario 1: HPR to a pest is coming from a species related to *G. hirsutum* and is going to take 15+ years to incorporate the resistance and develop an elite line. The pest is seasonally variable, and in seasons where there is high pest abundance, chemistry is used as a control method. At worst, a 5% yield penalty occurs. In this situation, it would be exceptionally hard for a breeding program to justify the ROI for this trait under the current cotton production system.
- Scenario 2: HPR to a pest has been identified in a *G. hirsutum* landrace cotton and will take 5–8 years to incorporate the resistance and develop an elite line. The pest is consistently present in most seasons and the chemistry used to control it is expensive and unsustainable for growers. Yield penalties of up to 15% have been reported. In this situation, due to the shorter timeline and cost of other control methods, it would be likely that a breeding program could justify the ROI for this trait under the current cotton production system.
- Scenario 3: HPR to a pest has been identified in a species related to *G. hirsutum* and will take 15+ years to incorporate the resistance and develop an elite line. The pest is devastating to the majority of cotton production areas in a large geographic area and has started to become resistant to chemistries used as a control method. Yield penalties of up to 40% are common. In this situation, the development and deployment of the HPR trait are critical for the survival of the industry and would heavily justify the ROI for this trait under the current cotton production system.

However, the suggested outcome of all the above-listed scenarios is dependent on the current control method and the social perception of the control method. Therefore, in some situations, environmental and social pressures may influence the justification of the HPR trait deployment.

The loss of an industry is the biggest imperative for breeding companies to pursue HPR traits. If a pest is damaging enough to collapse an industry, whether it be because the pest has become resistant to chemistries normally used to control them or there are no other available control options, the financial incentive for breeding companies to invest money into identifying and incorporating HPR traits is critical. Potential changes in legislation on the use of certain chemistries could result in an HPR trait becoming easier to justify the ROI. However, it is unlikely that a chemistry would be completely banned, rather restrictions on the timing and conditions of when it could be sprayed would be imposed largely for environmental/sustainability reasons.

3.5 Conclusions

The success of cotton production since the inception of the global industry has been laced with the parallel success of pests. As the effects of climate change alter the environments that we currently grow cotton and previously deemed unsuitable areas to grow cotton become suitable, new pests will consequently move into areas where they were not seen before.

HPR is commonly agreed upon as the most desirable means to control insect pests in cotton crops. HPR will remain a critical component of the production system for cotton, and resistant cultivars will remain the foundation upon which all other components of the crop production and pest control systems operate. Historical breeding efforts have identified HPR traits in cotton germplasm but very few have been advanced through to be incorporated into a commercial cultivar. The deployment of Bt cotton is one of the great GM success stories but does indicate both the frailties of relying on native HPRs, i.e. it was not successful, and the risks associated if Bt resistance occurs.

The continued success of HPR breeding in cotton will rely heavily on the continued screening of germplasm and the parallel development of biotechnological tools that can increase accuracy and speed of breeding decisions. However, the enormity of the work for HPR breeding should not be underestimated, and the adaptation of the pest to resistant breeding lines and the changing seasonal environment must always be revisited and evaluated. Furthermore, there is a never-ending list of insects that cause economic damage to cotton, and it is often questioned why there are not pest-resistant cultivars to all critical insects. Often it is overlooked that an increase in yield potential is a pest tolerance mechanism, and it is commonly forgotten about that current cotton cultivars carry a significant increase in resistance to many of the critical cotton pests.

The ever-advancing molecular tools are making breeding activities easier to deploy HPR through marker-assisted trait introgression to avoid time-consuming, and sometimes not so accurate, phenotyping-based trait selection. Once the causative mechanism is known, the HPR trait can be replicated in the elite genetic background by genotype-independent genetic engineering. The bottleneck limiting the adoption of molecular tools in breeding HPR traits in cotton is lack of target genes, so screening all three gene pools (primary, secondary, and tertiary) for pest resistance is highly encouraged.

To help aid in accommodating for HPR trait development, it is critical that the economic value of the HPR trait is estimated under both infested and non-infested environments so that growers can see the economic benefit. Loss of an industry, pest resistance to chemistries or previously resistant cultivars, and to a lesser extent, legislation that restricts chemical application are the three largest incentives to validate a ROI for an HPR trait.

References

Aboughanem-Sabanadzovic N, Allen T, Wilkerson T, Conner K, Sikora E, Nichols R et al (2019) First report of cotton leafroll dwarf virus in upland cotton (*Gossypium hirsutum*) in Mississippi. Plant Dis 103(7):1798. https://doi.org/10.1094/PDIS-01-19-0017-PDN

Adamczyk JJ Jr, Sumerford DV (2001) Potential factors impacting season-long expression of Cry1Ac in 13 commercial varieties of Bollgard® cotton. J Insect Sci 1(1):13. https://doi.org/10.1093/jis/1.1.13

Adamczyk J Jr, Holloway J, Church G, Leonard B, Graves J (1998) Larval survival and development of the fall armyworm (Lepidoptera: Noctuidae) on normal and transgenic cotton expressing the *Bacillus thuringiensis* CryIA (c) δ-endotoxin. J Econ Entomol 91(2):539–545. https://doi.org/10.1093/jee/91.2.539

Adjei-Maafo I, Wilson L, Thomson N, Blood P (1983) Effect of pest damage intensity on the growth, maturation, and yield of nectaried and nectariless cotton. Environ Entomol 12(2): 353–358. https://doi.org/10.1093/ee/12.2.353

Atak A, Göksel Z, Yılmaz Y (2021) Changes in major phenolic compounds of seeds, skins, and pulps from various *Vitis* spp. and the effect of powdery and downy mildew diseases on their levels in grape leaves. Plants 10(12):2554. https://doi.org/10.3390/plants10122554

Bag S, Roberts PM, Kemerait RC (2021) Cotton leafroll dwarf disease: an emerging virus disease on cotton in the US. Crops Soils 54(2):18–22. https://doi.org/10.1002/crso.20105

Bailey J (1981) Growth comparison of *Heliothis virescens* (F.) larvae fed white, yellow and orange cotton pollen. In: Proceedings of Beltwide cotton production research conference, 6–8 January, New Orleans, Louisiana

Bailey J, Scales A, Meredith W Jr (1984) Tarnished plant bug (Heteroptera: Miridae) nymph numbers decreased on caged nectariless cottons. J Econ Entomol 77(1):68–69. https://doi.org/10.1093/jee/77.1.68

Bally J, Fishilevich E, Doran RL, Lee K, de Campos SB, German MA et al (2020) Plin-amiR, a pre-microRNA-based technology for controlling herbivorous insect pests. Plant Biotechnol J 18(9): 1925–1932. https://doi.org/10.1111/pbi.13352

Balyan HS, Gupta PK, Kumar S, Dhariwal R, Jaiswal V, Tyagi S et al (2013) Genetic improvement of grain protein content and other health-related constituents of wheat grain. Plant Breed 132(5): 446–457. https://doi.org/10.1111/pbr.12047

Barros EM, Torres JB, Ruberson JR, Oliveira MD (2010) Development of *Spodoptera frugiperda* on different hosts and damage to reproductive structures in cotton. Entomol Exp Appl 137(3): 237–245. https://doi.org/10.1111/j.1570-7458.2010.01058.x

Batra G, Gupta D (1970) Screening of varieties of cotton for resistance to jassid. Cotton Grow Rev 47:285–291

Bayer (2023) *ThryvOn*. https://traits.bayer.com/Pages/Thryvon.aspx. Accessed 6 Feb 2023

Benedict J, Leigh T, Hyer A (1983) *Lygus hesperus* (Heteroptera: Miridae) oviposition behavior, growth, and survival in relation to cotton trichome density. Environ Entomol 12(2):331–335. https://doi.org/10.1093/ee/12.2.331

Bhattacharyya A, Leighton SM, Babu C (2007) Bioinsecticidal activity of *Archidendron ellipticum* trypsin inhibitor on growth and serine digestive enzymes during larval development of *Spodoptera litura*. Comp Biochem Physiol Part C Toxicol Pharmacol 145(4):669–677. https://doi.org/10.1016/j.cbpc.2007.03.003

Bi JL, Felton GW, Murphy JB (1997) Biochemical aspects of induced resistance in cotton to the cotton bollworm. In: Beltwide cotton conference, 6–10 January, New Orleans, Louisiana

Bommireddy PL, Leonard BR, Temple J, Price P, Emfinger K, Cook D et al (2011) Field performance and seasonal efficacy profiles of transgenic cotton lines expressing Vip3A and VipCot against *Helicoverpa zea* (Boddie) and *Heliothis virescens* (F.). J Cotton Sci 15(3): 251–259

Bottger G, Sheehan ET, Lukefahr M (1964) Relation of gossypol content of cotton plants to insect resistance. J Econ Entomol 57(2):283–285. https://doi.org/10.1093/jee/57.2.283

Bourland FM, Jones DC (2020) Registration of 'UA212ne', a nectariless cotton cultivar. J Plant Regist 14(3):266–272. https://doi.org/10.1002/plr2.20015

Butter N, Vir B (1989) Morphological basis of resistance in cotton to the whitefly *Bemisia tabaci*. Phytoparasitica 17(4):251

Byrd SA, Collins GD, Edmisten KL, Roberts PM, Snider JL, Spivey TA et al (2016) Leaf pubescence and defoliation strategy influence on cotton defoliation and fiber quality. J Cotton Sci 20(4):280–293. https://doi.org/10.56454/HXYT5212

Chen WB, Lu GQ, Cheng HM, Liu CX, Xiao YT, Xu C et al (2017) Transgenic cotton co-expressing chimeric Vip3AcAa and Cry1Ac confers effective protection against Cry1Ac-resistant cotton bollworm. Transgenic Res 26(6):763–774. https://doi.org/10.1007/s11248-017-0048-8

Chen W, Liu C, Lu G, Cheng H, Shen Z, Wu K (2018) Effects of Vip3AcAa+ Cry1Ac cotton on midgut tissue in *Helicoverpa armigera* (Lepidoptera: Noctuidae). J Insect Sci 18(4):13. https://doi.org/10.1093/jisesa/iey075

Chen CY, Liu YQ, Song WM, Chen DY, Chen FY, Chen XY et al (2019) An effector from cotton bollworm oral secretion impairs host plant defense signaling. Proc Natl Acad Sci 116(28): 14331–14338. https://doi.org/10.1073/pnas.1905471116

Chen X, Liu YQ, Wu MN, Yan L, Chen CY, Mu YP et al (2023) A highly accumulated secretory protein from cotton bollworm interacts with basic helix–loop–helix transcription factors to dampen plant defense. New Phytol 237(1):265–278. https://doi.org/10.1111/nph.18507

Constable G, Bange M (2015) The yield potential of cotton (*Gossypium hirsutum* L.). Field Crop Res 182:98–106. https://doi.org/10.1016/j.fcr.2015.07.017

Constable G, Llewellyn D, Wilson L, Stiller W (2011) An industry transformed-the impact of GM technology on Australian cotton production. Australian Farm Institute

Constable G, Llewellyn D, Walford SA, Clement JD (2015) Cotton breeding for fiber quality improvement. In: Cruz VMV, Dierig DA (eds) Industrial crops. Handbook of plant breeding. Springer, New York, pp 191–232

Conzemius SR, Reay-Jones FP, Greene JK, Campbell BT, Reisig DD, Wang H et al (2023) Field screening of wild cotton, *Gossypium hirsutum*, landraces for resistance to thrips (Thysanoptera: Thripidae). Crop Prot 163:106113. https://doi.org/10.1016/j.cropro.2022.106113

Cordeiro EM, Pantoja-Gomez LM, de Paiva JB, Nascimento AR, Omoto C, Michel AP et al (2020) Hybridization and introgression between *Helicoverpa armigera* and *H. zea*: an adaptational bridge. BMC Evol Biol 20(1):1–12. https://doi.org/10.1186/s12862-020-01621-8

Cotton Australia (2023a) Biotechnology and cotton [online]. https://cottonaustralia.com.au/fact-sheet. Accessed 6 Feb 2023

Cotton Australia (2023b) Crop protection [online]. Australia. https://cottonaustralia.com.au/crop-protection. Accessed 31 Jan 2023

Cui J, Luo J, Van Der Werf W, Ma Y, Xia J (2011) Effect of pyramiding Bt and CpTI genes on resistance of cotton to *Helicoverpa armigera* (Lepidoptera: Noctuidae) under laboratory and field conditions. J Econ Entomol 104(2):673–684. https://doi.org/10.1603/ec09228

de Oliveira RS, Oliveira-Neto OB, Moura HF, de Macedo LL, Arraes FB, Lucena WA et al (2016) Transgenic cotton plants expressing Cry1Ia12 toxin confer resistance to fall armyworm (*Spodoptera frugiperda*) and cotton boll weevil (*Anthonomus grandis*). Front Plant Sci 7:165. https://doi.org/10.3389/fpls.2016.00165

Dhivya K, Sathish S, Balakrishnan N, Udayasuriyan V, Sudhakar D (2016) Genetic engineering of cotton with a novel *cry2AX1* gene to impart insect resistance against *Helicoverpa armigera*. Crop Breed Appl Biotechnol 16(3):205–212. https://doi.org/10.1590/1984-70332016v16n3a31

Din SU, Azam S, Rao AQ, Shad M, Ahmed M, Gul A et al (2021) Development of broad-spectrum and sustainable resistance in cotton against major insects through the combination of *Bt* and plant lectin genes. Plant Cell Rep 40:707–721. https://doi.org/10.1007/s00299-021-02669-6

Downes S, Kriticos D, Parry H, Paull C, Schellhorn N, Zalucki MP (2017) A perspective on management of *Helicoverpa armigera*: transgenic Bt cotton, IPM, and landscapes. Pest Manag Sci 73(3):485–492. https://doi.org/10.1002/ps.4461

Dunnam EW, Clark JC (1939) The cotton aphid in relation to the pilosity of cotton leaves. J Econ Entomol 31:663–666. https://doi.org/10.1093/jee/31.6.663

Dunse K, Stevens J, Lay F, Gaspar Y, Heath R, Anderson M (2010) Coexpression of potato type I and II proteinase inhibitors gives cotton plants protection against insect damage in the field. Proc Natl Acad Sci 107(34):15011–15015. https://doi.org/10.1073/pnas.1009241107

Egan LM, Stiller WN (2022) The past, present, and future of host plant resistance in cotton: an Australian perspective. Front Plant Sci 13:5877. https://doi.org/10.3389/fpls.2022.895877

Ellis M, Silva T, Stiller W, Wilson L, Vaslin M, Sharman M et al (2013) Identification of a new Polerovirus (family *Luteoviridae*) associated with cotton bunchy top disease in Australia. Australas Plant Pathol 42:261–269. https://doi.org/10.1007/s13313-012-0177-8

Ellis MH, Stiller WN, Phongkham T, Tate WA, Gillespie VJ, Gapare WJ et al (2016) Molecular mapping of bunchy top disease resistance in *Gossypium hirsutum* L. Euphytica 210:135–142. https://doi.org/10.1007/s10681-016-1713-3

Feng H, Chen W, Hussain S, Shakir S, Tzin V, Adegbayi F et al (2022) Horizontally transferred genes as RNA interference targets for aphid and whitefly control. Plant Biotechnol J 21:754. https://doi.org/10.1111/pbi.13992

Fitt GP, Mares CL, Wilson LJ, Thomson NJ (1992) Development of resistance to insects in Australian cotton varieties. In: Proceedings of the 6th Australian cotton growers research conference, surfers paradise, Gold Coast, Australia, pp 307–322

Gowda A, Rydel TJ, Wollacott AM, Brown RS, Akbar W, Clark TL et al (2016) A transgenic approach for controlling Lygus in cotton. Nat Commun 7:12213. https://doi.org/10.1038/ncomms12213

Hagenbucher S, Olson DM, Ruberson JR, Wäckers FL, Romeis J (2013) Resistance mechanisms against arthropod herbivores in cotton and their interactions with natural enemies. Crit Rev Plant Sci 32(6):458–482. https://doi.org/10.1080/07352689.2013.809293

Han Q, Wang Z, He Y, Xiong Y, Lv S, Li S et al (2017) Transgenic cotton plants expressing the *HaHR3* gene conferred enhanced resistance to *Helicoverpa armigera* and improved cotton yield. Int J Mol Sci 18(9):1874. https://doi.org/10.3390/ijms18091874

Hanny B, Bailey JC, Meredith W Jr (1979) Yellow cotton pollen suppresses growth of larvae of tobacco budworm. Environ Entomol 8(4):706–707. https://doi.org/10.1093/ee/8.4.706

Head GP, Carroll MW, Evans SP, Rule DM, Willse AR, Clark TL et al (2017) Evaluation of SmartStax and SmartStax PRO maize against western corn rootworm and northern corn rootworm: efficacy and resistance management. Pest Manag Sci 73(9):1883–1899. https://doi.org/10.1002/ps.4554

Hequet E, Abidi N (2002) Processing sticky cotton: implication of trehalulose in residue build-up. J Cotton Sci 6(1):77–90

Hunter RC, Leigh TF, Lincoln C, Waddle BA, Bariola LA (1965) Evaluation of a selected cross-section of cottons for resistance to the boll weevil. In: Bulletin No. 700. Arkansas Experiment Station

ISAAA (2019) Biotech crops drive socio-economic development and sustainable environment in the new frontier. In: Global status of commercialized biotech/GM crops in 2019. SEAsiaCenter, Manila

Jadhav MS, Rathnasamy SA, Natarajan B, Duraialagaraja S, Varatharajalu U (2020) Study of expression of indigenous Bt cry2AX1 gene in T_3 progeny of cotton and its efficacy against *Helicoverpa armigera* (Hubner). Braz Arch Biol Technol 63:e20180428. https://doi.org/10.1590/1678-4324-2020180428

Jenkins JN, Wilson FD (1996) Host plant resistance. In: King EG, Phillips JR, Coleman RJ (eds) Cotton insects and mites: characterization and management. The Cotton Foundation, pp 563–597

Jesus FGD, Junior B, Leal A, Silva Alves GC, Busoli AC, Zanuncio JC (2014) Resistance of cotton varieties to *Spodoptera frugiperda* (Lepidoptera: Noctuidae). Revista Colomb Entomol 40(2): 158–163

Jin S, Singh ND, Li L, Zhang X, Daniell H (2015) Engineered chloroplast dsRNA silences cytochrome p450 monooxygenase, V-ATPase and chitin synthase genes in the insect gut and disrupts *Helicoverpa armigera* larval development and pupation. Plant Biotechnol J 13(3): 435–446. https://doi.org/10.1111/pbi.12355

Johnson M, Toscano N, Reynolds H, Sylvester E, Kido K, Natwick E (1982) Whiteflies cause problems for southern California growers. Calif Agric 36(9):24–26

Jones J, Sloane L, Phillips S (1970) Isogenic studies of red plant color in upland cotton. CABI, Wallingford

Jones JE, Dickson JI, Graves JB, Pavloff AM, Leonard BR, Burris E, et al (1989) Agronomically enhanced insect-resistant cottons. In: Proceedings of the Beltwide cotton production research conference, pp 135–137

Joußen N, Agnolet S, Lorenz S, Schöne SE, Ellinger R, Schneider B et al (2012) Resistance of Australian *Helicoverpa armigera* to fenvalerate is due to the chimeric P450 enzyme CYP337B3. Proc Natl Acad Sci 109(38):15206–15211. https://doi.org/10.1073/pnas.1202047109

Kamminga K, Koppel AL, Herbert DA Jr, Kuhar TP (2012) Biology and management of the green stink bug. J Integr Pest Manag 3(1):C1–C8. https://doi.org/10.1603/IPM12006

Karthik K, Negi J, Rathinam M, Saini N, Sreevathsa R (2021) Exploitation of novel *Bt* ICPs for the management of *Helicoverpa armigera* (Hübner) in cotton (*Gossypium hirsutum* L.): a transgenic approach. Front Microbiol 12:661212. https://doi.org/10.3389/fmicb.2021.661212

Karthik K, Hada A, Bajpai A, Patil BL, Paraselli B, Rao U et al (2023) A novel tasi RNA-based micro RNA-induced gene silencing strategy to tackle multiple pests and pathogens in cotton (*Gossypium hirsutum* L.). Planta 257(1):20. https://doi.org/10.1007/s00425-022-04055-2

Katta S, Talakayala A, Reddy MK, Addepally U, Garladinne M (2020) Development of transgenic cotton (Narasimha) using triple gene *Cry2Ab–Cry1F–Cry1Ac* construct conferring resistance to lepidopteran pest. J Biosci 45:1–11. https://doi.org/10.1007/s12038-020-0006-0

Kaushik P, Andújar I, Vilanova S, Plazas M, Gramazio P, Herraiz FJ et al (2015) Breeding vegetables with increased content in bioactive phenolic acids. Molecules 20(10): 18464–18481. https://doi.org/10.3390/molecules201018464

Khabbazi SD, Khabbazi AD, Özcan SF, Bakhsh A, Başalma D, Özcan S (2018) Expression of GNA and biting site-restricted cry1Ac in cotton; an efficient attribution to insect pest management strategies. Plant Biotechnol Rep 12:273–282. https://doi.org/10.1007/s11816-018-0493-8

Khalil H, Raza ABM, Afzal M, Aqueel MA, Khalil MS, Mansoor MM (2017) Effects of plant morphology on the incidence of sucking insect pests complex in few genotypes of cotton. J Saudi Soc Agric Sci 16(4):344–349. https://doi.org/10.1016/j.jssas.2015.11.003

Koch KG, Chapman K, Louis J, Heng-Moss T, Sarath G (2016) Plant tolerance: a unique approach to control hemipteran pests. Front Plant Sci 7:1363. https://doi.org/10.3389/fpls.2016.01363

Layton M (2000) Biology and damage of the tarnished plant bug, *Lygus lineolaris*, in cotton. Southwest Entomol 23:7–20

Leff B, Ramankutty N, Foley JA (2004) Geographic distribution of major crops across the world. Glob Biogeochem Cycles 18(1):2108. https://doi.org/10.1029/2003GB002108

Li L, Zhu Y, Jin S, Zhang X (2014) Pyramiding *B*t genes for increasing resistance of cotton to two major lepidopteran pests: *Spodoptera litura* and *Heliothis armigera*. Acta Physiol Plant 36: 2717–2727. https://doi.org/10.1007/s11738-014-1642-5

Li Y, Hallerman EM, Wu K, Peng Y (2020) Insect-resistant genetically engineered crops in China: development, application, and prospects for use. Annu Rev Entomol 65:273–292. https://doi.org/10.1146/annurev-ento-011019-025039

Liang S, Luo J, Alariqi M, Xu Z, Wang A, Zafar MN et al (2021) Silencing of a LIM gene in cotton exhibits enhanced resistance against *Apolygus lucorum*. J Cell Physiol 236(8):5921–5936. https://doi.org/10.1002/jcp.30281

Lincoln C, Dean G, Waddle B, Yearian V, Phillips J, Roberts L (1971) Resistance of frego-type cotton to boll weevil and bollworm. J Econ Entomol 64(5):1326–1327. https://doi.org/10.1093/jee/64.5.1326

Liu S, Constable G, Reid P, Stiller W, Cullis B (2013) The interaction between breeding and crop management in improved cotton yield. Field Crop Res 148:49–60. https://doi.org/10.1016/j.fcr.2013.04.006

Llandres AL, Verdeny-Vilalta O, Jean J, Goebel FR, Seydi O, Brévault T (2019) Cotton extrafloral nectaries as indirect defence against insect pests. Basic Appl Ecol 37:24–34. https://doi.org/10.1016/j.baae.2019.05.001

Lubbers EL, Chee PW (2009) The worldwide gene pool of *Gossypium hirsutum* and its improvement. In: Paterson A (ed) Genetics and genomics of cotton. Springer, New York, pp 23–52

Lukefahr M, Houghtaling J (1969) Resistance of cotton strains with high gossypol content to *Heliothis* spp. J Econ Entomol 62(3):588–591. https://doi.org/10.1093/jee/62.3.588

Lukefahr M, Martin D, Meyer J (1965) Plant resistance to five *Lepidoptera* attacking cotton. J Econ Entomol 58(3):516–518

Lukefahr M, Hovghtaling J, Graham H (1971) Suppression of *Heliothis* populations with glabrous cotton strains. J Econ Entomol 64(2):486–488. https://doi.org/10.1093/jee/64.2.486

Luttrell R, Fitt G, Ramalho F, Sugonyaev E (1994) Cotton pest management: part 1. A worldwide perspective. Annu Rev Entomol 39(1):517–526

Maes K (2014) *Spodoptera litura* (taro caterpillar). CABI Compendium. https://doi.org/10.1079/cabicompendium.44520

Mao YB, Tao XY, Xue XY, Wang LJ, Chen XY (2011) Cotton plants expressing CYP6AE14 double-stranded RNA show enhanced resistance to bollworms. Transgenic Res 20:665–673. https://doi.org/10.1007/s11248-010-9450-1

Mao YB, Xue XY, Tao XY, Yang CQ, Wang LJ, Chen XY (2013) Cysteine protease enhances plant-mediated bollworm RNA interference. Plant Mol Biol 83:119–129. https://doi.org/10.1007/s11103-013-0030-7

Maxwell F (1985) Utilization of host plant resistance in pest management. Int J Trop Insect Sci 6(3): 437–442. https://doi.org/10.1017/S1742758400004768

Maxwell FG, Jenkins JN, Parrott W, Buford W (1969) Factors contributing to resistance and susceptibility of cotton and other hosts to the boll weevil, *Anthonomus grandis*. Entomol Exp Appl 12(5):801–810. https://doi.org/10.1111/j.1570-7458.1969.tb02573.x

McCarty J Jr, Meredith W, Jenkins J, Parrott W, Bailey J (1983) Genotype × environment interaction of cottons varying in insect resistance. Crop Sci 23(5):970–973. https://doi.org/10.2135/cropsci1983.0011183X002300050037x

Meade T, Narva K, Storer NP, Sheets JJ, Burton SL, Woosley AT (2017) US Patent No 9:556,453

Mekala DK (2004) Screening upland cotton for resistance to cotton fleahopper (Heteroptera: Miridae). Texas A&M University, College Station, pp 1–75

Meredith WR Jr, Bridge R (1977) Registration of nine germplasm lines of nectariless cotton. Crop Sci 17(1):189

Milam M, Jenkins J, McCart J Jr, Parrott W (1985) Combining tarnished plant bug resistance with Frego bract, vol 264. Mississippi State University, Starkville

Miyazaki J, Stiller WN, Wilson LJ (2013) Identification of host plant resistance to silverleaf whitefly in cotton: implications for breeding. Field Crop Res 154:145–152. https://doi.org/10.1016/j.fcr.2013.08.001

Miyazaki J, Stiller WN, Wilson LJ (2017) Sources of plant resistance to thrips: a potential core component in cotton IPM. Entomol Exp Appl 162(1):30–40. https://doi.org/10.1111/eea.12501

Munro J (1994) Cotton and its production. In: Matthews G, Tunstall J (eds) Insect pests of cotton. CABI, Wallingford, pp 3–26

Ni M, Ma W, Wang X, Gao M, Dai Y, Wei X et al (2017) Next-generation transgenic cotton: pyramiding RNAi and Bt counters insect resistance. Plant Biotechnol J 15(9):1204–1213. https://doi.org/10.1111/pbi.12709

Niles G (1980) Plant breeding and improvement of the cotton plant. Outlook Agric 10(4):152–158. https://doi.org/10.1177/003072708001000401

Nuclear Institute for Agriculture and Biology (2022) Research divisions [online]. Faisalabad, Pakistan

Parkash V, Sharma DB, Snider J, Bag S, Roberts P, Tabassum A et al (2021) Effect of cotton leafroll dwarf virus on physiological processes and yield of individual cotton plants. Front Plant Sci 12:734386. https://doi.org/10.3389/fpls.2021.734386

Parrott WL (1990) Plant resistance to insects in cotton. Fla Entomol 73:392–396. https://doi.org/10.2307/3495457

Perveen SS, Qaisrani T, Siddiqui F, Perveen R, Naqvi S (2001) Cotton plant volatiles and insect's behavior. Pak J Biol Sci 4(5):554–558. https://doi.org/10.3923/pjbs.2001.554.558

Rao NV, Reddy A, Ankaiah R, Mukundan S (1989) Effects of whitefly, *Bemisia tabaci* (Genn.) on cotton yield and associated components. Int J Trop Insect Sci 10(5):685–690

Razzaq A, Ali A, Zafar MM, Nawaz A, Xiaoying D, Pengtao L et al (2021) Pyramiding of cry toxins and methanol producing genes to increase insect resistance in cotton. GM Crops Food 12(1):382–395. https://doi.org/10.1080/21645698.2021.1944013

Reddall A, Ali A, Able J, Stonor J, Tesoriero L, Wright P et al (2004) Cotton bunchy top: an aphid and graft transmitted cotton disease. Australas Plant Pathol 33:197–202. https://doi.org/10.1071/AP03094

Reid P (1998) Siokra V-16. Plant Var J 11:20

Riaz S, Johnson JB, Ahmad M, Fitt GP, Naiker M (2021) A review on biological interactions and management of the cotton bollworm, *Helicoverpa armigera* (Lepidoptera: Noctuidae). J Appl Entomol 145(6):467–498. https://doi.org/10.1111/jen.12880

Ribeiro TP, Arraes FBM, Lourenco-Tessutti IT, Silva MS, Lisei-de-Sa ME, Lucena WA et al (2017) Transgenic cotton expressing Cry10Aa toxin confers high resistance to the cotton boll weevil. Plant Biotechnol J 15:997–1009. https://doi.org/10.1111/pbi.12694

Rios DA, Specht A, Roque-Specht VF, Sosa-Gómez DR, Fochezato J, Malaquias JV et al (2022) *Helicoverpa armigera* and *Helicoverpa zea* hybridization: constraints, heterosis, and implications for pest management. Pest Manag Sci 78(3):955–964. https://doi.org/10.1002/ps.6705

Rizwan M, Abro S, Asif MU, Hameed A, Mahboob W, Deho ZA et al (2021) Evaluation of cotton germplasm for morphological and biochemical host plant resistance traits against sucking insect pests complex. J Cotton Res 4(1):1–8. https://doi.org/10.1186/s42397-021-00093-5

Scheffler JA, Romano GB, Blanco CA (2012) Evaluating host plant resistance in cotton (*Gossypium hirsutum* L.) with varying gland densities to tobacco budworm (*Heliothis virescens* F.) and bollworm (*Helicoverpa* zea Boddie) in the field and laboratory. Agric Sci 3:14–23. https://doi.org/10.4236/as.2012.31004

Schuster MF, Frazier JL (1976) Mechanisms of resistance to *Lygus* spp. in *Gossypium hirsutum* L. In: EUCARPIA/OILB host plant resistance proceedings, pp 129–135

Schuster M, Lukefahr M, Maxwell F (1976) Impact of nectariless cotton on plant bugs and natural enemies. J Econ Entomol 69(3):400–402. https://doi.org/10.1093/jee/69.3.400

Shad M, Yasmeen A, Azam S, Bakhsh A, Latif A, Shahid N et al (2021) Enhancing the resilience of transgenic cotton for insect resistance. Mol Biol Rep 49:5315–5323. https://doi.org/10.1007/s11033-021-06972-z

Shahzad K, Mubeen I, Zhang M, Zhang X, Wu J, Xing C (2022) Progress and perspective on cotton breeding in Pakistan. J Cotton Res 5(1):1–17. https://doi.org/10.1186/s42397-022-00137-4

Shen GM, Song CG, Ao YQY, Xiao YH, Zhang YJ, Pan Y et al (2017) Transgenic cotton expressing CYP392A4 double-stranded RNA decreases the reproductive ability of *Tetranychus cinnabarinus*. Insect Sci 24(4):559–568. https://doi.org/10.1111/1744-7917.12346

Shukla AK, Upadhyay SK, Mishra M, Saurabh S, Singh R, Singh H et al (2016) Expression of an insecticidal fern protein in cotton protects against whitefly. Nat Biotechnol 34(10):1046–1051. https://doi.org/10.1038/nbt.3665

Siddiqui HA, Asif M, Asad S, Naqvi RZ, Ajaz S, Umer N et al (2019) Development and evaluation of double gene transgenic cotton lines expressing Cry toxins for protection against chewing insect pests. Sci Rep 9(1):11774. https://doi.org/10.1038/s41598-019-48188-z

Singh AK, Paritosh K, Kant U, Burma PK, Pental D (2016) High expression of Cry1Ac protein in cotton (*Gossypium hirsutum*) by combining independent transgenic events that target the protein to cytoplasm and plastids. PLoS One 11(7):e0158603. https://doi.org/10.1371/journal.pone.0158603

Stiller WN, Wilson IW (2014) Australian cotton germplasm resources. In: Abdurakhmonov IY (ed) World cotton germplasm resources. IntechOpen, Croatia, pp 1–34

Stout MJ (2014) Host-plant resistance in pest management. In: Abrol DP (ed) Integrated pest management. Academic Press, pp 1–21

Sukhija H, Butter N, Singh J (1986) Determination of the economic threshold of whitefly, *Bemisia tabaci* Genn., on American cotton in the Punjab. Int J Pest Manag 32(2):134–136. https://doi.org/10.1080/09670878609371048

Tabashnik BE, Carrière Y (2019) Global patterns of resistance to Bt crops highlighting pink bollworm in the United States, China, and India. J Econ Entomol 112(6):2513–2523. https://doi.org/10.1093/jee/toz173

Tabashnik BE, Liesner LR, Ellsworth PC, Unnithan GC, Fabrick JA, Naranjo SE et al (2021) Transgenic cotton and sterile insect releases synergize eradication of pink bollworm a century after it invaded the United States. Proc Natl Acad Sci 118(1):e2019115118. https://doi.org/10.1073/pnas.2019115118

Tarazi R, Vaslin MF (2022) The viral threat in cotton: how new and emerging technologies accelerate virus identification and virus resistance breeding. Front Plant Sci 13:851939. https://doi.org/10.3389/fpls.2022.851939

Tariq M, Tabassum B, Bakhsh A, Farooq AM, Qamar Z, Akram F et al (2022) Heterologous expression of cry1Ia12 insecticidal gene in cotton encodes resistance against pink bollworm, *Pectinophora gossypiella* (Lepidoptera: Gelechiidae); an alternate insecticidal gene for insect pest management. Mol Biol Rep 49(11):10557–10564. https://doi.org/10.1007/s11033-022-07824-0

Thomson N (1984) Commercial growing of Sicot 3. In: Proceedings from the 1984 Australian cotton growers research conference. Toowoomba, Australia, pp 66–69

Thomson N (1994) Commercial utilisation of the okra leaf mutant of cotton—the Australian experience. In: Constable GA, Forrester NW (eds) Challenging the future: proceedings of the world cotton research conference, Brisbane, Australia. CSIRO, Melbourne, pp 393–401

Thomson N, Reid P, Williams E (1987) Effects of the okra leaf, nectariless, frego bract and glabrous conditions on yield and quality of cotton lines. Euphytica 36(2):545–553. https://doi.org/10.1007/BF00041501

Tian G, Cheng L, Qi X, Ge Z, Niu C, Zhang X et al (2015) Transgenic cotton plants expressing double-stranded RNAs target HMG-CoA reductase (HMGR) gene inhibits the growth, development and survival of cotton bollworms. Int J Biol Sci 11(11):1296. https://doi.org/10.7150/ijbs.12463

Tingey WM, Leigh TF, Hyer AH (1975) *Lygus hesperus*: growth, survival, and egg laying resistance of cotton genotypes. J Econ Entomol 68(1):28–30. https://doi.org/10.1093/jee/68.1.28

Trapero C, Wilson IW, Stiller WN, Wilson LJ (2016) Enhancing integrated pest management in GM cotton systems using host plant resistance. Front Plant Sci 7:500. https://doi.org/10.3389/fpls.2016.00500

USDA (2016) Plant variety protection office. United States Department of Agriculture, Washington, DC

USDA (2022) Cotton and products annual. Office of Agricultural Affairs, New Delhi, pp 1–9

Vajhala CS, Sadumpati VK, Nunna HR, Puligundla SK, Vudem DR, Khareedu VR (2013) Development of transgenic cotton lines expressing *Allium sativum* agglutinin (ASAL) for enhanced resistance against major sap-sucking pests. PLoS One 8(9):e72542. https://doi.org/10.1371/journal.pone.0072542

Vanti GL, Katageri IS, Inamdar SR, Hiremathada V, Swamy BM (2018) Potent insect gut binding lectin from *Sclerotium rolfsii* impart resistance to sucking and chewing type insects in cotton. J Biotechnol 278:20–27. https://doi.org/10.1016/j.jbiotec.2018.04.018

Voora V, Larrea C, Bermudez S (2020) Global market report: cotton. International Institute for Sustainable Development (IISD), Winnipeg

Vyavhare S (2017) Fall armyworm in cotton *Spodoptera frugiperda* (J.E. Smith). Texas A&M AgriLife Extension Service, College Station. http://lubbock.tamu.edu/files/2017/07/Fall-Armyworm-2017.pdf

Wamiq G, Khan JA (2018) Overexpression of ghr-miR166b generates resistance against *Bemisia tabaci* infestation in *Gossypium hirsutum* plants. Planta 247:1175–1189. https://doi.org/10.1007/s00425-018-2852-7

Wannamaker W (1957) The effect of plant hairiness of cotton strains on boll weevil attack. J Econ Entomol 50(4):418–423

Wendel JF, Brubaker C, Alvarez I, Cronn R, Stewart JM (2009) Evolution and natural history of the cotton genus. In: Paterson AH (ed) Genetics and genomics of cotton. Plant genetics and genomics: crops and models. Springer, New York, pp 3–22

Wilson FD, Flint HM (1987) Pink bollworm resistance and lint yield of a nectariless, okra-leaf germplasm line. The University of Arizona, Tucson, pp 157–158

Wilson F, George B (1981) Breeding cotton for resistance to pink bollworm. In: Proceedings of the Beltwide cotton research conference, pp 63–65

Wilson LJ, Whitehouse ME, Herron GA (2018) The management of insect pests in Australian cotton: an evolving story. Annu Rev Entomol 63:215–237. https://doi.org/10.1146/annurev-ento-020117-043432

Wink M (1988) Plant breeding: importance of plant secondary metabolites for protection against pathogens and herbivores. Theor Appl Genet 75:225–233. https://doi.org/10.1007/BF00303957

Wu J, Tian Y (2019) Development of insect-resistant transgenic cotton with chimeric TVip3A* accumulating in chloroplasts. In: Transgenic cotton methods protocols. Humana Press, New York, pp 281–292. https://doi.org/10.1007/978-1-4939-8952-2_23

Wu XM, Yang CQ, Mao YB, Wang LJ, Shangguan XX, Chen XY (2016) Targeting insect mitochondrial complex I for plant protection. Plant Biotechnol J 14(9):1925–1935. https://doi.org/10.1111/pbi.12553

Wu M, Dong Y, Zhang Q, Li S, Chang L, Loiacono FV et al (2022) Efficient control of western flower thrips by plastid-mediated RNA interference. Proc Natl Acad Sci 119(15):e2120081119. https://doi.org/10.1073/pnas.2120081119

Yasmin T, Tabbasam N, Ullah I, Asif M, Zafar Y (2008) Studying the extent of genetic diversity among *Gossypium arboreum* L. genotypes/cultivars using DNA fingerprinting. Genet Resour Crop Evol 55:331–339. https://doi.org/10.1007/s10722-007-9238-1

Zafar MM, Mustafa G, Shoukat F, Idrees A, Ali A, Sharif F et al (2022) Heterologous expression of *cry3Bb1* and *cry3* genes for enhanced resistance against insect pests in cotton. Sci Rep 12(1): 10878. https://doi.org/10.1038/s41598-022-13295-x

Zhang L, Wei Y, Wei L, Liu X, Liu N (2023) Effects of transgenic cotton lines expressing dsAgCYP6CY3-P1 on the growth and detoxification ability of *Aphis gossypii* Glover. Pest Manag Sci 79(1):481–488. https://doi.org/10.1002/ps.7220

Zhao X, Meng Z, Wang Y, Chen W, Sun C, Cui B et al (2017) Pollen magnetofection for genetic modification with magnetic nanoparticles as gene carriers. Nature Plants 3(12):956–964. https://doi.org/10.1038/s41477-017-0063-z

Zhao S, Wang F, Zhang Q, Zou J, Xie Z, Li K et al (2021) Transformation and functional verification of *Cry5Aa* in cotton. Sci Rep 11(1):2805. https://doi.org/10.1038/s41598-021-82495-8

Zhu Q-H, Wilson I, Llewellyn D (2017) Mapping-by-sequencing enabled fast forward genetics in crops with complex genomes. CABI Rev 2017:1–12. https://doi.org/10.1079/PAVSNNR201712016

Current Status of Host Plant Resistance to Insects in Rice and Future Perspectives

4

Ayyagari Phani Padmakumari, Suneetha Kota, and Raman Meenakshi Sundaram

Abstract

Rice is grown in diverse ecologies, and the crop is attacked by large number of insect pests of which planthoppers, stem borers, leaf folders and gall midge cause considerable yield losses, which may vary from 10% to 90%. This chapter reports the current status of research in host plant resistance (HPR) to major rice pests, as HPR is considered as the most economical and eco-friendly component of insect pest management. The journey from various phenotyping techniques to identification of resistant sources from diverse gene pools, through precise studies on mechanisms and genetics of resistance to genes, etc., is discussed at length. For pests like yellow stem borer, where very few sources of tolerance are available, novel strategies that have been developed for tackling this insect pest, like exploitation of host plant susceptibility, induced resistance, Bt transgenic and use of RNAi tools for pest suppression are discussed. In pests like gall midge and brown planthopper, HPR has an influence on the evolution of biotypes and on the symbionts that are harboured. To address the problem of multiple biotic stresses, marker-assisted backcross breeding is considered a potent tool, which helps in introgression of known resistance genes in the desired varietal background. With the availability of genomics resources for both rice host and insect pests like brown planthopper, white-backed planthopper, gall midge, leaf folder, yellow and striped stem borer, novel techniques like genomic selection, gene editing to address the issue of pest resistance in some of the incalcitrant traits are also discussed. Though development and deployment of a durable resistant variety with phenotypically acceptable traits are the need of the hour, the dire need for precision phenotyping keeping in view the pest behaviour and the target traits to

A. P. Padmakumari (✉) · S. Kota · R. M. Sundaram
ICAR-Indian Institute of Rice Research, Rajendranagar, Hyderabad, India
e-mail: Padma.APK@icar.gov.in; suneetha.kota@icar.gov.in; r.sundaram@icar.gov.in

S. Kumar, M. Furlong (eds.), *Plant Resistance to Insects in Major Field Crops*,
https://doi.org/10.1007/978-981-99-7520-4_4

address the location-specific needs is emphasized. Further, the need for a designer rice, addressing the location-specific needs, which involves both development of a resistant variety on one hand followed by strategic deployment of resistance genes as per the pest population prevalent in a location, consumer preference and market demands are emphasized in this chapter.

Keywords

Rice · Insect pests · Damage · Host plant resistance · Screening protocols · Molecular tools · Breeding strategies · Resistant varieties · IPM

4.1 Introduction

Rice is grown in several countries across the globe and is the staple cereal for teeming billions, and for many Asian countries including India, it is a way of life. With the rapid increase in the population growth, and the responsibility to feed the mankind with nutritious diet, there lies an unprecedented challenge for the rice workers to meet the demands for higher production, productivity and nutritious diets to cater to the needs of the increasing needs of the poverty stricken, malnourished mankind. Achieving food security through rice is challenged by insect pests, which reduce yields by up to 90%.

4.1.1 Insect Pests of Rice

Among the various pests attacking the rice plant, namely planthoppers, stem borers, leaf folders are universal and infest the rice crop at all stages of crop growth, i.e., from nursery to grain-filling stage causing significant economic loss to the crop. There are other pests like rice gall midge, thrips, leaf mites, which cause economic loss at vegetative phase (Plate 4.1). Some of the pests like gundhi bug, ear-cutting caterpillar attack the crop at reproductive phase. Of the many pests that infest the rice crop, most important are stem borers: yellow stem borer (YSB), *Scirpophaga incertulas* (Walker); striped stem borer (SSB), *Chilo suppressalis* (Walker); and pink stem borer (PSB), *Sesamia inferens* (Walker); planthoppers: brown planthopper (BPH), *Nilaparvata lugens* Stål; white-backed planthopper (WBPH), *Sogatella furcifera* Horvath; and small brown planthopper (SBPH), *Laodelphax striatellus*; leafhoppers: green leafhopper (GLH), *Nephottetix virescens* (Distant); green rice leafhopper (GRL/GRH), *Nephotettix cincticeps* Uhler and zigzag leafhopper (ZLH), *Recilia dorsalis* (Motchulsky); gall midges (GM): Asian rice gall midge (ARGM), *Orseolia oryzae* (Wood–Mason); and African rice gall midge (AfRGM), *Orseolia oryzivora* Harris & Gagné; and leaffolders: *Cnaphalocrocis medinalis* (Guenée) and *Marasmia* spp. In the USA, the fall armyworm, *Spodoptera frugiperda* (J.E. Smith), rice water weevil, *Lissorhoptrus oryzophilus* (Kuschel) and Mexican rice borer, *Eoreuma loftini* (Dyar) are the serious pests.

Plate 4.1 Major insect pests of rice crop

Despite the dynamic nature of pest infestations, stem borers and rice brown planthopper continue to be the two regular and major threats limiting rice production across the Asian countries. Both these pests inflict damage to the crop at all crop growth stages. Yellow stem borer (YSB), *Scirpophaga incertulas* (Walker), a monophagous pest with exclusive host specificity towards rice, is considered to be a dominant pest across the rice-growing regions. Some of the other stem borers like striped stem borer, *Chilo suppressalis* Walker, dark headed striped stem borer (DHSB), *Chilo polychrysa* Meyrick; gold fringed borer (GFB) *Chilo auricilius* Dudgeon; spotted stem borer, *Chilo partellus* Swinhoe and other *Scirpophaga* spp. are either specific to particular crop stage or are reported from rice and rice-based cropping systems. Apart from these two, there are pests like leaffolders, gall midge, rice hispa, caseworm, thrips and gundhi bugs which do cause economic losses but are endemic to certain pockets. Planthoppers and leafhoppers are also vectors of many plant viruses.

4.1.2 Losses Caused by Insect Pests in Rice

Oerke (2006) estimated 37.5% losses at global level due to animals, weeds, pathogens and viruses in rice crop. Average annual losses to rice borers in China, India, Bangladesh and Southeast Asia were approximately 5–10%, though losses in individual fields may reach 50–60% (Rahman et al. 2004). The losses in the rice crop due to animal pests increased from 15% in pre-green revolution era to 25% in the post-green revolution era in India, though at global level there was a decrease in the crop losses (Dhaliwal et al. 2007). Dhaliwal et al. 2010 estimated 25% crop loss due to insect pests in rice crop. The yield losses vary from one region to another, but range from 1.2 to 2.2 tons/ha due to the combined attack of diseases, insects and weeds in Asia (Savary et al. 2012). It is common to note that though infestation/

damage to rice appears high, the losses may be limited and control measures, sometime would be uneconomical. The intensity and effect of damage depend on the stage of the crop and confluence of the many biotic and abiotic factors that influence crop growth. Pest profile varies with the location, cultivation practices, nutritional profile of the crop, adopted management practices and the prevailing cropping system. Hence, insect pest management should be addressed based on the location-specific requirements. The losses caused by insect pests are very high when the yield components like productive tiller or grain yield are affected as in the case of BPH, WBPH, SB or GM or when the quality of the grain is affected as in the case of gundhi bug or stored grain pests. Nymphs and adults of planthoppers suck phloem sap (Auclair and Baldos 1982; Khan and Saxena 1984) from leaves and leaf sheaths turning them yellow and reducing plant height, tillering and filled grains. Plants infested by the BPH before maximum tillering usually have fewer panicles per unit area and fewer grains per panicle; a planthopper attack after the heading stage affects the percentage of ripened grain and grain weight. The time of insect attack in relation to plant growth, intensity of injury (or the population density of insects), duration of the attack and environmental factors affecting both insect activities and plant growth influence the relationship between an insect infestation and its effect on yield. Severe infestation of the insect leads to hopper-burn resulting in complete drying and death of the crop. Based on 770 experimental units from 28 years' data (All India Coordinated Rice Improvement Project from 1965 to 1992), empirical yield loss estimates caused by stem borers over various rice ecosystems due to 1% dead heart or white ear head or to both phases of damage were 2.5% (or 108 kg/ha), 4.0% (174 kg/ha) and 6.4% (278 kg/ha), respectively (Muralidharan and Pasalu 2006). White earhead damage had a much greater impact on rice yield in the irrigated ecosystem than due to dead heart, as it occurs later in the crop stage when no compensation is possible, thus resulting in direct loss of a yielding panicle. The grain yield loss from damage at the two phases, namely, dead heart and white earhead, is more than additive. On the other hand, potential yield gains of at least 10–20% of the current yields may be achieved through effective pest management techniques (Oerke 2006; Savary et al. 2012). Puri and Mote (2003) predicted that brown planthopper and gall midge in rice and pink stem borer in sorghum would assume serious proportions in future due to changes in the ecosystems and habitats. Deutsch et al. (2018) predicted that with every degree Celsius of increase in temperature, the loss caused by insects would increase by 10–25% which is more pronounced in temperate climates. This is imminent and evident with the impact of climate change on the crop growth life cycles and availability of the host plants beyond the traditional seasons, thus ensuring survival and multiplication of the pests.

4.1.3 Status of Host Plant Resistance in Rice IPM

In an agro-ecological system, the growth of insect pest population depends on the initial colonization, natural population growth rate and the impacts of control measures. Pest management strategies should focus on limiting the initial pest recruitment, colonization and build-up of the population (Cheng et al. 2008) before reaching the economic threshold level. Pest management strategies may include genetic resistance, acquired resistance during cultivation practices, improved environmental resistance, augmentative and inundative release of selected natural enemies. Host plant resistance is a built-in approach which is inherently the genetic makeup of the plant. It is environmentally safe, mostly compatible with other components of pest management and does not require additional equipment or cost except for the initial cost of development of the variety/breeding line. Large variation exists in the reaction of the rice pests to germplasm, breeding lines and varieties. A wealth of literature and information are available on the resistance mechanisms, genes conferring resistance to BPH, WBPH, GLH, ARGM, AfRGM, leaf folder and to some extent for pests like rice hispa, gundhi bug and thrips. Thus, integrated pest management through deployment of host plant resistance should be thought of within the context of crop management. Litsinger (2005) opined that with the advances in the breeding methodology and change in plant type with high tillering capacity and improved yield potential as compared to traditional varieties, the capacity for compensation from insect pest damage is enhanced. Appropriate crop management practices also supplement the compensatory ability. Prior knowledge on the crop compensatory ability and severity of crop stress acting on it would help the farmers in accurate decision-making to take up control measures.

4.2 Phenotyping for Insect Resistance

Accurate phenotyping for insect resistance, elucidating the mechanisms and genetics, serves as a back bone on which all the modern biotechnological tools and breeding programmes depend upon. So, a precise phenotyping of a resistant source is very much essential and is the base on which all the further studies and results rely upon.

4.2.1 Screening Methodology

Germplasms from various sources, breeding material at different stages of development with known sources of resistance, are evaluated to identify resistance for various pests. Standard screening protocols are available for various rice pests. Protocols for screening and identifying sources of resistance and studying various mechanisms of resistance were reviewed at length for all the rice pests, namely plant and leafhoppers, stem borers, gall midge, leaf folders, rice thrips, gundhi bug, case worm, whorl maggot, etc., by many rice workers (Heinrichs et al. 1985; Panda and Khush 1995; Bentur et al. 2011), and the protocol for screening and scoring for damage by rice pests was summarized in IRRI (2013) which till date is followed by rice workers all over the world. Later on, based on the need, many protocols were designed, developed, revised and improvised for deeper understanding of rice–pest interactions and studying their mechanisms of resistance. Some of these techniques are detailed below:

4.2.1.1 Field Screening Techniques

Screening of large quantum of material can be taken up at "hot spots" or endemic areas where occurrence of natural populations is regular. Infestation by natural populations helps to identify the promising lines which are further retested under stringent conditions to identify the resistant cultures (Plate 4.2a). Field screening requires thorough knowledge of the pest incidence pattern so as to adjust the time of sowing/planting in order to ensure pest infestation at the right crop stage. The major limitations of evaluating germplasm in the field are seasonality, unpredictability with reference to incidence of the insect pests, uneven distribution of the pest, asynchrony between insect pest incidence and the stage of the crop for effective screening, interference from other biotic and abiotic factors rendering field screening unreliable at times. Under most of the field conditions in production systems, it is common to have two or more biotic stresses at the same time. For example, WBPH colonizes the plant first, but as the crop advances, BPH takes over. Therefore, in some of the locations there would be a mixture of both BPH and WBPH, and the resultant phenotype would be due to damage by both the insect pests. To address the issue of uneven infestation, staggered sowing to coincide with the time of the specific pest incidence in the field, followed by supplementing the infestation, is suggested.

4.2.1.1.1 Augmentative Releases

Augmentative release of insects at specific stage can be carried out so as to ensure reliable screening of the material. These techniques are currently available for BPH, gall midge, yellow stem borer, leaffolder and hispa (Plate 4.2b). To develop screening techniques, a deeper understanding of the insect biology, behaviour and its interaction with the host is very much essential. Under All India Coordinated Research Programme on Rice (AICRPR) in NARS, India, multi-location evaluation of rice lines is carried out for major pests like planthoppers, GM, YSB, LF and rice hispa. Details of phenotypic screening protocols for the major rice pests are described below:

BPH

A susceptible variety is raised as infestor rows around the test material. A polythene sheet barrier of 2.5 ft height is erected all around the planting area within 15 days after planting. For better results, it is desirable to plant test entries in longitudinal strips not wider than 2 m and each strip separately covered around with polythene sheet. Currently, TN1 is used as the susceptible check. PTB 33 and RP2068-18-3-5 are the resistant checks for BPH and MO1 for WBPH screening. Adults and nymphs of planthoppers collected from adjacent areas or greenhouse culture are released to build up the pest pressure. A spray of Deltamethrin (0.002%) is given on infestor/feeder rows to ensure further build-up of the pest population. Field screening can identify field tolerance of the genotypes.

Plate 4.2 (**a**) Field screening of rice lines based on natural pest incidence at hot spots. Damage by (A) Gall midge (courtesy: Dr Atanu Seni); (B) Planthoppers (courtesy: Dr Mallikarjuna Rao); (C) Stem borer—dead heart (Dr AP Padmakumari); (D) Stem borer (white ear: Dr Sanjay Sharma). (**b**) Screening of rice lines through insect augmentation. (A) Screening against BPH (photo courtesy: Dr JS Bentur); (B) Release of GM adults in cages at IIRR (photo courtesy: Dr AP Padmakumari); (C) Release of leaffolder adults (photo courtesy: Dr Ch. Padmavathi); (D) Augmenting with YSB egg mass (photo courtesy); (E) Release of two neonate YSB larvae per tiller to augment the infestation, E-1 in pot, E-2 in field (Dr AP Padmakumari); (F) Screening in net house against adult rice hispa (F2)-F1-hispa damage (photo courtesy: Dr A Srivastava)

b

Plate 4.2 (continued)

Yellow Stem Borer

Augmentative release is carried either through pinning of the egg mass at black head stage or release of two neonate larvae per tiller at maximum tillering stage in vegetative phase for dead heart damage and at booting stage in reproductive phase for white ear damage as the damage by the pest differs between crop phases. The material thus infested is scored for dead heart, white ear damage and grain yield in the infested plants (Bentur et al. 2011; Padmakumari and Ram 2012; Padmakumari and Katti 2018; Jeer et al. 2018). Pusa Basmati 1 is the susceptible check; TKM6, Sasyasree and W1263 are the resistant checks used for comparison. The methodology is highly dependent on availability of adult YSB in the field. In order to synchronize the infestation at appropriate crop stage, staggered sowings are suggested. It should be noted that the resistance/tolerance to YSB can be at both vegetative and reproductive phase or only at vegetative phase and susceptible at reproductive phase or vice versa. This protocol is useful to screen any material for yellow stem borer.

Gall Midge

The material to be screened can be raised in two rows. Generally, late planting is preferred in hot spot areas to ensure sufficient infestation and catch up with the pest incidence. For phototropic insects like gall midge, it is suggested to deploy incandescent bulb around the field to attract adults for oviposition. TN1 is the susceptible check.

Leaffolder

In case of leaffolder, field-collected adults are released on to the screening material covered by nylon net at 40–60 days after transplanting (DAT). The damaged leaves are counted at 60–80 DAT (in the reproductive phase) or 20 DAT after release of the adults. TN1 is the susceptible check, and W1263 is used as a resistant check (Bentur et al. 2011). Bentur and Kalode (1990) studied the mechanism of resistance through a feeding test which involved caging of individual V instar larvae for 48 h on 30–45-day-old plants of test varieties with PTB12 as resistant check and recording area of leaf damage when the leaf nitrogen content varied from 2% to 3.4%. As leaffolder has a behaviour to undergo precocious pupation when exposed to nutritional stress (resistant variety), a rapid screening test was developed wherein single third instar leaffolder larva reared on TN1 in greenhouse was released on to the leaves of test entries and allowed to feed for 48 h (Chintalapati et al. 2017). After 48 h of confinement, the larvae were collected back and the leaf damage was scored for extent of feeding. Using this technique, promising RILs developed from W1263/TN1 were screened and identified for leaffolder resistance (Chintalapati et al. 2019). A study using a doubled haploid population of CJ06/TN1 uncovered five QTLs for rice leaffolder (RLF) resistance on chromosomes 1, 2, 3, 4, and 8. The effect of a single locus is limited, but QTLs pyramiding markedly improved leaffolder resistance in rice (Rao et al. 2010).

Rice Hispa

The adult and grubs of this pest cause extensive damage. Srivastava and Kamini (2020) identified resistant lines by collecting adults from field and releasing @40 adults/m^2 on the test material at maximum tillering phase and scoring for leaf damage by the adults.

4.2.1.2 Greenhouse Techniques

Screening for BPH, WBPH and GLH is carried out as per the Standard Seed box Screening Technique (SSST) (Heinrichs et al. 1985). It is to be noted that the status of resistance is dependent on crop age. In order to characterize resistance for a particular pest, it is most appropriate to go in for screening of the shortlisted entries under greenhouse conditions through modified seed box screening technique (MSST). MSST provides a reliable method of identifying field resistance under greenhouse conditions and is expected to be a useful tool in breeding high-yielding rice cultivars with durable resistance to BPH (Velusamy et al. 1986). A rapid and quick mass screening technique at seedling stage against brown planthopper by following the standard evaluation system (SES) was employed by Soundararajan and Jeyaprakash (2019). Through this technique all the susceptible lines can be identified so that large germplasm can be screened within a short time. Gall midge is maintained on susceptible rice cv. TN1 plants, and screening is carried out as per Vijayalakshmi et al. (2006). Cheng et al. (2021) reported that 24 seedlings treated with at least five female ARGM adults were efficacious and could offer adequate samples for insect development observation or molecular biological studies.

Though greenhouse screening is a fool-proof method, it is highly labour-intensive and laborious. The basic requirement for greenhouse screening is the continuous maintenance of insect population conditioned on a susceptible check entry and infestation of the test material at the right stage of the plants with right stage of the insect. Therefore, depending on the need of experimentation, the desired number of insects can be maintained.

Kobayashi (2015) emphasized the role of development and maintenance of a virulent population on susceptible host from single progeny in characterizing resistance so that the base line data related to identification of resistant sources would be accurate. This rule would apply to all insects in the screening programme and caution should be exercised in the maintenance and release of the avirulent test insect populations as the reaction to a particular genotype would vary with the virulence of the insect. Authenticity of the variations exhibited is highly dependent on the different seed sources, their purity and/or accessions used in these evaluations.

4.2.2 Mechanisms of Resistance

Host plant resistance mechanisms are primarily categorized as antixenosis, antibiosis and tolerance (Painter 1968). Antixenosis is mainly due to host-plant characters which are responsible for non-preference by the insects for shelter, oviposition and/or feeding because of the presence of morphological or chemical factors. Antibiosis denotes the adverse effect of the biochemical and biophysical factors of the host plant on the biology (survival, development and reproduction) of the insect pests and their progeny. Both non-preference and antibiosis interfere with insect behaviour and metabolism.

Tolerance is the ability of the host plant to grow and produce satisfactory yield in the presence of pests that would cause loss in a susceptible plant. The phenomenon of tolerance is generally cumulative and a result of interacting with the host plant. These include general vigour, inter- and intra-plant compensatory growth, wound compensation, mechanical strength of tissues and organs and nutrient and growth regulator partitioning (Velusamy and Heinrichs 1986). This mechanism does not provide any selection pressure on the insect pests that can lead to the development of insect biotype. Tolerant varieties have higher decision thresholds than susceptible varieties resulting in reduction of insecticide usage and enhanced natural enemy activity (Panda and Heinrichs 1983). Usually, vertical resistance governed by single/major genes leads to development of biotypes. If this first line of defence provided by major genes breaks down, the tolerance mechanism becomes a secondary line of defence and will continue to function while efforts can be directed towards the release of a new resistant variety (Litsinger 2005).

4.2.2.1 Planthopppers

The predominant mechanisms operating for planthoppers are non-preference for oviposition; for feeding quantified through honey dew tests, days to wilt and functional plant loss index and the tolerance mechanism (Heinrichs et al. 1985; Panda and Khush 1995) which operates with age and plant damage. Biochemical factors like phenols, salicylic acid, shaftosides and sucking inhibitors like oxalic acid, soluble silicic acid, were attributed to lowering the damage in rice by BPH and WBPH. Various studies pointed out the significance of feeding behaviour of planthoppers in quantifying resistance. Use of the electrical penetration graph (EPG) technique has revealed details of host plant suitability and feeding behaviour, including probing frequency, duration of stylet insertion, duration of phloem or xylem feeding and honeydew secretion (Velusamy et al. 1986; Walling and Thompson 2013). Host searching and feeding behaviour of BPHs on rice were implicated in host resistance together with flavonoid glycosides, which are phloem secondary metabolites that have anti-feeding activity against BPH (Grayer et al. 1994), the resistance factors are thought to occur within the phloem and in the pathways to the vascular system (Walling and Thompson 2013). BPH feeding up-regulated callose synthase genes and induced callose deposition in the sieve tubes at the point where the stylets were inserted. In the resistant plants, the compact callose remained intact while in susceptible plants, genes encoding β-1,3-glucanases were activated, causing unplugging of the sieve tube occlusions (Hao et al. 2008). Shen et al. (2022) reported that the plant elicitor peptide (Pep), rice PEP RECEPTORs (PEPRs) and PRECURSORs of PEP (PROPEPs), particularly *OsPROPEP*, were transcriptionally induced in leaf sheaths upon BPH infestation conferring resistance to BPH. Knockout of *OsPEPRs* impaired rice resistance to BPH, whereas exogenous application of *OsPep3* improved the resistance. Hormone measurement and co-profiling of transcriptomics and metabolomics in *OsPep3*-treated rice leaf sheaths suggested potential contributions of jasmonic acid biosynthesis, lipid metabolism and phenylpropanoid metabolism to OsPep3-induced rice immunity. In case of BPH, primary tiller not infested by the brown planthopper translocates nutrients and assimilates into the main shoot, to reduce the effects of brown planthopper feeding on the main shoot (Rubia-Sanchez et al. 1999). Our own studies identified that some of the lines damaged by brown planthopper at reproductive phase recover back, but continue to remain in the vegetative phase without putting forth any grain. This phenotype which appears to be resistant in terms of plant stand is not useful for the farmers. Application of nitrogenous fertilizers did not confer or improve plant tolerance against BPH attack as hoppers being phloem feeders utilized the extra nutrients before they are used by host plant (Horgan et al. 2016).

4.2.2.2 Gall Midge

It is well established that nature of resistance to ARGM is antibiosis and is governed by a single gene^{-}. The death of the maggot occurs due to feeding on the meristem which is often associated with the induction of a hypersensitive reaction where there is browning of the tissue due to phenols (i.e. HR^{+}) or only dead maggots without browning (HR^{-}) which is mostly constitutive in nature. Amudhan et al. (1999) implicated the role of phenols in conferring HR^{+} resistance. Agarrwal et al. (2014) identified biomarkers for gall midge infestation. Trehalose was classified as an infestation feature and linoleic acid, as a resistance feature accumulated in higher levels in infested tissues as compared to uninfested tissues of Kavya (*Gm1*) but not in RP2068 (*gm3*). Levels of glutamine were significantly lower in infested tissues during compatible interaction as compared to its levels in uninfested tissues, thus suggesting that depleted levels of glutamine in gall midge infested tissues as compared to uninfested played a role in the plant's susceptibility to gall midge. Furthermore, Agarrwal et al. (2016) studied the transcriptome and metabolome of RP 2068-18-3-5 (*gm3* and HR^+) infested with gall midge biotype1 (GMB 1) and implicated the involvement of reactive oxygen species, i.e. singlet oxygen and accumulation of azelaic acid, a marker metabolite of lipid peroxidation (LPO)-mediated cell death leading to induced HR^+ reaction. This coincided with a greater accumulation of GABA (a neurotransmitter and an insect antifeedant) at the feeding site which could have resulted in maggot morality. Microarray data revealed deregulation of carbon (C) and nitrogen (N) metabolism causing a C/N shift; up-regulation of tetrapyrrole synthesis and down-regulation of chlorophyll synthesis and photosynthesis which manifests as tissue necrosis.

4.2.2.3 YSB

YSB moths do not exhibit any ovipositional preference. TKM6 was reported to confer non-preference mechanism of resistance. Padhi (2004) worked on the biochemical basis of resistance in rice to *S. incertulas* and reported that rice varieties TKM 6 and PTB 18 recorded lower borer incidence, larval survival and sugar content, but higher quantity of total phenols, ortho-dehydroxy phenol and silica indicating their resistance to yellow stem borer as compared to Jaya, the susceptible check. Rubia et al. (1996) through simulation studies reported that dead heart damage by stem borer could be compensated up to 10%. The response of varieties to YSB damage is diverse. The response of a variety to stem borer damage was calculated from the observed dead hearts and their regeneration capacity. In majority of the varieties, the first and fourth tillers were damaged. Regeneration capacity varied from 4% to 80% in the varieties studied (Padmakumari et al. 2008). The study show that compensation through the recovery resistance is the major mechanism operating in rice against yellow stem borer at the vegetative phase. Varieties/breeding lines, wherein panicles were produced without any delay soon after damage by YSB and among those wherein grain filling is not affected despite borer damage, can be graded as tolerant lines. Recovery resistance at vegetative phase and low level of antibiosis with tolerance could be the possible mechanisms of resistance in IR64/*O. glaberimma* lines, Swarna$*^2$//*O. longistaminata* introgressed lines (Padmakumari and Ram 2012; Padmakumari et al. 2015). Selvi et al. (2002) identified four phenotype-specific RAPD markers through Random Amplified Polymorphic DNA (RAPD) analysis, in conjunction with bulked segregant analysis in an F_2 mapping population of Co43 (susceptible parent)/W1263 (resistant parent). The markers $C1_{320}$ and $K6_{695}$ (at 12.8 cM) were linked with resistance and $AH5_{660}$ and $C41_{300}$ with susceptibility. However, the marker $K6_{695}$ was consistently found to be associated with resistance. Bose et al. (2015) introgressed yellow stem borer (YSB) resistant traits like short stature, increased tiller number and reduced stem diameter from the wild rice species *Oryza brachyantha* and BPH resistance from *O. rufipogon* to the high-yielding durable BPH resistant varieties.

4.2.2.4 Leaffolder

In the case of the rice leaffolder, both antixenosis and antibiosis as mechanisms of resistance were reported (Khan et al. 1989; Ramachandran and Khan 1991; Dakshayani et al. 1993; Xu et al. 2010; Punithavalli et al. 2014). Bio-physical plant factors like trichomes, leaf length are associated with resistance to leaffolder. Different genotypes had varying effects on leaf folding behaviour by the larvae (Punithavalli et al. 2013). The time taken by the larva to fold the leaves or developmental time for the larvae to complete one generation is one of the criteria developed based on the insect behaviour to evaluate the resistance (Javvaji et al. 2021).

4.2.3 Sources, Genes and Genetics of Resistance

Heinrichs et al. (1985) reported various sources of resistance to BPH, WPH, GRH, GLH, YSB, LF and other minor pests like hispa from land races, wild rices, breeding lines and cultivars. However, the levels of resistance were not good enough to confer resistance to all the major pests in one varietal background. With the improvement of techniques to identify the resistance, developments in sources of tolerance have been reported for many pests which were earlier reported that no sources of resistance were available. The details of sources of resistance, markers and genes identified against BPH, WBPH. GLH, GRH and GM are detailed in Table 4.1.

BPH

Around 40 BPH resistance genes have been identified so far in indica, African rice varieties along with wild germplasms such as *O. nivara* (AA), *O. officinalis* (CC), *O. minuta*, *O. punctata* (BB and BBCC), *O. rufipogon*, *O. australiensis* (EE) and *O. latifolia* (CCDD) (Jena and Kim 2010; Muduli et al. 2021). Genes/QTLs involved in BPH resistance, including *Bph1*, *bph2/BPH26*, *Bph3*, *bph4*, *Bph6*, *bph7*, *BPH9*, *Bph12*, *Bph14*, *Bph15*, *Bph17*, *BPH18*, *bph19*, *Bph20*, *Bph21(t)*, *Bph27*, *Bph27(t)*, *Bph28(t)*, *BPH29*, *QBph3*, *QBph4*, *QBph4.2*, *Bph30*, *Bph32*, *Bph33*, *Bph35 and Bph36*, have been fine-mapped by different researchers across the globe. The majority of genes/QTLs are located on rice chromosomes 1, 3, 4, 6, 11 and 12 (Sarao et al. 2016; Muduli et al. 2021). However, there are inconsistencies in assigning gene number and locating their position on the chromosome. Fujita et al. (2013) detailed on the confusions and anomalies in the naming of the *Bph* genes and suggested that CGSNL convention (McCouch and CGSNL 2008) of naming may be adopted to avoid confusions in future. Of the total BPH genes identified for resistance, 14 genes have been identified through map-based cloning (Yang et al. 2020; Du et al. 2020). *BPH1*, *bph2*, *bph7*, *BPH 9*, *BPH10*, *BPH18*, *BPH21*, *BPH26* encoded for CC-NB-NB-LRR protein localized in the endomembrane bundles. *BPH3* and *BPH15* encoded for leptin receptor kinases localized in plasma membrane; BPH6 for atypical LRR protein; *BPH 14* encoded for CC-NB-LRR and localized in LRR nucleus and cytoplasm; *bph29* encoded for DNA-binding domain, and *BPH32* for unknown SCR domain though localized in nucleus and plasma membrane, respectively. It is interesting to note that despite their subcellular

Table 4.1 List of insect resistance genes identified in rice germplasm against major pests

Pest	Source of resistance	*Gene/QTL*	Chromosome	Associated marker(s)	Reference
Brown planthopper (BPH)	Mudgo, TKM6	*Bph1*[a]	12	BpE18-3	Kim and Sohn (2005)
	ASD7	*bph2*[a]	12	RM463, RM7102	Sun et al. (2007)
	Rathu Heenati	*Bph3*[a]	6	RM469, RM588	Jairin et al. (2007a)
	Babawee	*bph4*	6	RM190, C76A	Kawaguchi et al. (2001)
	ARC10550	*bph5*	–	–	Khush et al. (1985)
	Swarnalata	*Bph6*[a]	4	Y19, RM119	Qiu et al. (2010)
	T12	*bph7*[a]	12	RM3448, RM313	Qiu et al. (2014)
	Chinsaba (Acc. no. 33016)	*bph8*	–	–	Nemoto et al. (1989)
	Kaharamana	*BPH9*	12	RM463, RM5341	Su et al. (2006)
	Pokkali	*BPH9*[a]	12	OPR04, S2545	Murata et al. (2001)
	O. australiensis	*BPH10*[a]	12	RM260, RM313	Ishii et al. (1994)
	O. officinalis	*bph11*	3	–	Hirabayashi et al. (1998)
	B14 (*O. latifolia*)	*Bph12*	4	RM16459, RM1305	Qiu et al. (2012)
	IR54745-2-21 (*O. officinalis*)	*Bph13(t)*	3	AJ09b230, AJ09c	Renganayaki et al. (2002)
	B5 (*O. officinalis*)	*BPH14*[a]	14	–	Huang et al. (2001)
	B5 (*O. officinalis*)	*Bph15*[a]	4	RM261, S16	Lü et al. (2014)
	M1635–7	*Bph16*	12	RM6732-R10289	Hirabayashi et al. (2004)
	RathuHeenati	*Bph17*	4	RM8213, RM5953	Sun et al. (2005)
	IR65482-7-216 (*O. australiensis*)	*BPH18(t)*[a]	12	RM463, S15552	Jena et al. (2006)
	AS20-1	*bph19*	3	RM6308, RM3134	Chen et al. (2006)
	IR71033-121-15 (*O. minuta*)	*Bph20(t)*	4	MS10, RM5953	Rahman et al. (2009)

	IR71033-121-15 (*O. minuta*)	*Bph21(t)*[a]	12	RM3726, RM5479	Rahman et al. (2009)
	IR 75870-5-8-5-B-1-B (*O. glaberrima*)	*Bph22(t)*	–	–	Ram et al. (2010)
	IR 71033-121-15 (*O. minuta*)	*Bph23(t)*	–	–	Ram et al. (2010)
	IR 73678-6-9-B (*O. rufipogon* (Acc. 106412))	*Bph24(t)*	–	–	Deen et al. (2010)
	ADR52	*BPH25*	6	S00310	Myint et al. (2012)
	IR1154-243	*bph2/Bph26*[a]	12	KAM4	Murai et al. (2001)
	GX2183 (*O. rufipogon*)	*Bph27*	4	RM273, RM471	Huang et al. (2013)
	Balamawee	*Bph27(t)*	4	RM471, RM5742	He et al. (2013)
	DV85	*Bph28(t)*	11	Indel55, Indel66	Wu et al. (2014)
	RBPH54 (*O. rufipogon*)	*BPH29*[a]	6	BYL7, BID2	Wang et al. (2015)
	AC-1613 (*O. indica*)	*Bph30*	4	RM16294, RM16299	Wang et al. (2018)
	CR2711-76 (indica)	*BPH31*	3	PA26, RM2334	Prahalada et al. (2017)
	Ptb33	*Bph32*[a]	6	RM19291, RM8072	Ren et al. (2016)
	RP2068	*Bph33(t)*	1	RM488, RM11522	Naik et al. (2018)
	Kolayal and Poliyal	*Bph33*	4	H99, H101	Hu et al. (2018a)
	IRGC104646 (*O. nivara*)	*Bph34*	4	RM16994, RM17007	Kumar et al. (2018)
	RBPH660	*Bph35*	4	PSM16, RM413	Zhang et al. (2020)
	O. rufipogon Griff	*Bph36*	4	S13, X48	Li et al. (2019)
	IR64	*BPH37*	1	RM302	Yang et al. (2019)
	Khazar	*BPH38(t)*	1	SNPs 693, 369	Balachiranjeevi et al. (2019)

(continued)

Table 4.1 (continued)

Pest	Source of resistance	*Gene/QTL*	Chromosome	Associated marker(s)	Reference
	[RPBio4918-230S line-(IRGC81848 × Swarna)] × Swarna	bph39(t) and bph40(t)	–	–	Akanksha et al. (2019)
	RathuHeenati	*qBph3*	3	RM3180, RM2453	Kumari et al. (2010)
	IR02W101 (*O. officinalis*)	*QBph4.1*	4	P17, xc4-27	Hu et al. (2015b)
	IR65482-17-511 (*O. australiensis*)	*QBph4.2*	4	RM261, S1	Hu et al. (2015a)
White backed planthopper (WBPH)	N22 (IRRI Acc. no. 4819)	*Wbph1*		Genetic Analysis	Sidhu et al. (1979)
	ARC10239	*Wbph2*		Genetic Analysis	Angeles et al. (1981)
	ADR52	*Wbph3*		Genetic Analysis	Hernandez and Khush (1981)
	Podiwi A8	*wbph 4*		Genetic Analysis	Hernandez and Khush (1981)
	N′ Diang Marie	*Wbph 5*		Genetic Analysis	Wu and Khush (1985)
	Giu-yi-gu	*Wbph 6(t)*		Genetic Analysis	Brar and Khush (1991)
	B5 (*O. officinalis*)	*Wbph 7(t)*	3	R1925, G1318	Tan et al. (2002)
	B5 (*O. officinalis*)	*Wbph 8(t)*	4	R288, S11182	Tan et al. (2002)
	SinnaSivappu	*wbph 9(t)*	6	RM589-RM539	Ramesh et al. (2014)
	SinnaSivappu	*wbph10(t)*	12	SSR12-17.2-RM28487	Ramesh et al. (2014)
	SinnaSivappu	*wbph11(t)*	4	RM3643-RM1223	Ramesh et al. (2014)
	SinnaSivappu	*Wbph12(t)*	4	RM16913-RM471	Ramesh et al. (2014)
	Asonimori	Ovc	6	S1520, L688	Yamasaki et al. (2003)
	Asonimori	Ovc	6	R2373-C946	Yamasaki et al. (2003)
	Asonimori		1	XNpb346-C112	Yamasaki et al. (2003)

		qOVA-1-3 Asominori 1			
	Asonimori	*qOVA-5-1*	5	XNpb251-R331	Yamasaki et al. (2003)
	Asonimori	qOVA-5-2	5	C1268	Yamasaki et al. (2003)
	Asominori	Ovc	6	R2373-C946	Yamasaki et al. (2003)
	Asominori	qOVA-1-3	1	XNpb346-C112	Yamasaki et al. (2003)
	Asominori	qOVA-4	4	R1854	Yamasaki et al. (2003)
	Asominori	qOVA-5-1	5	XNpb251-R3313	Yamasaki et al. (2003)
	Asominori	qOVA-5-2	5	C1268	Yamasaki et al. (2003)
	O. rufipogon	qWPH2	2	RM1285-RM555	Chen et al. (2010)
	O. rufipogon	qWBPH5	5	RM3870-RZ70	Chen et al. (2010)
	O. rufipogon	qWBPH9	9	RG451-RM245	Chen et al. (2010)
	Chunjiang 06	qWL6	6	M3-M5	Yang et al. (2014)
	IR54751	qWBPH3.2	3	InDel3-23-InDel3-26	Fan et al. (2018)
	IR54751	qWBPH11	11	DJ53973-SNP56	Fan et al. (2018)
	Mudgo	WBPHM1, WBPHM2	–	–	Sidhu et al. (2005)
	MO1	*WBPH0*	–	–	Sidhu et al. (2005)
	NCS2041	*WBPHN*	–	–	Sidhu et al. (2005)
Green rice leafhopper (GRLH), *N. cincticeps*	Norin-PL2 (Pe-bi-hun), IR24	*Grh1*	5	XNpb260R2558, C309, R566	Tamura et al. (1999), Kadowaki et al. (2003)
	Lepedumai	*Grh2*	11	S723, G1465	Fukuta et al. (1998), Kadowaki et al. (2003)
	Aichi42	*Grh3(t)*	6	C288B, C133A, C81	Saka et al. (2006)
	DV85	*Grh4*	3	XNpb144	Kadowaki et al. (2003)
	W1962 (*Oryza rufipogon*)	*Grh5*	8		Fujita et al. (2006)

(continued)

Table 4.1 (continued)

Pest	Source of resistance	*Gene/QTL*	Chromosome	Associated marker(s)	Reference
				RM3754, RM3761	
	SML17	*Grh6*	4	RM5414, C60248	Fujita et al. (2004), Phi et al. (2019)
	IRGC104038 O. glaberrima	*qGRH9*	9	RM201, RM205	Fujita et al. (2010)
Green leaf hopper, *N. virescens*	*GLH1*	Pankahari 203	5	–	Athwal et al. (1971)
	GLH 2	ASD7	11	–	Athwal et al. (1971)
	GLH 3	IR8	6	–	Athwal et al. (1971)
	glh4	PTB8	3	–	Siwi and Khush (1977)
	GLH 5	ASD8	8	–	Siwi and Khush (1977)
	GLH 6 *GLH 7*	TAPL 796 MaddaniKaruppan	5 –	– –	Rezaul Karim and Pathak (1982)
	glh8	DV85	–	–	Ghani and Khush (1988)
	GLH 9, glh10t	IR28, IR36	–	–	Angeles and Khush (2000a)
	GLH 11t	IR20965-11-3-3	–	–	Angeles and Khush (2000a)
	GLH 12t	Hashikalmi, ARC10313, Ghaiya, Garia	–	–	Angeles and Khush (2000b)
	GLH 13	Asmaita	–	–	Angeles and Khush (2000b)
	GLH 14	Asmaita	–	–	Sebastian et al. (1996)
Gall midge (ARGM), *Orseolia oryzae*	W1263, Kavya	*Gm1*	9	RM23941, RM23956	Reddy et al. (1997), Sundaram (2007)
	Phalguna	*Gm2*	4	RM17473, RM17503	Mohan et al. (1994), Sundaram (2007)

	RP2068-18-3-5[a]	*gm3*	4	RM17480, gm3SSR4	Kumar et al. (1998a), Sama et al. (2014)
	Abhaya[a]	*Gm4*	8	RM22551, RM22562	Srivastava et al. (1993), Divya et al. (2015)
	ARC 5984	*Gm5*			Kumar et al. (1998b)
	Dukong 1	*Gm6*	4	OPM061400, RG2014, RG476	Katiyar et al. (2001)
	RP2333-156-8	*Gm7*	4	SA 598, F8	Kumar et al. (1999), Sardesai et al. (2002)
	Jhitpiti/Aganni[a]/INRC 3021	*Gm8*	8	RM22685, RM22709	Kumar et al. (2000), Sama et al. (2012)
	Madhuri-L9	*Gm9*	7	NA	Shrivastava et al. (2003)
	BG308	*Gm10*		NA	Kumar et al. (2005)
	CR57-MR1523	*Gm11*	12	NA	Himabindu et al. (2010)
	MN 62M	*gm12*		RM3340, S2-76222	Leelagud et al. (2020)

NA = marker details not available in the publication cited
[a] Cloned gene

variations, all these genes are expressed in vascular bundles, thus affecting the insect which is a phloem feeder. Ishwarya Lakshmi et al. (2022) validated *Bph17* and *Bph32* gene-specific SNPs and identified six SNPs, namely snpOS00912, snpOS00915, snpOS00922, snpOS00923, snpOS00927 and snpOS00929 through Kompetitive allele-specific PCR (KASP) assay and distinguished BPH resistant genotypes. This would help in the detection of SNPs with high precision and has potential usefulness in MAS related to many breeding programmes.

WBPH

Till date, 13 genes, 10 QTLs were identified to confer resistance to WBPH. Studies on the inheritance of WBPH resistance revealed eight major genes, i.e. *Wbphl* in N22 (Sidhu et al. 1979), *Wbph2* in ARC 10239 (Angeles et al. 1981), *Wbph3* in ADR 52 (Hernandez and Khush 1981), *wbph 4* in Podiwi A8 (Hernandez and Khush 1981), *Wbph 5* in N'Diang Marie (Wu and Khush 1985), *Wbph 6(t)* in Giu-yi-gu (Brar and Khush 1991), *Wbph 7(t)* and Wbph 8(t) in B5 (Tan et al. 2002), *wbph 9t, 10t, 11t,* and *WBPH 12t* from Sinnasivappu (Ramesh et al. 2014); Ovc gene, qOVA-4, qOVA-5-1, qOVA-5-2 from Asominori (Yamasaki et al. 2003); *qWPH2, qWBPH5, qWBPH9* from *O. rufipogon* (Chen et al. 2010); *qWL6* from Chunjiang 06; M3–M5 (Yang et al. 2014); qWBPH3.2 and qWBPH11 in IR54751 (Fan et al. 2018). The other sources of resistance include a single dominant gene in ADR 52; a single recessive gene in ARC 6650, ARC 5984 and Podiwi A8 (all allelic to each other); two dominant genes in Velluthecherra; and one dominant and one recessive gene in MO 1 though the dominant genes present in ADR 52 and Velluthacherra are different. The recessive gene of MO1 was non-allelic to that of Podiwi A8 (*wbph 4*). *Wbph 3* and *wbp4* can be effectively utilized in the breeding programme for development of resistant cultures (Padmavathi et al. 2007).

Leaf Hoppers

Leaf hoppers play a major role as vectors of viral diseases rather than causing damage as insect pests. Six dominant genes (GRH1–GRH6) and one minor QTL were identified for GRH. Against GLH, 14 genes were identified of which three were recessive (*glh* 4, 8, 10).

Gall Midge

Of the 12 gall midge resistance genes identified thus far, 10 (*Gm1, Gm2, gm3, Gm4, Gm5, Gm6, Gm7, Gm8, Gm11* and *gm12*) have been tagged and mapped with reported linked markers (Bentur et al. 2021). Among the 12 genes, only two genes *gm3* and *gm12* are recessive in nature. The recessive resistance gene, *gm3*, which displays HR^+, encodes an NB-ARC (NBS-LRR) domain-containing protein, while the dominant gene, *Gm4*, which also displays HR^+, encodes a leucine-rich repeat (LRR) protein, and *Gm8*, displaying HR^-, encodes a proline-rich protein (PRP). Of the 12 genes, *Gm2, gm3, Gm6* and *Gm7* are located on chr 4, *Gm4* and *Gm8* on chr 8, *Gm1* on Chr 9 and *Gm11* on Chr 12. So far, three gall midge resistance genes, *gm3* (Sama et al. 2014), *Gm4* (Divya et al. 2015) and *Gm8* (Divya et al. 2018b), have been cloned and characterized. *Gm2* has been reported to be allelic to *gm3*

(Sundaram 2007; Sama et al. 2014). Since resistance against gall midge is single gene-governed conferring antibiosis, it can break down within a short time. Hence, the strategy of pyramiding two or more genes with divergent mechanisms of resistance (i.e. HR^+ and HR^-) has been advocated for durable resistance against this insect pest (Sundaram et al. 2013).

YSB

Yellow stem borer, *S. incertulas* is the most predominant stem borer species across rice ecosystems in India and infests rice crop from nursery to grain-filling stage. Conventional breeding has led to development of rice varieties like Ratna, Sasyasree and Vikas which derived moderate level of resistance from the donor source TKM6. Very limited information is available on the donors for stem borer resistance in the *sativa* germplasm. The progress in resistance breeding for yellow stem borer has been limited for most of the ecosystems due to polygenic nature of resistance. The wild species are a reservoir of useful genes for rice improvement. Wide hybridization and chromosome manipulation are the important techniques to transfer useful genes from wild to cultivated rice leading to rapid genomic changes like chromosomal rearrangements, genome expansion, differential gene expression and gene silencing. The tertiary gene pool was exploited for identification of resistant sources to BPH, gall midge and YSB. Wild rice germplasm has been screened against yellow stem borer. Protocols for extensive screening of material against yellow stem borer were standardized through utilization of field populations (Bentur et al. 2011; Padmakumari et al. 2015; Padmakumari and Katti 2018). *O. brachyantha, O. officinalis, O. ridleyi* and *Porteresia coarctata* were found to be resistant/tolerant against yellow stem borer (Padhi and Sen 2002) and the genes have been introgressed from *O. brachyantha* (Thapa 2008; Panda et al. 2011). *O. glaberrima, O. longistaminata* and *O. rufipogon* as sources of tolerance to yellow stem borer were identified (Padmakumari and Ram 2012) as a good source of resistance/tolerance to yellow stem borer. Cut stem assay of these tolerant lines with neonate larvae also revealed 30–100% larval mortality and delay in the larval growth as compared to 10% mortality in the recurrent parent (RP) IR 64 (Padmakumari, unpublished). In a novel method to address the issue of stem borer tolerance, intensive field screening of 4500 lines of EMS mutants of BPT 5204 against yellow stem borer at both phases of crop growth was carried out. This led to identification of 8 lines as tolerant to YSB with low level of antibiosis and recovery resistance (Padmakumari et al. 2017b; Potupureddi et al. 2021). Since stem borer is an internal feeder, the mechanism of resistance is always either low level of antibiosis or coupled with recovery resistance due to compensation. Hence, the yield would be almost at par with the national check variety or less than the wild type (Padmakumari and Ram 2016). Hence, while evaluating resistant varieties, it is pertinent to note that the economics need to be worked based on the ecosystem services, cost involved in raising the variety and cost towards the reduction in insecticide sprays so that proper deployment can be ensured.

4.2.3.1 Multiple Pest Resistance

Many donors were identified to possess multiple pest resistance to two or more pests. Some of the identified donors are listed in Table 4.2. These donors can be effectively utilized in the resistance breeding programmes.

4.2.3.2 Insect Virulence, Biotypes and Effective Genes

Here we discuss two important insects of rice BPH and GM where the concept of biotype with reference to the reaction of insect population to a set of host plant differentials is well established and would play a crucial role in the designing of location-specific breeding programmes and gene deployment for durable resistance.

4.2.3.2.1 BPH

Four biotypes were identified for BPH based on the virulence to the host plant differentials. Biotypes 1 and 2 are widely distributed in Southeast and East Asia, whereas biotype 3 was developed in the laboratory by rearing the insects on the resistant variety ASD7 which has the *bph2* gene for resistance (Panda and Heinrichs 1983) while biotype 4 or the South Asian biotype which is most destructive in nature occurs in the Indian subcontinent. Ferrater et al. (2013) reviewed the status of virulence in BPH populations. Using genome-wide scanning and interval mapping, the Qhp7 locus that governs preference for Bph1 plants was mapped to a 0.1 cM region of chromosome 7. In addition, two major QTLs that govern the rate of insect growth on resistant rice plants were identified on chromosomes 5 (Qgr5) and 14 (Qgr14) (Jing et al. 2014). Many salivary proteins have been identified using transcriptome and proteome techniques. The genetic loci of virulence were mapped in BPH genome based on the linkage map (Jing et al. 2017). Widespread virulence in South and SE Asia for *Bph1*, *Bph3* and *bph4*—India (Verma et al. 1979), Thailand (Thanysiriwat et al. 2009), *bph4* and *bph8*—China and *BPH26*—widespread in SE Asia (Myint et al. 2009b); *Bph5*, *Bph6*, Japan (Myint et al. 2009a); *BPH25* and *Bph7*, *bph8*, *Bph9*, *Bph10*, *Bph18*, *Bph20 and Bph21*—Philippines (Ferrater et al. 2013) was reported. From current literature on rice resistance to the brown

Table 4.2 Donors with multiple pest resistance

Insect pests	Donors/cross combination
BPH and WBPH	Mudgo, MO1, Podiwi A8
BPH and ARGM	RP 2068-18-3-5 (Swarnadhan × Velluthacheera)
BPH and GLH, LF	ASD7
BPH and ZLH	PTB33
GM and BPH, MR to SB	PTB21 (IRRI Acc. 6113)
GM and SB (MR)	PTB18
GM, SB (MR), LF, BPH (biotype1 & 3)	W1263 (MTU15 × Eswarakorra (*Gm1*))
GM, BPH (biotype1 & 3)	Suraksha
BPH and ARGM	Sinnasivappu (IRRI Acc. 15444)
BPH and ARGM	SuduHondrawala (IRRI Acc. 15541)
BPH and SB (MR)	RathuHeenati

Modified from Heinrichs et al. (1985)

planthopper, it is evident that newly identified resistant varieties and genes appear less effective than the varieties and genes that were identified in the 1970s and 1980s (Horgan et al. 2015). In a study on 39 rice accessions with 12 populations of BPH from South and South East Asia, only 6 varieties were less damaged than the susceptible check Taichung Native 1 (TN1). *Bph1, bph2, bph5, bph7, bph8, Bph9, Bph10* and *Bph18* were ineffective against most planthopper populations. *Bph20, Bph21* and *Bph17* have the potential to be used in resistance breeding in both South and South East Asia, whereas BPH25 and BPH26 have potential for use in South Asia (Horgan et al. 2015) suggesting pyramiding of two or more genes with different mechanisms could improve resistance level and durability. The variations in the reactions could be attributed to different seed sources or accessions used in these evaluations or the preconditioning of the insect used in evaluations. Horgan et al. (2015) reported that IR71033-121-15 (IRTP23991, a line introgressed with *Oryza minuta* from Philippines with *Bph20, Bph21* and MO1 (*WbphN, WbphO*) were apparently resistant to all South Asian populations and South East Asian populations of BPH though RathuHeenati (*Bph3*), Balamawee, Swarnalata and PTB33 showed resistance/tolerance across 4–6 populations tested. It is interesting to note that MO1 identified as resistant donor for WBPH is also tolerant to BPH. These can be used as differentials to check/monitor for the variation in virulence of the BPH populations. Genetic analysis revealed that BPH resistant genes are either linked with other gene or clustered together. BPH resistant genes are localized on 7 of the 12 chromosomes (1, 2, 3, 4, 6, 11 and 12) with several of the genes reported to be on chromosome 4 followed by chromosome 12. Kuang et al. (2021) in their studies on effect of climate change on the efficacy of BPH resistance genes in IR24-NILs reported that NIL-BPH17 has a strong inhibitory effect on BPH feeding on phloem and would be unaffected by environmental changes, while NIL-BPH20 would lose its ability during the environmental changes.

4.2.3.2.2 GM

Asian rice gall midge (GM), *O. oryzae* (Wood-Mason) is a serious pest in rice-growing countries, especially in China, India, Vietnam and Sri Lanka. Based on the reaction of the populations from endemic locations against a set of host plant differentials grouped into 4 categories, seven biotypes of rice gall midge were reported from India (Vijayalakshmi et al. 2006). Recent studies indicate that most of the gall midge populations in India are avirulent on *Gm1* and *Gm8* both with HR^{-} mechanism of resistance followed by *Gm4* and *virulent* on *Gm11, gm3* and *Gm4* (ICAR-IIRR, 2018–2021). Studies on inheritance of virulence in gall midge suggest that virulence against plant resistance gene *Gm2* is conferred by a single recessive gene (v*Gm2*) and is inherited in a sex-linked manner (Bentur et al. 1992), whereas virulence against *Gm1* gene is also conferred by a recessive gene (*vGm1*) but shows autosomal pattern of inheritance predicting the durability of *vGm1* as compared to *vGm2*. A simple modified F_2 screen method was developed and tested for estimating frequency of rare recessive alleles in pedigreed crosses of natural populations of gall midge (Bentur et al. 2008; Andow and Bentur 2010). Hence, it is very much essential to monitor the virulence of the major insect pests and generate more information on the insect virulence with respect to the effective plant resistance genes. This prior knowledge would aid in effective deployment of the plant resistance genes, thus aiding in durable resistance.

4.3 Breeding for Insect Resistance

4.3.1 Conventional Approaches

Owing to the prevalence and economic losses caused by insect pests across the world, concerted efforts were directed towards development of a resistant variety against either planthoppers/leaf hoppers, gall midge or stem borer which is the ultimate goal of host plant resistance in the service of mankind. Mudgo (*Bph1*), ASD 7 (*bph2*), RathuHeenati (*Bph3*) and Babawee (*Bph4*) were extensively utilized in breeding programmes as donors to develop BPH resistant varieties in Southeast Asia (Jairin et al. 2007b). Some of them were mega varieties for more than a decade before their breakdown due to development of new biotypes. Because of the high variability of BPH and the apparent specificity of BPH resistance, an understanding of the linkage and allelic relationship of resistance genes is particularly important for the development of BPH-resistant cultivars (Jena and Kim 2010). It is interesting to note that some of the cultures bred for resistance through conventional approaches are not under cultivation, but few varieties are still in the seed production chain for more than a decade in India (Table 4.3). This is mainly due to the patronage by the consumers in certain areas for their grain and cooking quality. But no information is available on the current status of pest incidence in these varieties.

4.3.2 Molecular approaches

4.3.2.1 Application of Molecular Markers in Breeding for Insect Resistance

Biotic stresses like stem borer, BPH, leaffolder, gall midge and among the diseases, blast, bacterial blight, sheath blight, false smut and rice tungro disease reduce rice

Table 4.3 Resistant varieties in seed production chain in India

Resistant variety	Seed production (no. of years)	Cultivation in India	Resistance trait
Jyothi	21	Kerala	Gall midge
Uma	15	Kuttanad area, Kerala	Gall midge
Phalguna	14	Andhra Pradesh	Gall midge
Danteshwari	14	Chhattisgarh	Gall midge
Ratna	18	Maharashtra	Stem borer
Erramallelu	11	Telangana	Gall midge
Surekha	11	Grown in Telangana but exported to Kerala	Gall midge
Karma Mahsuri	9	Chhattisgarh	Gall midge

production in India significantly. As chemical control of pests and diseases has several limitations as a strategy, and with the phasing out of many broad spectrum insecticides, deployment of host plant resistance has been advocated. BPH, gall midge, blast and bacterial blight can be effectively managed through deployment of resistant varieties. Availability of molecular markers facilitates easy identification of resistance genes and aids in the selection of good and relevant donors, thus saving the cost, time and energy involved in the rigorous phenotyping. Bentur et al. (2021) had reviewed at length the literature on the identified genes conferring resistance to BPH, WBPH, GLH, GRH and GM, the chromosome number, primer identity along with the forward and reverse sequences for the benefit of the molecular breeders. With the basic knowledge available on the mechanisms of resistance to all the effective genes, it is easy to breed a resistant/tolerant variety with desired genes. Hence, development of an insect resistant variety through any breeding programme for a specific location would involve selection of a good donor wherein these molecular markers can be effectively used to develop gene pyramids for wide spectrum and durable resistance. However, from the stand point of acceptability, when the same rice variety may not be grown across the length and breadth of the country, a single variety need not be pyramided to provide resistance to all the biotypes of gall midge. Instead, locally adopted rice varieties can be used to introgress two or more effective genes that will confer resistance against all the prevalent biotypes/pests in that area. Durability of resistance is influenced by initial frequency of alleles in the pest population that confer adaptation against the plant gene, nature of inheritance of this allele and possible cross-virulence (Cohen et al. 2004). Hence, it is suggested that deployment of two or more undeployed genes that differ in their mechanism of resistance is the ideal option. Figure 4.1 depicts the schematic representation of the work flow for the development of a durable pest resistant rice variety.

4.3.2.2 Gene Pyramiding Strategies for Durable Resistance

Pyramiding resistance genes is predicted to increase the durability of resistant rice varieties against phloem feeding herbivores and other rice pests. Attempts to pyramid insect resistance genes were initiated with GM genes as resistance is governed by a single gene with the availability of closely linked markers/functional markers for developing durable pest-resistant rice cultivars/hybrids. Sama et al. (2014) introduced the recessive gene *gm3* into the genetic background of elite rice variety improved Samba Mahsuri with the help of markers. A focussed molecular breeding programme, which is still ongoing, was initiated in 1999–2000 at IIRR, Hyderabad, for introgression of several genes conferring resistance against bacterial blight, blast, gall midge and BPH into elite Indica rice varieties and hybrid rice parental lines with the help of molecular markers. Samba Mahsuri, Swarna, Akshayadhan, few elite *Basmati* varieties along with the hybrid rice parental lines KMR3R, RPHR-1005R, IR58025 and DRR17B served as recipient varieties while several breeding lines/ genotypes possessing one or more genes conferring resistance against BB, blast, gall midge and BPH served as donors. The genes considered for introgression included *Xa21*, *xa13*, *xa5*, *Xa33*, *Xa38* for BB resistance, *Pi1*, *Pi2* and Pi54 for blast resistance, *Gm1*, *gm3*, *Gm4 and Gm8* for gall midge resistance and *Bph18* and

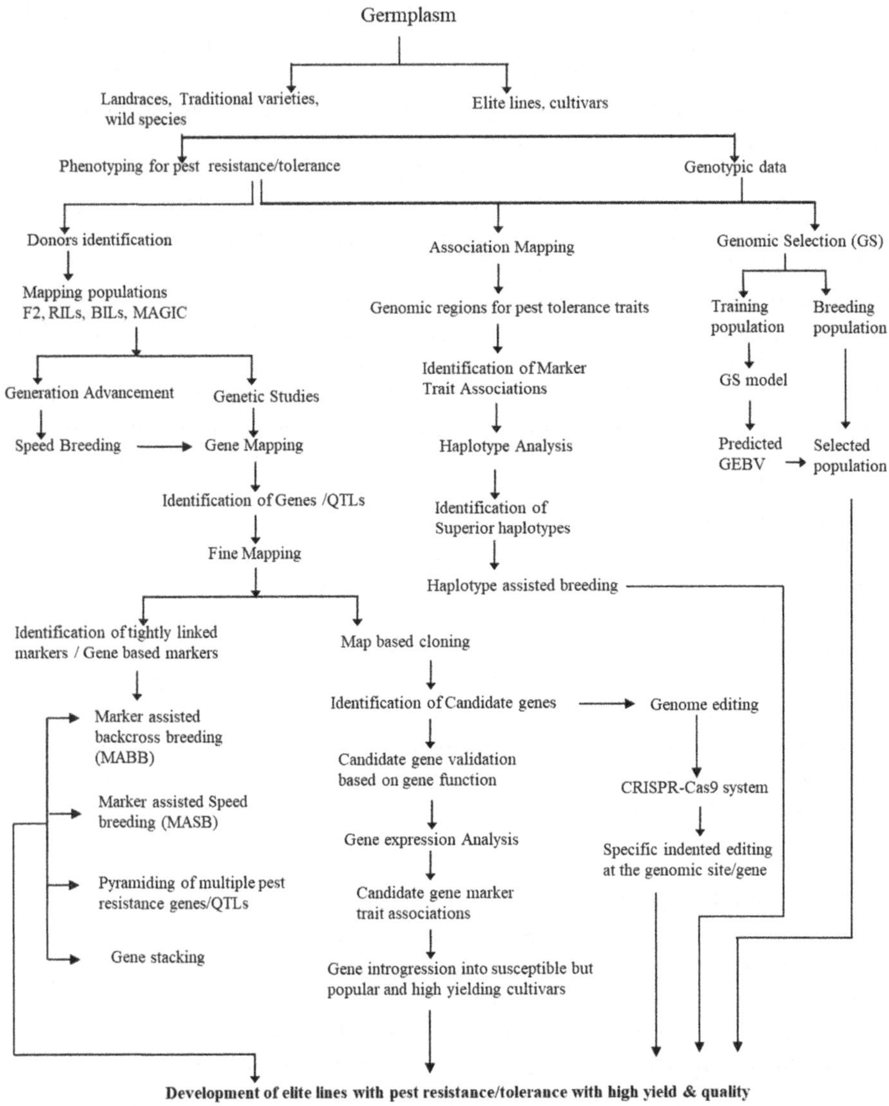

Fig. 4.1 Schematic representation depicting integration of breeding, molecular and genomic approaches for developing elite pest resistant lines. *RIL* recombinant inbred lines, *BIL* backcross inbred lines, *GEBV* genomic estimated breeding value, *GS* genomic selection, *MAGIC* multiparent advanced generation intercross, *QTLs* quantitative trait loci, *CRISPR* clustered regularly interspaced short palindromic repeats

Bph33 for BPH resistance. Two approaches, namely marker-assisted backcross breeding (MABB; i.e., 3–4 backcrosses), limited MABB (i.e. a single backcross) in conjunction with pedigree breeding were deployed for targeted resistance gene transfer. Recently, a novel BPH resistance gene, named *Bph33* was transferred along with the gall midge resistance genes, *gm3*, *Gm4 and Gm8* into improved Samba

Mahsuri through MABB (Naik et al. 2018). In another project funded by DBT, Govt. of India, initiated in the year 2009, a total of two genes each for resistance against BB (*Xa21* and*Xa33*), blast (*Pi2* and *Pi54*), gall midge (*Gm4* and *Gm8*) and BPH (*Bph18*) have been transferred into the high-yielding rice variety Akshayadhan and hybrid rice parental line DRR17B. Backcross derived breeding lines possessing four, five and six resistance genes conferring tolerance to multiple biotic stresses have been developed and screened for their resistance against multiple biotic stresses.

Using *Gm4* and *Gm8*, the research group at ICAR-IIRR has developed gene-pyramided lines (PYL) in the genetic background of the elite restorer line RPHR1005R (restorer line for the popular rice hybrid DRRH3) (Abhilash Kumar et al. 2017) and into a high-yielding rice variety Akshayadhan through marker-assisted breeding. In another such effort, two resistance genes, *gm3* and *Gm8*, have been pyramided in the genetic background of the fine-grain-type rice variety Kavya, which is known to have *Gm1* (Venkanna et al. 2018). Multilocation studies on evaluation of gene PYL for gall midge resistance in India indicated that pyramided lines with two effective genes *Gm4* and *Gm8* were conferring resistance across 7–8 populations of gall midge (ICAR-IIRR, 2018–2021). But one of the major problems encountered in the development of rice varieties resistant to gall midge is that over a period of time, gall midge develops resistance against the deployed resistance genes, through the evolution of new biotypes. But later, it has been recorded from the research stations that only *Gm8* and *Gm1*, both with HR^- reaction, are holding promise across locations against gall midge populations in India (ICAR-IIRR 2018–2021).In another study from the Philippines, it was recorded that pre-conditioning of GLH, *N. virescens* on a pyramided line (PYL) with GRH2 and GRH 4-PYL gained partial virulence (feeding and development equal to that on T65) to the pyramided line within 10 generations and complete virulence (egg-laying equal to that on T65) within 20 generations. Studies with various rice pests suggest that previous long-term exposure to ineffective genes (including unperceived resistance genes) could dramatically reduce the durability of pyramided resistance (Horgan 2018). However, Gupta and Nair (2021) in her laboratory studies on BPH exposed to resistant rice variety, Rathu Heenati (*Bph 3*), observed that the epigenetic changes were induced in the context of DNA methylation levels. These epigenetic changes were reversible, i.e. the methylation levels, especially in CHG and CHH contexts reverted to its initial state once the stress was removed. The time taken for these epigenetic marks to "fade away" was shorter in case nutritional stress (resistant variety) as compared to pesticide stress. This information can be utilized in the effective deployment of a cafeteria of resistant genes.

4.3.2.3 Genomics-Assisted Breeding for Resistance/Tolerance Trait

It is interesting to note that though resistance breeding programme was in vogue since time immemorial, not many resistance varieties are under large cultivation. Of the many factors, it is pertinent that the final product/resistant variety will find a place in cultivation if it caters to the regional requirements of the farmers/consumers apart from possessing reasonable level of resistance to a particular pest(s). It is well known that resistance to insect pests is often associated with low yields. In this direction to address the issue, many of the BPH resistance genes have been introduced alone or in combination into modern rice varieties/parental lines of hybrids by marker-assisted selection. Wang et al. (2017a, 2017b) pyramided *Bph6* and *Bph9* into elite restorer line 9311, while Fan et al. (2017) developed three broad-spectrum BPH-resistant restorer lines by pyramiding big-panicle gene *Gn8.1*, BPH resistance genes *Bph6* and *Bph9* and fertility restorer genes *Rf3*, *Rf4*, *Rf5* and *Rf6* through molecular marker-assisted breeding. *Gm4* and *Gm8*, the effective genes, for conferring resistance to gall midge were incorporated into major mega varieties like MTU1010 (Duriseti et al. 2018), Swarna (Dixit et al. 2020), WGL 44 (Jai Vidhya et al. 2018); *Gm8* for gall midge (GM) in the background of Naveen (Janaki Ramayya et al. 2021); *Bph*20/21 into Krishna Hamsa (Badri et al. 2022) along with other genes for biotic (blast and BLB) and abiotic stresses (drought). To address this issue of both resistance and yield, care was taken to design the final product which has both insect resistance genes along with yield genes. Nagamani (2022) developed pyramided lines in the background of the elite variety MTU1010 through MABB conferring resistance to both blast and virulent population of gall midge and possessing high yield (*Ospl*4 and *Gna* genes). This involved intensive phenotyping under field conditions and selection for both resistance and yield in each generation of development for various stresses and advancement of promising material. In this endeavour, vast genetic resources with various gene combinations and different levels of expression were generated out of which pyramided lines with 3–7 genes under homozygous condition for gall midge resistance along with BLB genes and good grain yield ready for deployment are currently available. In this method, genotyping or presence of marker was correlated with the phenotypic data so as to ensure the presence of genes. However, the expression of genes under infested conditions in the finished product is not quantified. It is to be noted that the type of yield genes to be chosen would depend on the insect damage so that the gene would negate/compensate for the effect of insect feeding.

4.3.2.4 Molecular Cross Talks Between Insect Resistance/Susceptibility Pathways and Other Pathways

Plants are able to activate different types of induced resistance, depending on the organism that interacts with the plant. Well-studied examples of induced resistance are systemic acquired resistance, which is triggered by pathogens causing limited infection, such as hypersensitive necrosis. Wound-induced resistance, which is typically elicited upon tissue damage such as that caused by insect feeding (Kessler and Baldwin 2002; Howe 2004). Salicylic acid (SA), jasmonic acid (JA) and ethylene (ET) are recognized as key players in the regulation of the signalling pathways involved in conferring resistance (Howe 2004). It is generally stated that pathogens with a biotrophic lifestyle are more sensitive to SA-mediated induced defences, whereas necrotrophic pathogens and herbivorous insects are resisted more through JA/ET-mediated defences (Kessler and Baldwin 2002). In nature, however, plants often deal with simultaneous or subsequent invasion by multiple aggressors, which can influence the primary induced defence response of the host plant. Cross talk is the interaction between the signalling pathways which helps the plant to minimize energy costs and create a flexible signalling network that allows the plant to finely tune its defence response to the invaders encountered (Reymond and Farmer 1998; Pieterse et al. 2001; Bostock 2005). SA plays an important role in the regulation of systemic acquired resistance (SAR). Transduction of the signal leads to activation of the genes encoding pathogenesis related proteins. The regulatory protein Non expressor of PR genes 1 (NPR1), WRKY transcription factors are important factors in SA-dependent signalling. MAP kinases and Gluaredoxin GRX 480 are implicated in SA/JA cross talk. Highlight diversity in resistance pathways in rice against the gall midge triggered by different R genes has been reported by Divya et al. (2018a). Ample evidence is available on the interaction of SA, JA and ET pathways which could be either positive or negative. Although most reports indicate a mutually antagonistic interaction between SA and JA-dependent signalling, synergistic interactions have been described as well (Mur et al. 2006). As a result of negative cross talk between SA and JA, activation of the SA response should render a plant more susceptible to attackers that are resisted via JA-dependent defences and vice versa.

Usually under field conditions, the rice crop is challenged by many biotic stresses like insect pests, diseases or weeds. The challenge may vary. It could be one insect or many insects, or a disease and an insect at a particular crop stage. So a cautious deployment of a rice culture resistant to a single biotic stress/multiple biotic stresses is necessary. Though progress in tagging, mapping and cloning of several resistance (R) genes against various biotic stresses has led to marker-assisted multigene introgression into elite cultivars for multiple and durable resistance, no detailed studies on possible interactions among these genes when expressed simultaneously under combined stresses are available. Divya et al. (2018c) reported induction of several defence genes in response to attack by more than one biotic stress under controlled conditions through expression profiles of 14 defence-related genes in 11 rice breeding lines derived from Samba Mahsuri (BPT 5204), an elite cultivar with different combination of R genes against bacterial blight (BB), blast (BL) and ARGM. Cross response of *Pi54* and *Gm4* was observed, and inhibition of expression of *Pi54* by other target genes was indicated. In another study, marker-assisted intercross (IC) breeding involving multiple donors was used to combine three BB (bacterial leaf blight) resistance genes—*xa5*, *xa13* and *Xa21*, two blast resistance genes—*Pi9* and *Pi54*, two BPH resistance genes—*Bph20* and *Bph21* and four drought-tolerant quantitative trait loci (QTL)—qDTY1.1, qDTY2.1, qDTY3.1 and qDTY12.1—in the genetic background of the elite Indian rice cultivar 'Krishna Hamsa'. It was observed that though introgressed lines (ILs) were marker-positive, there was a selective exclusion of BPH resistance genes especially when BB genes were present. However, such interactions were not observed with *Pi* genes (blast), and this requires further validation in different backgrounds to confirm the results (Badri et al. 2022). Multiple genes introgression in single background may result in the selective combination of compatible genes or selective exclusion of some combinations due to recombination events or chromosome-related factors (Haque et al. 2021). Significant variation was observed in the phenotypic response of the ILs in the background of Swarna despite the presence/absence of corresponding R genes and BPH susceptibility in ILs with *Bph3* and *Bph17* (Dixit et al. 2020) which could be due to the difference in the genome recovery and background interaction of genes/QTLs of the ILs. The study also reports that mere presence of markers would not ensure a positive phenotype. The negative phenotype in marker-positive plants can be the effect of varying expression of a specific gene in different background genomes. This could be the trade-off between the different signalling pathways associated with the genes that are operating. Expression profiling of stress-inducible genes was carried out in rice varieties and reported variability in gene expression patterns indicating the complex network of pathways for regulation of multiple stresses (Basu and Roychoudhury 2014). Detailed studies on cross talk between defence pathways are essential (Sharoni et al. 2011) when genes are pyramided to engineer plants resistant to multiple stress conditions (Saidi and Hajibarat 2019). So, it is important to know the pathways which are operating prior to the deployment of the varieties with various defence genes. To understand the functioning of the complex defence signalling network in nature, molecular biologists and ecologists should work in tandem to study molecular mechanisms of induced plant defences in an ecological perspective (Koornneef et al.2008).

4.4 Novel Methods to Identify and Characterize Resistance to Insect Pests

4.4.1 Susceptible Variety as Trap Crop for Stem Borer Management

It is an eco-friendly pest management strategy for small farmers where a susceptible cultivar was used to attract YSB larvae and minimize YSB damage. Growing one row of a susceptible aromatic variety (Pusa basmati 1) as an inter-crop/trap crop for every 9 rows of main crop (2.5–3 m) in east-west direction helps in reducing the damage by stem borer in the main crop by increasing the damage in the trap crop. The main crop variety can be of farmers' choice. The sowing date of aromatic variety has to be adjusted so as to flower a week before the main crop (Plate 4.3) (Padmakumari and Pasalu 2003). Under All India Coordinated Crop Improvement Programme for Rice, this concept was successfully tested in different regions for 3 years with a cafeteria of varieties like Swarna, MTU1010, BPT5204, HMT Sona, Tellahamsa, Krishna Hamsa, Jyoti, Rajendra Kasturi, TPS3, Karjat2, Karjat3 and ADT36 as main crop varieties in both wet and dry seasons (DRR 2009–2011). The grain yield in the treatments with main crop + trap crop was more than that in the treatment with main crop alone and the B C: ratio varied from 1.18 to 4.17 across locations and the monetary gain obtained varied from 3.27% to 8.25%. The damage in the main crop is reduced to half of that in the trap crop, thus avoiding one impulsive spraying of insecticide at the vegetative phase in a stem borer endemic area. No insecticide spray is required on main crop up to 25% dead heart damage in the trap crop (Padmakumari et al. 2017a).

Plate 4.3 Host plant susceptibility for yellow stem borer management

4.4.2 Exploitation of Susceptibility Genes

Besides plant R-genes, plant susceptibility genes can be very useful in pest management. Some herbivorous insects have the ability to suppress effective defence responses of plants to their own advantage. It is hypothesized that these insects activate certain plant genes, the so-called susceptibility genes, to suppress or avoid the defence that antagonizes insect performance resulting in the susceptibility of the host plant (Walling 2008; Giordanengo et al. 2010). Knockdown of such susceptibility genes may be particularly interesting to control plant-manipulating insects. Therefore, investigating the potential to use susceptibility genes to control insect pests deserves to be explored though it was reported to confer resistance to certain pathogens (Lorang et al. 2007). It would be more appropriate to explore the mutagenic resources where both the reference genome and the lines susceptible for a particular trait are available. These lines can be further utilized to study the genetics, molecular and biochemical factors that confer susceptibility to a biotic stress and manipulate through gene editing to identify tolerant lines.

4.4.3 Induced Resistance

During the coevolution of plant–insect relationships, plants developed defensive strategies to minimize insect herbivory, which in turn led to the development of counter adaptations by herbivores. The plant hormone jasmonate (JA) plays a central role in regulating induced defence responses to a broad spectrum of herbivores.

4.4.3.1 Herbivore- and Elicitor-Induced Resistance

Bentur and Kalode (1996) observed that infestation of primary tiller induced HR^+ in resistant rice var. Phalguna, but mortality of gall midge maggots was induced in the secondary tiller without HR^+. The effects of prior herbivory by the fall armyworm, *Spodoptera frugiperda* and of exogenous applications of jasmonic acid (JA) on the resistance of rice plants to the rice water weevil, *L. oryzophilus*, manifested as reduced numbers of eggs and first instars associated with armyworm-injured or JA-treated plants relative to control plants (Hamm et al. 2010). In field experiments, there was a transient but significant reduction in the number of immature *L. oryzophilus* on JA-treated plants relative to untreated plants. The duration of effectiveness of such induced resistance varied with the sequence and time lag between infestations. Such studies are essential where induced resistance to same insect or different insects, to know the effectiveness of the genes under real-time situations where the sequence and intensity of insect infestations vary and are not under our control under field situations and also the exogenous application of methyl salicylate/jasmonate for pest management.

4.4.3.2 Silicon as Source of Resistance

Many elements that are considered as essential nutrients play a significant role in host plant–pest interactions. But silicon (Si) though not identified as an essential element was found to play a significant role in imparting biotic and abiotic stress resistance and had a predominant negative impact on herbivore performance (Ma 2004). The application of Si to crops is a viable component of an integrated management programme for insect pests and diseases as it is relatively less expensive, leaves no pesticide residue in food or the environment and could easily be integrated with other pest management practices (Laing et al. 2006). Extensive work on carrier-induced silicon transportation into rice in relation to yellow stem borer (Ranganathan et al. 2006) identified imidazole as the best carrier cum solubilizer for silicon of the many amino acids studied. The effect of imidazole, when applied in split doses at 1500 mmol of the carrier X four times in the field in lowering yellow stem borer damage in rice varieties (Voleti et al. 2008) due to accumulation of Si. Study by Jeer et al. (2017), the first study of its kind, demonstrated that application of imidazole, a solubilizer cum carrier, enhanced the uptake of Si, deposited in the stem region (2.1–5.3-fold), thereby increasing the abrasion of the larval incisors (mandibular wearing) and effected the YSB larval growth, larval recovery through effect on mandible wearing and inner lining of the midgut. The work has led to the identification of a novel class of bio-compatible molecules, which exhibited remarkable resistance to damage by yellow stem borer. Combined application of rice husk ash (a cheap source of Si) and imidazole reduced YSB damage with a benefit–cost ratio (1.16–1.31) which was at par with one granular insecticide application (Jeer et al. 2018). In BPH, Si acts as a sucking inhibitor (Yoshihara et al. 1979). Studies on BPH proved that Si amendment adds to fast and strong callose deposition in

BPH-infested plants, resulting in low feeding by BPH on Si-supplemented plants. Si amendment is involved in modulation of the gene expression of callose synthase and hydrolase triggered by BPH infestation and that high expression of the synthase gene *OsGSL1* and low expression of the hydrolase gene *Gns5* were recorded in Si-amended plants during the initial stages of BPH infestation (Yang et al. 2018). The reported relationships between soluble silicon and the jasmonic acid (JA) defence pathway and herbivore-induced plant volatiles (HIPVs) suggest that soluble silicon may enhance the production of herbivore-induced plant volatiles (HIPVs) (Reynolds et al. 2016).

4.4.4 Transgenics

Research in late 1990s explored various options for development of resistance to incalcitrant pests like stem borers where adequate sources of resistance in the gene pool were not available, and genetic engineering as a tool was explored and utilized to develop resistant lines. Among the widely used genes encoding insecticidal proteins/molecules against rice insect pests, endotoxin crystal proteins of *Bacillus thuringiensis*, digestive enzyme-specific protease inhibitors, plant lectins, α amylase inhibitors, insect chitinases and insecticidal viruses are important. Of these, Bt toxin genes (*cry1A*, *cry1B*, *cry1C*, *Cry1Aa*, *cry1Ab*, *cry1AC*, *cry1Aa*, *cry1Ac*, *cry2A*, *cry1C*, *cry1c* and *cry2A*) and protease inhibitors PINII (potato proteinase inhibitor) (cowpea serine P1) *Cry1AB*, *Cri Ac*, *Cry1A*, *Cry2Aa*, with CaMV 35S promoters were tested. Chimeric Bt gene, *cry1Ab*; *cry1Ab*/*cry1Ac* fusion gene against stem borer and lectin protein gene (gna, asa lectin, snowdrop lectin) against hoppers have been reported to be effective. Nayak et al. (1997) and Datta et al. (1998) were the first to report transformation of rice with the Bt gene against yellow stem borer. Research spanning more than two decades had resulted in many genes effective against stem borers, YSB and SSB which has been extensively reviewed (Makkar and Bentur 2017; Bentur et al. 2021).

Apart from YSB, genetic engineering approaches were also attempted for the most destructive pest brown planthopper in south and south-east Asia. Plant-derived lectins were utilized for engineering resistance to plant and leafhoppers. The snowdrop lectin gene, *Galanthus nivalis* agglutinin (*GNA*), conferred partial to complete resistance to planthoppers and leafhoppers in the rice varieties. Partial resistance to leafhoppers and planthoppers was demonstrated by rice transformation with a lectin gene from garlic (*Allium sativum* leaf agglutinin gene, *ASAL*; Rao et al. 1998; Saha et al. 2006). Bala et al. (2013) reported that *ASAL* interacts with NADH quinine oxidoreductase (NQO), a key player in the electron transport chain, and results in toxicity and increased mortality of BPH in transgenic rice lines. This study also demonstrated that, among all the transgenes available for control of sucking pests, *ASAL* holds significant promise, particularly against BPH. Yoshimura et al. (2012) developed transgenic rice possessing lectin1 gene from *Dioscorea batatas* under the control of a phloem-specific promoter (i.e. promoter of sucrose synthase-1gene) that showed a 30% decrease in the survival rate of BPH. Even though, in general, it is known that Cry proteins are ineffective against sucking pests, through loop replacements with gut-binding peptides in *Cry1AB* domain II, enhanced toxicity against BPH has been demonstrated (Shao et al. 2016). Liu et al. (2018) have shown the efficacy of Cry64Ba and Cry64Ca, two ETX/MTX2-type *Bt* proteins, against hemipteran pests. Boddupally et al. (2018) recently demonstrated that the expression of hybrid fusion protein (Cry1Ac::ASAL) in transgenic rice plants imparted resistance against multiple insect pests: BPH, stem borer and leaffolder.

Bt rice lines Huahui No. 1 (CMS restorer line) and *Bt* Shanyou 63 (a hybrid of Huahui No. 1 and Zhenshan97A, aCMS line), expressing *Cry1Ab/Ac* fusion gene, were developed in China (Chen et al. 2011) and earned biosafety certificates in 2009. However, because of political reasons, its commercialization in China has been postponed, and, to date, Bt rice is not grown in China. Some of the reasons for the delay in commercialization are (a) Bt rice was developed only for stem borer resistance and not any other rice pests. So, there is a need to develop insect resistance management strategies, (b) low level of understanding and acceptance of GM crops in China (Li et al. 2016), (c) resolving trade policy impediments (High et al. 2004; Liu et al. 2016). Jin et al. (2019) estimated the opportunity cost of postponement of Bt rice between 2009 and 2019 to be 12 billion US dollars per year.

4.4.5 RNA Interference (RNAi) for Plant Resistance to Insects

RNAi also known as post-transcriptional gene silencing (PTGS) is a novel method wherein gene expression of specific sequences is suppressed. Gene silencing mechanism is triggered by double-stranded RNA at the cellular level. The RNAi strategy for pest control is based on ingestion of double-stranded RNA (dsRNA) into the target pest system. After ingestion, dsRNA expresses either through hairpin or by other different means and spreads throughout the insect system (Katoch et al. 2013). Resistance to *C. suppressalis* was provided by rice knockdown lines TT51 (*cryAb* and *cry1Ac*) and T1C-19 (*cry1Ac*) with two aminopeptidase N genes (APN1 and APN2) (Qiu et al. 2017). Kola et al. (2016) demonstrated the silencing of CYP and APN genes of *S. incertulas* and knockdown of acetylcholine esterase gene (Ache) in rice lines which resulted in reduced larval length and weight of yellow stem borer within 15 days after exposure to dsRNA (Kola et al. 2019). However, they are still considered as genetically modified crops and more data need to be generated on specificity of dsRNA, fate of nanoparticle/dsRNA formulation in the environment, effects of RNAi-based products on non-target organisms, etc.

4.4.6 Genome Editing for Developing Resistance to Insect Pests in Elite Background

CRISPR gene editing is gaining prominence as an effective tool to tackle insect pest problems as it has the capacity to alter the specific gene of interest. In rice, the cytochrome P450 gene CYP71A1 encodes tryptamine 5-hydroxylase, which catalyses conversion of tryptamine to serotonin. In susceptible wild-type rice, planthopper feeding induces biosynthesis of serotonin and salicylic acid, whereas in mutants with an inactivated CYP71A1 gene, no serotonin is produced, but salicylic acid levels are higher. Insect-resistant rice plants with mutations in the cytochrome P450 gene accumulated high levels of salicylic acid but lacked serotonin. Suppression of serotonin confers resistance to planthoppers and stem borers, the two most destructive pests of rice (Lu et al. 2018). However, commercial use of CRISPR/Cas9 in insect pest management is still in its infancy. With the governments favouring SD2 genome-edited events, this method of genome editing is more appealing as compared to transgenic approach for the production of next-generation insect-resistant crops.

4.5 Role of Genomics in Insect Pest Resistance

4.5.1 Rice Genome

The 3k rice genome panel is a valuable resource which can be exploited for identification of genotypes with similar haplotypes. Once the haplotypes are identified, they may be evaluated for their phenotype. This will reduce the drudgery of screening large germplasm against the target pests. The work flow propounded by Varshney (2016) can be applied to the generation of re-sequencing and phenotyping data on 3000 rice genotypes. The re-sequencing data will be analysed for better understanding the rice genome and identification of structural variations in the germplasm collection. These structural variations together with phenotyping data will be analysed for establishing marker–trait associations for resistance traits of importance. The associated structural variations can be utilized for accelerating development of improved cultivars using genomics-assisted breeding. The availability of the rice genome sequence in conjunction with cutting-edge genomic technologies, including transcriptomics, proteomics and metabolomics, provides scientists an unprecedented opportunity to unravel the functions of rice defence genes against insects. The knowledge and molecular tools developed through these studies will undoubtedly facilitate the development of novel rice varieties with enhanced insect resistance (Chen et al. 2012).

4.5.2 Insect Genome and Transcriptome

The genome of BPH and its endosymbionts have been sequenced (Xue et al. 2014). It is a large genome (1141 Mb) with 27,571 protein-coding genes, of which 16,330 are specific to this species. In comparison, the WBPH genome is relatively smaller (720 Mb) with 21,254 protein-coding genes (Wang et al. 2017a), while the SBPH genome size is 541 Mb with 17,736 protein-coding genes (Zhu et al. 2017). Mitochondrial (mt) genomes of these three planthopper species have also been sequenced (Zhang et al. 2013, 2014). Draft genome (Kattupalli et al. 2021) and larval transcriptome (Renuka et al. 2017) of *S. incertulas*; chromosome level genome assembly of *C suppressalis* (Ma et al. 2020) and genome of *C. medinalis* (Zhao et al. 2021) are currently available. Transcriptomes of individual tissues like secretory salivary glands of GM with details on SSGPs in resistant and susceptible varieties (Sinha et al. 2012a, 2012b) and BPH (Ji et al. 2013) which have major implications in pest–rice interactions have been studied. Transcriptome analysis provides valuable information regarding BPH development, wing dimorphism and sex differences (Xue et al. 2010), which could facilitate further investigations on the detailed physiological mechanisms of BPH. Comparative analyses of the methylation patterns of Tf2 elements in BPH feeding on resistant- and susceptible-rice varieties confirmed that methylation, particularly in non-CG context, is involved in TE regulation and dynamics under nutritional stress which is reversible (Gupta and Nair 2022). These changes likely account for its capacity to thrive under unstable environments and its remarkable resilience to several stresses. Information on genes involved in sensory and chemoreception, those regulating metabolic pathways, detoxification genes, RNAi machinery in rice pests, can be further explored for developing pest-resistant crops.

4.5.3 Application of Structural and Functional Genomics in Genomics-Assisted Breeding

The insect genome and transcriptome databases of rice, the host plant and the insects have opened new vistas in the understanding of pest behaviour and can be exploited in future for developing pest-resistant lines. Deeper understanding of the pest–host interaction would help in manipulation of the signalling or metabolic pathways for conferring resistance.

4.6 HPR as a Component of Rice IPM

4.6.1 HPR with Biological Control

Earlier studies suggest that insect feeding on resistant varieties has delayed development and is more exposed to natural enemies. There are also some volatile organic compounds [*S*-linalool, caryophyllene, green leaf volatile (GLV) etc.] whose expression level in response to BPH attack determines the plant–pest interaction. *S*-linalool is one such volatile which is strongly induced by BPH attack (Cheng et al. 2007). Inducible *S*-linalool attracts predators and parasitoids but repels BPH. Similarly, GLV encoded by a gene HPL3 positively modulates resistance to BPH by modulating oxylipin pathway (Tong et al. 2012). Compared with rice plants exposed to the volatiles from uninfested plants, plants exposed to SSB-induced volatiles showed enhanced direct and indirect resistance to SSB. When subjected to caterpillar damage, the HIPV-exposed plants showed increased expression of jasmonic acid (JA) signalling genes, resulting in JA accumulation and higher levels of defensive proteinase inhibitors. Plants exposed to *C. suppressalis*-induced volatiles emitted larger amounts of inducible volatiles and were more attractive to the parasitoid, *Cotesia chilonis* (Yao et al. 2023) implicating the role of proteinase inhibitors. However, in-depth studies are essential to prove the effect and role of damage-associated molecular patterns (DAMP), HIV, green leaf volatiles (GLVs) in pest resistance and their utility in pest management.

4.6.2 HPR with Insecticides

Horgan et al. (2020) reported that BPH populations selectively reared on IR65482-4-136-2-2 (*Bph10* gene) had increased susceptibility to imidacloprid and fipronil, representing a possible trade-off with virulence. In contrast, a population with virulence to the highly resistant variety PTB33 was 4.88 times more resistant to imidacloprid and 3.18 times more resistant to BPMC compared to planthoppers of the same origin but reared on the susceptible variety IR22. Their experimental results suggest complex relations between insecticide resistance and virulence that vary according to insecticidal toxins and resistance genes and include potentially increased insecticide susceptibility (a trade-off) as well as common detoxification mechanisms (a benefit).

4.6.3 HPR and Endosymbionts

Symbionts influence the performance of herbivores on plants, to such an extent that they can ultimately determine whether a plant becomes a suitable host for an insect or not. In BPH, until 2000, yeast-like symbionts (YLS) were reported to be associated. But of late, other eukaryotic and bacterial symbionts have been reported to be associated with *N. lugens* and *N. cincticeps*. Ferrater et al. (2013) reviewed the complexity of the rice hopper symbiont interactions. He detailed on the nature of symbiont communities and their functions in planthoppers and leafhoppers—focusing on their likely roles in mediating adaptation to plant resistance. Evidence from a small number of experimental studies suggested that bacterial and eukaryotic (including yeast-like) symbionts can determine or mediate hopper virulence on rice plants and that symbiont functions could change over successive generations of selection on both resistant and susceptible plants. The role of symbionts in mediating virulence could differ depending on the strength of resistance in rice and the extent of planthopper exposure to the resistant variety or to varieties with similar resistance genes. Yeast-like symbionts (YLS) may mediate virulence adaptation in early generations of selection with the planthoppers themselves ultimately adapting to the novel resistance after several generations.

Preliminary mating studies with *N. lugens* have indicated that YLS did contribute to virulence on a resistant variety (IR62—*Bph3* gene), the male parent also influenced fitness on the resistant host, suggesting that other mechanisms (which may include bacterial symbionts) also played a role in virulence adaptation. Gupta et al. (2022) reported that *Pseudomonas* sp. in BPH microbiome varied with seasons and geographical locations and its composition and abundance correlated with BPH survivability. Environment-guided microbial shifts drive rapid stress adaptations in BPH.

Sinha et al. (2022) in their studies on analysis of rice tissues at the site of feeding of ARGM maggots revealed differential representation of the phylum Proteobacteria in the GM-infested and uninfested rice tissues. Analysis of the species diversity of *Pseudomonas* and *Wolbachia* supergroups at the feeding sites indicated the exchange of bacterial species between ARGM and its host, the rice upon infestation. In another study, sequencing of 16S rRNA bacterial gene (V3-V4 region) revealed differences in the microflora of the ARGM maggots feeding on susceptible (var. TN1) or resistant rice (RP 2068-18-3-5) hosts though *Wolbachia* sp. and *Pseudomonas* sp. formed the major constituents of the Proteobacteria in all the ARGM samples analysed. *Wolbachia* was the predominant bacterium in pupae and adults. *Pseudomonas* sp. was predominant in maggots. High species diversity in maggots isolated from susceptible rice and high representation of unclassified bacteria in maggots isolated from resistant rice was observed (Ojha et al. 2017). During the different stages of metamorphosis of an insect, there is a sterilization process at each stage, and insects acquire new set of microbes at each stage.

Symbionts do not work or have no role to play on mechanisms of resistance which are associated with plant surfaces, mechanical barriers, volatiles or chemicals that work on sensory features of the insect. Studies till date suggest an influence of endosymbionts on insect adaptation and virulence, but the mechanisms operating the adaptation are unclear. However, bacterial functions or the nature of the association between the bacteria and the insect pest (i.e. primary, secondary, internal or external) has not been established. The relative role of these symbionts in each of the insect pest–rice interactions needs more in-depth studies to actually define their role in insect survival and conferring virulence. Gupta and Nair (2020) opined that the gut microbiome can dramatically influence the physiology, behaviour and genetics of its insect host, and therefore, targeting the microbiome could be counted as an effective approach for developing an integrated, environment-friendly and a sustainable pest-management strategy.

4.6.4 Conclusion

IPM is the careful selection and intelligent deployment of the various proven and available strategies from the basket of options that are available for managing insect pests. It is solely based on location-specific needs of the farmers and not prescription-based. The current rice varieties developed and deployed for cultivation possess high tillering capacity with improvement in various yield components than traditional rice for compensation from insect pest damage and that capacity is enhanced by agronomic practices, thus integrated pest management should be thought of within the context of crop management (Litsinger 2005). Systems biology-guided analysis plays a major role in understanding the needs of plant–

insect interaction. The order of recruitment/colonization of insects in a field and understanding the relative importance of biotic and abiotic factors that affect insect community assemblages and species co-occurrence on plants is thus central to our understanding of insect–plant interactions (Trivellone et al. 2017). More careful, targeted and intelligent deployment (both temporally and spatially) of resistance genes together with crop management practices that avoid pest outbreaks will reduce the emergence of further virulent populations especially in the case of planthoppers which are widespread and gall midge which are endemic to certain pockets. So a thorough knowledge on the crop management practices and deployment strategies should be developed for rice lines with pyramided resistance, and the farmers need to be informed and educated thoroughly so as to avoid the build-up of virulent herbivore populations and increase the durability of resistance. Another opinion is that resistant varieties can be deployed to delay the outbreaks of the pests but not as an integral part of IPM (Horgan 2018).

4.6.5 Way Forward

Last decade witnessed a tremendous shift from exclusive selection of plants based on phenotype to section of plants based on molecular markers. Availability of trait-linked molecular markers and reduction in genotyping costs have facilitated better selection of lines in the breeding programmes. Most of the genotyping depends on the precise, reliable and repeatable phenotyping protocols with specific insect population. More accurate and precise phenotyping strategies are necessary to empower high-resolution linkage mapping and genome-wide association studies (GWAS) and for training genomic selection models in crop improvement (Varshney 2016). It is anticipated that selection of lines in breeding programmes will be predominantly based on genotyping/sequencing in the framework of genomic selection (GS). Cobb et al. (2013) reiterated that phenotyping is emerging as the major operational bottleneck limiting the power of genetic analysis and genomic prediction. To ensure relevance, the implementation of efficient and informative phenotyping experiments also requires familiarity with diverse germplasm resources, population structures and target populations of environments. Therefore, they proposed to establish next-generation phenotyping to increase the accuracy, precision and throughput of phenotypic estimation at all levels of biological organization while reducing costs and minimizing labour through automation, remote sensing, improved data integration and experimental design. Robust and field-relevant trait phenotyping systems are needed to characterize the full suite of genetic factors that contribute to quantitative phenotypic variation across cells, organs and tissues, developmental stages, years, environments, species and research programmes. Today, with the impact of climate change evident, there is a need to look for climate-resilient lines tolerant to insect pests. Varshney (2014) proposed the utility of NGS technologies combined with precise phenotyping methods, as tools that are powerful and rapid for identifying the genetic basis of agriculturally important traits and for predicting the breeding value of individuals in a plant breeding population. Unlike MAS, GS does not necessarily need QTL information before selection. GS uses reference population data containing phenotype and high-density marker data to predict breeding values for all the markers. Based on the predicted values, the breeding population data will be analysed to select the individual that possesses the desirable phenotype (Perez-de-Castro et al. 2012; Hu et al. 2018b). In this way, it is possible to introgress even minor-effect QTLs efficiently, as there are no biased marker effects, unlike with MAS.

As the plants and insects co-evolve, evaluation, identification and characterization of genetic resources are a continuous process. Finally, the age at which plants are attacked, how resistance manifests vis-à-vis over various stages of crop growth and how farmers manage their crop (vis-à-vis fertilizer and pesticide inputs), the knowledge level of the farmers, access to various pest management options will be the key determinants of pest population growth and damage responses in fields of resistant rice.

References

Abhilash Kumar V, Balachiranjeevi CH, Bhaskar Naik S, Rekha G, Rambabu R, Harika G, Pranathi K, Hajira SK, Anila M, Kousik M, Kale R, Dilip Kumar T, Prasad MS, Hari Prasad AS, Padmakumari AP, Laha GS, Balachandran SM, Madhav MS, Senguttuve LP, Kemparajau KB, Fiyaz AR, Bentur JS, Viraktamath BC, Ravindra Babu V, Sundaram RM (2017) Marker-assisted pyramiding of bacterial blight and gall midge resistance genes into RPHR-1005, the restorer line of the popular rice hybrid DRRH-3. Mol Breed 37:86. https://doi.org/10.1007/s11032-017-0687-8

Agarrwal R, Bentur J, Nair S (2014) Gas chromatography mass spectrometry based metabolic profiling reveals biomarkers involved in rice-gall midge interactions. J Integr Plant Biol 56. https://doi.org/10.1111/jipb.12244

Agarrwal R, Padmakumari AP, Bentur JS, Nair S (2016) Metabolic and transcriptomic changes induced in host during hypersensitive response mediated resistance in rice against the Asian rice gall midge. Rice 9:5. https://doi.org/10.1186/s12284-016-0077-6

Akanksha S, Jhansi Lakshmi V, Singh AK, Deepthi Y et al (2019) Genetics of novel brown planthopper *Nilaparvata lugens* (Stål) resistance genes in derived introgression lines from the interspecific cross *O. sativa* var. Swarna × *O. nivara*. J Genet 98:113

Amudhan S, Prasad Rao U, Bentur JS (1999) Total phenol profile in some rice varieties in relation to infestation by Asian rice gall midge *Orseolia oryzae* (Wood-Mason). Curr Sci 76:1577–1580

Andow DA, Bentur JS (2010) Pedigreed crosses to estimate recessive virulence allele frequencies in natural populations of gall midges. Entomol Exp Appl 135:18–36

Angeles ER, Khush GS (2000a) Genetic analysis of resistance to green leafhopper, *Nephotettix virescens* (Distant), in three varieties. Plant Breed 119:446–448

Angeles ER, Khush GS (2000b) Genetics of resistance to green leafhopper in five cultivars of rice, *Oryza sativa* L. SABRAO J Breed Genet 32:1–4

Angeles ER, Khush GS, Heinrichs EA (1981) New genes for resistance to white-backed planthopper in rice. Crop Sci 21:47–50

Athwal DS, Pathak MD, Bacalangco E, Pura CD (1971) Genetics of resistance to brown planthoppers and green leaf hoppers in *Oryza sativa* L. Crop Sci 11:747–750

Auclair JL, Baldos E (1982) Feeding by the whitebacked planthopper, *Sogatella furcifera*, within susceptible and resistant rice varieties. Entomol Exp Appl 32:200–203

Badri J, Lakshmidevi G, Jai Vidhya LRK, Prasad MS, Laha GS, Lakshmi VJ, Isetty SR, Padmashree R, Balakrishnan D, Varanasi YVP et al (2022) Multi-parent derived, marker-assisted introgression lines of the elite Indian rice cultivar, 'Krishna Hamsa' show resistance against Bacterial Blight and Blast and tolerance to drought. Plants 11:622. https://doi.org/10.3390/plants11050622

Bala A, Roy A, Behura N, Hess D, Das S (2013) Insight to the mode of action of *Allium sativum* leaf agglutinin (ASAL) expressing in T3 rice lines on brown planthopper. Am J Plant Sci 4:400–407

Balachiranjeevi CH, Prahalada GD, Mahender A, Jamaloddin M, Sevilla MAL, Marfori-Nazarea CM, Vinarao R, Sushanto U, Baehaki SE, Li ZK, Ali J (2019) Identification of novel locus, *BPH38(t)*, conferring resistance to brown planthopper (*Nilaparvata lugens* Stål.) using early backcross population in rice (*Oryza sativa* L.). Euphytica 215(11):185. https://doi.org/10.1007/s10681-019-2506-2

Basu S, Roychoudhury A (2014) Expression profiling of abiotic stress-inducible genes in response to multiple stresses in rice (*Oryza sativa* L.) varieties with contrasting level of stress tolerance. Biomed Res Int 2014:1 12

Bentur JS, Kalode MB (1990) A feeding test to identify rice varieties resistant to the leaf folder, *Cnaphalocrocis medinalis* (Guenee). Proc Indian Acad Sci (Anim Sci) 99(6):483–491

Bentur JS, Kalode MB (1996) Hypersensitive reaction and induced resistance in rice against the Asian rice gall midge, *Orseolia oryzae*. Entomol Exp Appl 78. https://doi.org/10.1111/j.1570-7458.1996.tb00766.x

Bentur JS, Pasalu IC, Kalode MB (1992) Inheritance of virulence in rice gall midge, *Orseolia oryzae*. Indian J Agric Sci 62:492–493

Bentur JS, Cheralu C, Ram Mohan Rao P (2008) Monitoring virulence in Asian rice gall midge populations in India. Entomol Exp Appl 129:96–106

Bentur JS, Padmakumari AP, Jhansilakshmi V, Padmavathi Ch, Kondala Rao Y, Amudhan S, Pasalu IC (2011) Insect resistance in rice (2000-2009). Technical bulletin no. 59. Directorate of Rice Research, Rajendranagar, Hyderabad, p 89

Bentur JS, Sundaram RM, Mangrauthia S, Nair S (2021) Molecular approaches for insect pest management in rice. In: Ali J, Wani SH (eds) Rice improvement: physiological, molecular breeding and genetic perspectives, pp 379–424. https://doi.org/10.1007/978-3-030-66530-2_11

Boddupally D, Tamirisa S, Gundra SR, Vudem DR, Rao KV (2018) Expression of hybrid fusion protein (Cry1Ac::ASAL) in transgenic rice plants imparts resistance against multiple insect pests. Sci Rep 8:8458. https://doi.org/10.1038/s41598-018-26881-9

Bose LK, Jambhulkar NN, Jena M, Mohapatra SD, Roy S, Singh ON (2015) Broadening genetic base of rice against major pests through wide hybridization. IRS15–PP 054. Genetic gains. Management in IRS 2015 compendium of abstracts. Paper presented in the International rice symposium—"Rice science for global food security and nutritional security" IRS 2015, Nov 18–20, Hyderabad, organized by ICAR-Indian Institute of Rice Research in association with International Rice Research Institute

Bostock RM (2005) Signal crosstalk and induced resistance: straddling the line between cost and benefit. Annu Rev Phytopathol 43:545–580. https://doi.org/10.1146/annurev.phyto.41.052002.095505

Brar DS, Khush GS (1991) Genetics of resistance to insects in crop plants. Adv Agron 45:223–274

Chen JW, Wang L, Pang XF, Pan QH (2006) Genetic analysis and fine mapping of a rice brown planthopper (*Nilaparvata lugens* Stål) resistance gene bph19(t). Mol Gen Genomics 275(4):321–329

Chen J, Huang DR, Wang L, Liu GJ (2010) Identification of quantitative trait loci for resistance to whitebacked planthopper, *Sogatella furcifera*, from an interspecific cross *Oryza sativa* × *O. rufpogon*. Breed Sci 60:153–159

Chen M, Shelton A, Ye GY (2011) Insect-resistant genetically modified rice in China: from research to commercialization. Annu Rev Entomol 56:81–101. https://doi.org/10.1146/annurevento-120709-144810

Chen H, Stout MJ, Qian Q, Chen F (2012) Genetic, molecular and genomic basis of rice defense against insects. Crit Rev Plant Sci 31(1):74–91. https://doi.org/10.1080/07352689.2011.616052

Cheng AX, Xiang CY, Li JX, Yang CQ, Hu WL, Wang LJ, Lou YG, Chen XY (2007) The rice (E)-beta-caryophyllene synthase (OsTPS3) accounts for the major inducible volatile sesquiterpenes. Phytochemistry 68(12):1632–1641

Cheng JA, Zhu JL, Zhu ZR, Zhang LG (2008) Rice planthopper outbreak and environmental regulation. J Environ Entomol 30(2):177–183

Cheng L, Huang F, Jiang Z, Baiyi L, Zhong X, Qiu Y (2021) Improved phenotyping procedure for evaluating resistance in rice against gall midge (*Orseolia oryzae*, WoodMason). Plant Methods 17:121. https://doi.org/10.1186/s13007-021-00823-5

Chintalapati P, Javvaji S, Gururaj K (2017) A rapid field screening method for evaluation of resistance to leaffolder, *Cnaphalocrocis medinalis* Guenee in rice. J Rice Res 10(1):67

Chintalapati P, Balakrishnan D, Venu Gopal Nammi TV, Javvaji S, Muthusamy SK, Lella Venkata SR, Neelamraju S, Katti G (2019) Phenotyping and genotype × environment interaction of resistance to leaffolder, *Cnaphalocrocis medinalis* Guenee (Lepidoptera: Pyralidae) in rice. Front Plant Sci 10:49. https://doi.org/10.3389/fpls.2019.00049

Cobb JN, DeClerck G, Greenberg A, Clark R, McCouch S (2013) Next-generation phenotyping: requirements and strategies for enhancing our understanding of genotype–phenotype relationships and its relevance to crop improvement. Theor Appl Genet 126:867–887. https://doi.org/10.1007/s00122-013-2066-0

Cohen MB, Bentur JS, Gould F (2004) Durable deployment of gall midge-resistant varieties. In: Bennett J, Bentur JS, Pasalu IC, Krishnaiah K (eds) New approaches to gall midge resistance in rice. Proceedings of the international workshop, 22–24 Nov 1998, Hyderabad, India

Dakshayani K, Bentur S, Kalode MB (1993) Nature of resistance in rice varieties against leaffolder *Cnaphalocrocis medinalis* (Guenee). Int J Trop Insect Sci 14:107–114

Datta K, Vasquez A, Tu J, Torrizo L, Alam MF, Oliva N, Abrigo E, Khush GS, Datta SK (1998) Constitutive and tissue-specific differential expression of the *cryIA(b)* gene in transgenic rice plants conferring resistance to rice insect pest. Theor Appl Genet 97:20–30

Deen RK, Ramesh S, Gautam K, Rao YK, Lakshmi VJ, Viraktamath BC, Brar DS, Ram T (2010) Identification of new genes for brown planthopper resistance in rice introgressed from *O. glaberrima* and *O. minuta*. Rice Genet Newsl 25:70–71

Deutsch CA, Tewksbury JJ, Tigchelaar M et al (2018) Increase in crop losses to insect pests in a warming climate. Science 361:916–919

Dhaliwal GS, Dhawan AK, Singh R (2007) Biodiversity and ecological agriculture: issues and perspectives. Indian J Ecol 34(2):100–109

Dhaliwal GS, Jindal V, Dhawan AK (2010) Insect pest problems and crop losses: changing trends. Indian J Ecol 37(1):1–7

Divya D, Himabindu K, Nair S et al (2015) Cloning of a gene encoding LRR protein and its validation as candidate gall midge resistance gene, *Gm4*, in rice. Euphytica 203:185–195

Divya D, Nair S, Bentur JS (2018a) Expression profiles of key genes involved in rice gall midge interactions reveal diversity in resistance pathways. Curr Sci 115:74–82

Divya D, Sahu N, Nair S, Bentur JS (2018b) Map-based cloning and validation of a gall midge resistance gene, *Gm8*, encoding a proline-rich protein in the rice variety Aganni. Mol Biol Rep 45:2075. https://doi.org/10.1007/s11033-018-4364-8

Divya D, Ratna Madhavi K, Dass MA et al (2018c) Expression profile of defense genes in rice lines pyramided with resistance genes against bacterial blight, fungal blast and insect gall midge. Rice 11:40

Dixit S, Singh UM, Singh AK, Alam S, Venkateshwarlu C, Nachimuthu VV, Yadav S, Abbai R, Selvaraj R, Devi MN et al (2020) Marker assisted forward breeding to combine multiple biotic-abiotic stress resistance/tolerance in rice. Rice 13:29

DRR (2009–2011) Progress report. Crop protection: entomology and plant pathology, vol. 2. Directorate of Rice Research (DRR), Hyderabad. All India Coordinated Rice Improvement Project. Indian Council of Agricultural Research, India, p 17

Du B, Chen RZ, Guo JP, He GC (2020) Current understanding of the genomic, genetic, and molecular control of insect resistance in rice. Mol Breed 40(2):24

Duriseti M, Durga Rani Ch.V, Padmakumari AP, Yamini KN, Malathi S, Cheralu C, Jamaloddin Md., Swathi G, Vanisri S (2018) Marker assisted selection for durable gall midge (Orseolia oryzae)resistance in rice. Presented in the National symposium on Entomology 2018: Advances and challenges from 10–12, Dec 2018 organised by PPAI, ESI, ABF and EAI at PJTSAU, Hyderabad

Fan F, Li N, Chen Y, Liu X, Sun H, Wang J, He G, Zhu Y, Li S (2017) Development of elite BPH-resistant wide-spectrum restorer lines for three and two line hybrid rice. Front Plant Sci 8: 986. https://doi.org/10.3389/fpls.2017.00986

Fan D, Liu Y, Zhang H, He J, Huang F, Huang S, Wu B, Liu D, Wen P, Liu L, Jiang L, Cheng X, Wan J (2018) Identification and fine mapping of qWBPH11 conferring resistance to white backed planthopper (*Sogatella furcifera* Horvath) in rice (*Oryza sativa* L.). Mol Breed 38:96

Ferrater JB, Jong PW, Dicke M, Chen YH, Horgan FG (2013) Symbiont-mediated adaptation by planthoppers and leafhoppers to resistant rice varieties. Arthropod Plant Interact 7:591–605

Fujita D, Doi K, Yoshimura A, Yasui H (2004) Introgression of a resistance gene for green rice leafhopper from *Oryza nivara* into cultivated rice, *Oryza sativa* L. Rice Genet Newsl 21:64–66

Fujita D, Doi K, Yoshimura A, Yasui H (2006) Molecular mapping of a novel gene, *Grh5*, conferring resistance to green rice leafhopper (*Nephotettix cincticeps* Uhler) in rice, *Oryza sativa* L. Theor Appl Genet 113:567–573

Fujita D, Yoshimura A, Yasui H (2010) Development of near isogenic lines and pyramided lines carrying resistance genes to green rice leafhopper (*Nephotettix cincticeps* Uhler) with the Taichung 65 genetic background in rice (*Oryza sativa* L.). Breed Sci 60:18–27

Fujita D, Kohli A, Horgan FG (2013) Rice resistance to planthoppers and leafhoppers. Crit Rev Plant Sci 32(3):162–191. https://doi.org/10.1080/07352689.2012.735986

Fukuta Y, Tamura K, Hirae M, Oya S (1998) Genetic analysis of resistance to green rice leafhopper (*Nephotettix cincticeps* Uhler) in rice parental line, Norin-16, using RFLP markers. Breed Sci 48:243–249

Ghani MV, Khush GS (1988) A new gene for resistance to green leafhopper, Nephotettix virescens (Distant) in rice. J Genet 67:151–159

Giordanengo P, Brunissen L, Rusterucci C, Vincent C, van Bel A, Dinant S, Girousse C, Faucher M, Bonnemain JL (2010) Compatible plant-aphid interactions: how aphids manipulate plant responses. Crit Rev Biol 333:566–573

Grayer RJ, Kimmins FM, Stevenson PC, Stevenson JB, Wijayagunesekera HNP (1994) Phenolics in rice phloem sap as sucking deterrents to the brown planthopper (*Nilaparvata lugens*). Acta Hortic 381:391–394

Gupta A, Nair S (2020) Dynamics of insect–microbiome interaction influence host and microbial symbiont. Front Microbiol 11:1357. https://doi.org/10.3389/fmicb.2020.01357

Gupta A, Nair S (2021) Methylation patterns of Tf2 retrotransposons linked to rapid adaptive stress response in the brown planthopper (*Nilaparvata lugens*). Genomics 113:4214–4226

Gupta A, Nair S (2022) Heritable epigenomic modifications influence stress resilience and rapid adaptations in the brown planthopper (*Nilaparvata lugens*). Int J Mol Sci 2022(23):8728. https://doi.org/10.3390/ijms23158700028

Gupta A, Deepak S, Nair S (2022) Shifts in Pseudomonas species diversity influence adaptation of brown planthopper to changing climates and geographical locations. iScience 25:104550. https://doi.org/10.1016/j.isci.2022.104550

Hamm JC, Stout MJ, Riggio RM (2010) Herbivore- and elicitor-induced resistance in rice to the rice water weevil (*Lissorhoptrus oryzophilus* Kuschel) in the laboratory and field. J Chem Ecol 36: 192–199. https://doi.org/10.1007/s10886-010-9751-0

Hao P, Liu C, Wang Y, Chen R, Tang M, Du B, Zhu L, He G (2008) Herbivore-induced callose deposition on the sieve plates of rice: an important mechanism for host resistance. Plant Physiol 146(4):1810–1820

Haque A, Rafii MY, Yusoff MM, Ali NS, Yusuff O, Datta DR, Anisuzzaman M, Ikbal MF (2021) Recent advances in rice varietal development for durable resistance to biotic and abiotic stresses through marker-assisted gene pyramiding. Sustainability 13:10806

He J, Liu YQ, Liu YL, Jiang L, Wu H, Kang HY, Liu SJ, Chen LM, Liu X, Cheng XN, Wan JM (2013) High-resolution mapping of brown planthopper (BPH) resistance gene Bph27(t) in rice (*Oryza sativa* L.). Mol Breed 31(3):549–557

Heinrichs E, Medrano FG, Rapusas HR (1985) Genetic evaluation for insect resistance in rice. XF2006254048. 352 pp. http://books.irri.org/9711041103_content.pdf

Hernandez JE, Khush GS (1981) Genetics of resistance to white-backed planthopper in some rice (*Oryza sativa* L.) varieties. Oryza 18:44–50

High SM, Cohen MB, Shu Q, Altosaar I (2004) Achieving successful deployment of Bt rice. Trends Plant Sci 9(6):286–292. https://doi.org/10.1016/S1360-1385(04)00099-8

Himabindu K, Suneetha K, Sama VSAK et al (2010) A new rice gall midge resistance gene in the breeding line CR57-MR1523, mapping with flanking markers and development of NILs. Euphytica 174:179–187

Hirabayashi H, Angeles ER, Kaji R, Ogawa T, Brar DS, Khush GS (1998) Identification of a brown planthopper resistance gene derived from *O. officinalis* using molecular markers in rice. Breed Sci 48:82 (in Japanese)

Hirabayashi H, Ideta O, Sato H, Takeuchi Y, Ando I, Nemoto H, Imbe T, Brar DS, Ogawa T (2004) Identification of a resistance gene to brown planthopper derived from *Oryza minuta* in rice. Breed Res 6:285

Horgan FG (2018) Integrating gene deployment and crop management for improved rice resistance to Asian planthoppers. Crop Prot 110:21–33. https://doi.org/10.1016/j.cropro.2018.03.013

Horgan FG, Ramal AF, Bentur JS, Kumar R, VasantaBhanu K, Sarao PS, Iswanto EH, Chien HV, Phyu MH, Bernal CC, Almazan MLP, Alam MZ, Lu Z, Huang SH (2015) Virulence of brown planthopper (*Nilaparvata lugens*) populations from South and South East Asia against resistant rice varieties. Crop Prot 78:222–231

Horgan FG, Srinivasan TS, Naik BS, Ramal AF, Bernal C, Almazan MLP (2016) Effects of nitrogen on egg-laying inhibition and ovicidal response in planthopper-resistant rice varieties. Crop Prot 89:223–230

Horgan FG, Garcia CPF, Haverkort F, de Jong PW, Ferrater JB (2020) Changes in insecticide resistance and host range performance of planthoppers artificially selected to feed on resistant rice. Crop Prot 127:104963. https://doi.org/10.1016/j.cropro.2019.104963

Howe GA (2004) Jasmonates as signals in the wound response. J Plant Growth Regul 23:223–237. https://doi.org/10.1007/s00344-004-0030-6

Hu J, Xiao C, Cheng MX, Gao GJ, Zhang QL, He YQ (2015a) Fine mapping and pyramiding of brown planthopper resistance genes QBph3 and QBph4 in an introgression line from wild rice *Oryza officinalis*. Mol Breed 35(1):3

Hu J, Xiao C, Cheng MX, Gao GJ, Zhang QL, He YQ (2015b) A new finely mapped *Oryza australiensis*-derived QTL in rice confers resistance to brown planthopper. Gene 561(1): 132–137

Hu H, Scheben A, Edwards D (2018a) Advances in integrating genomics and bioinformatics in the plant breeding pipeline. Agriculture 8:75

Hu J, Chang XY, Zou L, Tang WQ, Wu WR (2018b) Identification and fine mapping of *Bph33*, a new brown planthopper resistance gene in rice (*Oryza sativa* L.). Rice 11(1):55

Huang Z, He GC, Shu LH, Li XH, Zhang Q (2001) Identification and mapping of two brown planthopper resistance genes in rice. Theor Appl Genet 102:923–934

Huang D, Qiu YF, Zhang Y, Huang F, Meng J, Wei S, Li R, Chen BS (2013) Fine mapping and characterization of BPH27, a brown planthopper resistance gene from wild rice (*Oryza rufipogon* Griff.). Theor Appl Genet 126(1):219–229

ICAR-IIRR (2018–2021) Progress reports. Crop protection. Entomology and plant pathology, vol II. All India Coordinated Rice Improvement Project. ICAR-Indian Institute of Rice Research, Indian Council of Agricultural Research, Rajendranagar, Hyderabad

IRRI (2013) Standard evaluation system (SES) for rice, 5th edn. IRRI, Manila, Philippines

Ishii T, Brar DS, Multani DS, Khush GS (1994) Molecular tagging of genes for brown planthopper resistance and earliness introgressed from *Oryza australiensis* into cultivated rice *O. sativa*. Genome 37(2):217–221

Ishwarya Lakshmi VG, Sreedhar M, Jhansi Lakshmi V, Gireesh C, Rathod S, Bohar R, Deshpande S, Laavanya R, Kiranmayee KN, Siddi S, Vanisri S (2022) Development and validation of diagnostic KASP markers for brown planthopper resistance in rice. Front Genet 13. https://doi.org/10.3389/fgene.2022.914131

Jai Vidhya LRK, Lakshmi Devi G, Lakshmi Bhavani IP, Swathi G, Badri J, Padmakumari AP, Prasad MS, Laha GS, Ram T (2018) Polymorphic survey in donors for bacterial leaf blight, blast, brown planthopper, gall midge and recurrent parent Krishnahamsa and WGL14 for background selection. Paper presented in the 5th international rice congress held at Singapore from 15–17 Oct 2018, organised by IRRI, Manila, Philippines

Jairin J, Phengrat K, Tcangdeerith S, Vanavichit A, Toojinda T (2007a) Mapping of a broad-spectrum brown planthopper resistance gene, *Bph3*, on rice chromosome 6. Mol Breed 19:35–44

Jairin J, Teangdeerith SN, Leelagud P, Phengrat K, Vanavichit A, Toojinda T (2007b) Physical mapping of Bph3, a brown planthopper resistance locus in rice. Maejo Int J Sci Technol 1(2): 166–177

Janaki Ramayya P, Vinukonda VP, Singh UM, Alam S, Venkateshwarlu C, Vipparla AK et al (2021) Marker-assisted forward and backcross breeding for improvement of elite Indian rice variety Naveen for multiple biotic and abiotic stress tolerance. PLoS One 16(9):e0256721. https://doi.org/10.1371/journal.pone.0256721

Javvaji S, Uma MT, Ramana DBV, Sheshu MM et al (2021) Characterization of resistance to rice leaf folder, *Cnaphalocrocis medinalis* in mutant SambaMahsuri rice lines. Entomol Exp Appl 169(9):859–875. https://doi.org/10.1111/eea.13082169

Jeer M, Telugu UM, Voleti SR, Padmakumari AP (2017) Soil application of silicon reduces yellow stem borer, *Scirpophaga incertulas* (Walker) damage in rice. J Appl Entomol 141:189–201. https://doi.org/10.1111/jen.12324

Jeer M, Suman K, Uma Maheswari T, Voleti SR, Padmakumari AP (2018) Rice husk ash and imidazole application enhances silicon availability to rice plants and reduces yellow stem borer damage. Field Crops Res 224:60–66

Jena KK, Kim SM (2010) Current status of brown planthopper (BPH) resistance and genetics. Rice 3:161–171. https://doi.org/10.1007/s12284-010-9050-y

Jena KK, Jeung JU, Lee JH, Choi HC, Brar DS (2006) High-resolution mapping of a new brown planthopper (BPH) resistance gene, *Bph18*(t), and marker-assisted selection for BPH resistance in rice (*Oryza sativa* L.). Theor Appl Genet 112:288–297

Ji R, Yu H, Fu Q, Chen H, Ye W, Li S, Lou Y (2013) Comparative transcriptome analysis of salivary glands of two populations of rice brown planthopper, *Nilaparvata lugens*, that differ in virulence. PLoS One 8:e79612

Jin Y, Drabik D, Heerink N, Wesseler J (2019) The cost of postponement of Bt Rice commercialization in China. Front Plant Sci 10:1226. https://doi.org/10.3389/fpls.2019.01226

Jing S, Zhang L, Ma Y, Liu B, Zhao Y, Yu H, Zhou X, Qin R, Zhu L, He G (2014) Genome-wide mapping of virulence in brown planthopper identifies Loci that break down host plant resistance. PLoS One 9:e98911

Jing S, Zhao Y, Du B, Chen R, Zhu L, He G (2017) Genomics of interaction between the brown planthopper and rice. Curr Opin Insect Sci 19:82–87

Kadowaki M, Yoshimura A, Yasui H (2003) RFLP mapping of antibiosis to rice green leafhopper. In: Khush GS, Brar DS, Hardy B (eds) Advances in rice genetics. International Rice Research Institute, Los Baños, pp 270–272

Katiyar SK, Tan Y, Huang B et al (2001) Molecular mapping of gene Gm-6(t) which confers resistance against four biotypes of Asian rice gall midge in China. Theor Appl Genet 103:953–961

Katoch R, Sethi A, Thakur N et al (2013) RNAi for insect control: current perspective and future challenges. Appl Biochem Biotechnol 171:847–873. https://doi.org/10.1007/s12010-013-0399-4

Kattupalli D, Barbadikar KM, Balija V, Ballichatla S, Athulya R, Padmakumari AP, Saxena S, Gaikwad K, Yerram S, Kokku P, Madhav MS (2021) The draft genome of yellow stem borer, an agriculturally important pest, provides molecular insights into its biology, development and specificity towards rice for infestation. Insects 12(6):563. https://doi.org/10.3390/insects12060563

Kawaguchi M, Murata K, Ishii T, Takumi S, Mori N, Nakamura C (2001) Assignment of a brown planthopper (*Nilaparvata lugens*, Stål) resistance gene *bph4* to the rice chromosome 6. Breed Sci 51(1):13–18

Kessler A, Baldwin IT (2002) Plant responses to insect herbivory: the emerging molecular analysis. Annu Rev Plant Biol 53:299–328

Khan ZR, Saxena RC (1984) Electronically recorded waveforms associated with the feeding behavior of *Sogatella furcifera* (Homoptera: Delphacidae) on susceptible and resistant rice varieties. J Econ Entomol 77:1479–1482

Khan ZR, Rueda BP, Caballero P (1989) Behavioral and physiological responses of rice leaffolder *Cnaphalocrocis medinalis* to selected wild rices. Entomol Exp Appl 52:7–13

Khush GS, Karim ANMR, Angeles ER (1985) Genetics of resistance of rice cultivar ARC10550 to Bangladesh brown planthopper teletype. J Genet 64:121–125

Kim SM, Sohn JK (2005) Identification of a rice gene (*Bph1*) conferring resistance to brown planthopper (*Nilaparvata lugens* Stål) using STS markers. Mol Cells 20(1):30–34

Kobayashi T (2015) Evolving ideas about genetics underlying insect virulence to plant resistance in rice-brown planthopper interactions. J Insect Physiol. https://doi.org/10.1016/j.insphys.2015.12.001

Kola VSR, Renuka P, Padmakumari AP, Mangrauthia SK, Balachandran SM, Babu VR, Madhav MS (2016) Silencing of CYP6 and APN genes affects the growth and development of rice yellow stem borer, *Scirpophaga incertulas*. Front Physiol 7:20. https://doi.org/10.3389/fphys.2016.00020

Kola VSR, Renuka P, Padmakumari AP, Mangrauthia SK, Balachandran SM, Madhav MS (2019) Knockdown of acetylcholinesterase (AChE) gene in rice yellow stem borer, *Scirpophaga incertulas* (Walker) through RNA interference. Agri Gene. https://doi.org/10.1016/j.aggene.2019.100081

Koornneef A, Leon-Reyes A, Ritsema T, Verhage A, Den Otter FC, Van Loon LC, Pieterse CM (2008) Kinetics of salicylate-mediated suppression of jasmonate signaling reveal a role for redox modulation. Plant Physiol 147(3):1358–1368. https://doi.org/10.1104/pp.108.121392. Epub 2008 Jun 6. PMID: 18539774; PMCID: PMC2442557

Kuang YH, Fang YF, Lin SC et al (2021) The impact of climate change on the resistance of rice near-isogenic lines with resistance genes against brown planthopper. Rice 14:64. https://doi.org/10.1186/s12284-021-00508-6

Kumar A, Shrivastava MN, Shukla BC (1998a) Inheritance and allelic relationship of gall midge biotype 1 resistance genes in some new donors. Oryza 35:70–73

Kumar A, Shrivastava MN, Sahu RK (1998b) Genetic analysis of ARC5984 for gall midge resistance—a reconsideration. Rice Genet Newsl 15:142–143

Kumar A, Shrivastava MN, Shukla BC (1999) A new gene for resistance to gall midge in rice cultivar RP2333-156-8. Rice Genet Newsl 16:85–87

Kumar A, Bhandarkar S, Pophlay DJ et al (2000) A new gene for gall midge resistance in rice accession Jhitpiti. Rice Genet Newsl 17:83–84

Kumar A, Jain A, Sahu RK et al (2005) Genetic analysis of resistance genes for the rice gall midge in two rice genotypes. Crop Sci 45:1631–1635

Kumar K, Sarao PS, Bhatia D, Neelam K, Kaur A, Mangat GS, Brar DS, Singh K (2018) High-resolution genetic mapping of a novel brown planthopper resistance locus, *Bph34* in *Oryza sativa* L. × *Oryza nivara* (Sharma & Shastry) derived interspecific F_2 population. Theor Appl Genet 131:1163–1171

Kumari S, Sheba JM, Marappan M, Ponnuswamy S, Seetharaman S, Pothi N, Subbarayalu M, Muthurajan R, Natesan S (2010) Screening of IR50 × RathuHeenati F_7 RILs and identification of SSR markers linked to brown planthopper (*Nilaparvata lugens* Stål) resistance in rice (*Oryza sativa* L.). Mol Biotechnol 46(1):63–71

Kumari P, Jasrotia P, Kumar D, Kashyap PL, Kumar S, Mishra CN, Kumar S, Singh GP (2022) Biotechnological approaches for host plant resistance to insect pests. Front Genet 13:914029

Laing MD, Gatarahiya MC, Adandonon A (2006) Silicon use for pest control in agriculture: a review. Proc S Afr Sugar Technol Assoc 80:278–286

Leelagud P, Kongsila S, Vejchasarn P et al (2020) Genetic diversity of Asian rice gall midge based on mtCOI gene sequences and identification of a novel resistance locus gm12 in rice cultivar MN62M. Mol Biol Rep 47.4273. https://doi.org/10.1007/s11033-020-05546-9

Li Y, Hallerman E, Liu Q, Wu K, Peng Y (2016) The development and status of Bt rice in China. Plant Biotechnol J 14(3):839–848. https://doi.org/10.1111/pbi.12464

Li ZH, Xue YX, Zhou HL, Li Y, Usman B, Jiao XZ, Wang XY, Liu F, Qin BX, Li RB, Qiu YF (2019) High-resolution mapping and breeding application of a novel brown planthopper resistance gene derived from wild rice (*Oryza rufipogon* Griff). Rice 12(1):41

Litsinger JA (2005) When is a rice insect a pest: yield loss and the green revolution. In: Peshin R, Dhawan AK (eds) Integrated pest management: innovation-development process. Springer Science+Business Media B.V., pp 391–498. https://doi.org/10.1007/978-1-4020-8992-3

Liu Q, Hallerman E, Peng Y, Li Y (2016) Development of Bt rice and Bt maize in China and their efficacy in target pest control. Int J Mol Sci 17(10):1561–1576. https://doi.org/10.3390/ijms17101561

Liu Y, Wang Y, Shu C, Lin K, Song F, Bravo A, Soberón M, Zhang J (2018) Cry64Ba and Cry64Ca, two ETX/MTX2-type Bacillus thuringiensis insecticidal proteins active against hemipteran pests. Appl Environ Microbiol 84(3):e01996-17. https://doi.org/10.1128/AEM.01996-17

Lorang JM, Sweat TA, Wolpert TJ (2007) Plant disease susceptibility conferred by a "resistance" gene. Biol Sci 104(37):14861–14866. https://doi.org/10.1073/pnas.0702572104

Lü WT, Du B, Shangguan XX, Zhao Y, Pan YF, Zhu LL, He YQ, He GC (2014) BAC and RNA sequencing reveal the brown planthopper resistance gene *Bph15* in a recombination cold spot that mediates a unique defense mechanism. BMC Genomics 15(1):674

Lu HP, Luo T, Fu HW, Wang L, Tan YY, Huang JZ et al (2018) Resistance of rice to insect pests mediated by suppression of serotonin biosynthesis. Nat Plants 4(6):338–344. https://doi.org/10.1038/s41477-018-0152-7

Ma JF (2004) Role of silicon in enhancing the resistance of plants to biotic and abiotic stresses. Soil Sci Plant Nutr 50:11–18

Ma W, Zhao X, Yin C, Jiang F, Du X, Chen T, Zhang Q, Qiu L, Xu H, Joe Hull J, Li G, Sung WK, Li F, Lin Y (2020) A chromosome-level genome assembly reveals the genetic basis of cold tolerance in a notorious rice insect pest, *Chilo suppressalis*. Mol Ecol Resour 20(1):268–282. https://doi.org/10.1111/1755-0998.13078

Makkar GS, Bentur JS (2017) Breeding for stem borer and gall midge resistance in rice. In: Breeding insect resistant crops for sustainable agriculture. Springer, Singapore, pp 323–352

McCouch SR, CGSNL (2008) Gene nomenclature system for rice. Rice 1:72–84

Mohan M, Nair S, Bentur JS et al (1994) RFLP and RAPD mapping of rice GM2 gene that confers resistance to biotype 1 of gall midge (*Orseoliaoryzae*). Theor Appl Genet 87:782–788

Muduli L, Pradhan SK, Mishra A, Bastia DN, Samal KC, Agrawal PK, Dash M (2021) Understanding brown planthopper resistance in rice: genetics, biochemical and molecular breeding approaches. Rice Sci 28(6):532–546. https://doi.org/10.1016/j.rsci.2021.05.013

Mur LAJ, Kenton P, Atzorn R, Miersch O, Wasternack C (2006) The outcomes of concentration-specific interactions between salicylate and cross talk in defense signaling: Jasmonate signaling include synergy, antagonism, and oxidative stress leading to cell death. Plant Physiol 140:249–262

Murai H, Hashimoto Z, Sharma PN, Shimizu T, Murata K, Takumi S, Mori N, Kawasaki S, Nakamura C (2001) Construction of a high-resolution linkage map of a rice brown planthopper (*Nilaparvata lugens* Stål) resistance gene bph2. Theor Appl Genet 103(4):526–532

Muralidharan K, Pasalu IC (2006) Assessments of crop losses in rice ecosystems due to stem borer damage (Lepidoptera: Pyralidae). Crop Prot 25(5):409–417. https://doi.org/10.1016/j.cropro.2005.06.007

Murata K, Fujiwara M, Murai H, Takumi S, Mori N, Nakamura C (2001) Mapping of a brown planthopper (*Nilaparvata lugens* Stål) resistance gene BPH9 on the long arm of rice chromosome 12. Cereal Res Commun 29:245–250

Myint KKM, Matsumura M, Takagi M, Yasui H (2009a) Demographic parameters of long-term laboratory strains of the brown planthopper, *Nilaparvata lugens* Stål (Homoptera: Delphacidae) on resistance genes bph20(t) and Bph21(t) in rice. J Faculty Agric Kyushu Univ 54:159–164

Myint KKM, Yasui H, Takagi M, Matsumura M (2009b) Virulence of long-term laboratory populations of the brown planthopper, *Nilaparvata lugens* (Stål), and whitebacked planthopper, *Sogatella furcifera* (Horvath; Homoptera: Delphacidae), on rice differential varieties. Appl Entomol Zool 44(1):149–153

Myint KKM, Fujita D, Matsumura M, Sonoda T, Yoshimura A, Yasui H (2012) Mapping and pyramiding of two major genes for resistance to the brown planthopper (*Nilaparvata lugens* [Stål]) in the rice cultivar ADR52. Theor Appl Genet 124(3):495–504

Nagamani G (2022) Gene pyrimiding for biotic stress resistance and high yield in rice. Thesis submitted to Prof. Jaya Shankar. Telangana State Agricultural University, Telangana

Naik SB, Divya D, Sahu N, Sundaram RM, Sarao PS, Singh K, Lakshmi VJ, Bentur JS (2018) A new gene Bph33(t) conferring resistance to brown planthopper (BPH), *Nilaparvata lugens* (Stål) in rice line RP2068-18-3-5. Euphytica 214(3):53

Nayak P, Basu D, Das S et al (1997) Transgenic elite indica rice plants expressing CryIAc-endotoxin of *Bacillus thuringiensis* are resistant against yellow stem borer (*Scirpophaga incertulas*). Proc Natl Acad Sci USA 94:2111–2116

Nemoto H, Ikeda R, Kaneda C (1989) New genes for resistance to brown planthopper *Nilaparvata lugens* Stål in rice. Jpn J Breed 39:23–28

Oerke EC (2006) Crop losses to pests. J Agric Sci 144:31–43

Ojha A, Sinha DK, Padmakumari AP et al (2017) Bacterial community structure in the Asian rice gall midge reveals a varied microbiome rich in proteobacteria. Sci Rep 7:9424. https://doi.org/10.1038/s41598-017-09791-0

Padhi G (2004) Biochemical basis of resistance in rice to yellow stem borer, Scirpophaga incertulas Wlk. Madras Agric J 91:253–256

Padhi G, Rao PSP (1979) Preliminary studies on reaction of some wild rices to infestation by the yellow rice stem borer *Tryporyza incertulas* Wlk. Oryza 15(1):99–101

Padhi G, Sen P (2002) Evaluation of wild species against yellow stem borer, *Scirpophagaincertulas* Walk. J Appl Zool Res 11(2 & 3):103–104

Padmakumari AP, Katti G (2018) Designing insect bioassays for studying components of insect pest management-rice yellow stem borer—as a case study. Extended summary# 054 page 117. Paper presented in National symposium entomology 2018: advances and challenges, 8–10 Dec 2018, Hyderabad, India

Padmakumari AP, Pasalu IC (2003) Influence of planting pattern of trap crops on yellow stem borer, *Scirpophaga incertulas* (Walker) damage in rice. Indian J Plant Prot 30(1):78–83

Padmakumari AP, Ram T (2012) Wild rice as source of tolerance to yellow stem borer, *Scirpophaga incertulas*, Walker in rice. In: International conference on plant health management for food security held at Hyderabad from 28–30 Nov 2012

Padmakumari AP, Ram T (2016) Tolerance of wild rice introgressed lines to yellow stem borer, *Scirpophaga incertulas* (Walker). IIRR Newsl 14(2&3)

Padmakumari AP, Padmavathi Ch, Band S, Pasalu IC (2008) Compensatory mechanism in rice against yellow stem borer damage. Poster presented in Dr. Leslie C. Coleman Memorial National Symposium on Plant Protection, 4–6 Dec 2008. University of Agricultural Sciences, GKVK, Bangalore

Padmakumari AP, Ram T, Suman K, Ram Deen (2015) Stem borer tolerance in IR 64 BILs developed from *O. glaberrima*. IRS15–OP417 in theme biotic stress management in *IRS 2015 compendium of abstracts*. Paper presented in the International rice symposium—"Rice science for global food security and nutritional security" IRS 2015, Nov 18–20, Hyderabad, organized by ICAR-IIRR in association with IRRI

Padmakumari AP, Katti G, Rajender Kumar Reddy V (2017a) Aromatic variety as a trap crop for stem borer management in rice. Agric Update 12(TECHSEAR-9):1–8. https://doi.org/10.15740/HAS/AU/12.TECHSEAR(9)2017/1-8

Padmakumari AP, Seshu Madhav M, Uma Kanth B, Sunil B, Gopi P, Karteek J, Laha GS, Sundaram RM, Subba Rao LV, Patel HK, Sonti RV (2017b) EMS mutants as a novel source of tolerance to yellow stem borer. IIRR Newsl 15(2&3):10–11

Padmavathi G, Ram T, Ramesh K, Kondala Rao Y, Pasalu IC, Viraktamath BC (2007) Genetics of whitebacked planthopper, *Sogatella furcifera* (Horváth), resistance in rice. SABRAO J Breed Genet 39(2):99–105

Painter RH (1968) Insect resistance in crop plants. The University Press of Kansas, Lawrence, KS, p 520

Panda N, Heinrichs E (1983) Levels of tolerance and antibiosis in rice varieties having moderate resistance to the brown planthopper, *Nilaparvata lugens* (Stål) (Hemiptera: Delphacidae). Environ Entomol 12(4):1204–1214

Panda N, Khush GS (1995) Host plant resistance to insects. International Rice Research Institute, CAB International, Wallingford, pp 151–523. 9780851989044. https://books.google.co.in/books?id=xe3wAAAAMAAJ

Panda B, Kar MK, Padhi G, Sen P (2011) Development of F_1 and BC_1F_1 interspecific hybrids of *O. sativa* cv. Savitri/*O.brachyantha* to introgress yellow stem borer resistance genes into cultivated rice. Oryza 48(3):195–199

Perez-de-Castro A, Vilanova S, Canizares J et al (2012) Application of genomic tools in plantbreeding. Curr Genomics 13:179–195. https://doi.org/10.2174/138920212800543084

Phi CN, Fujita D, Yamagata Y, Yoshimura A, Yasui H (2019) High-resolution mapping of GRH6, a gene from *Oryza nivara* (Sharma et Shastry) conferring resistance to green rice leafhopper (*Nephotettix cincticeps* Uhler). Breed Sci 69(3):439–446. https://doi.org/10.1270/jsbbs.19029

Pieterse CMJ, Ton J, Van Loon LC (2001) Cross-talk between plant defence signalling pathways: boost or burden? Agbiotechnet 3:1–8

Potupureddi G, Balija V, Ballichatla S, Gokulan CG, Awalellu K, Lekkala S, Jallipalli K, Gayathri MG, Mohammad E, Milton M, Arutla S, Burka R, Shankar LG, Phani PA, Lella Venkata S, Raman Meenakshi S, Viraktamath BC, Vemuri RB, Brahma K, Madnala R, Patel HK, Sonti RV, Madhav MS (2021) Mutation resource of Samba Mahsuri revealed the presence of high extent of variations among key traits for rice improvement. PLoS One 16(10):e0258816. https://doi.org/10.1371/journal.pone.0258816. ISSN: 1932-6203.

Prahalada GD, Shivakumar N, Lohithaswa HC, Gowda DKS, Ramkumar G, Kim SR, Ramachandra C, Hittalmani S, Mohapatra T, Jena KK (2017) Identification and fine mapping of a new gene, BPH31 conferring resistance to brown planthopper biotype 4 of India to improve rice, *Oryza sativa* L. Rice 10(1):41

Punithavalli M, Muthukrishnan N, Balaji Rajkumar M (2013) Influence of rice genotypes on folding and spinning behaviour of leaffolder (*Cnaphalocrocis medinalis*) and its interaction with leaf damage. Rice Sci 20:442–450

Punithavalli M, Muthukrishnan N, Balaji Rajkumar M (2014) Impact of rice genotypes on the developmental characteristics of rice leaf folder, *Cnaphalocrocis medinalis*. Indian J Plant Prot 42:38–43

Puri SN, Mote UN (2003) Emerging pest problems of India and critical issues in their management. In: Subrahmanyam B, Ramamurthy VV, Singh VS (eds) Proceedings of national symposium on frontier areas of entomological research. Entomological Society of India, New Delhi, pp 13–24

Qiu YF, Guo JP, Jing SL, Zhu LL, He GC (2010) High resolution mapping of the brown planthopper resistance gene Bph6 in rice and characterizing its resistance in the 9311 and Nipponbare near isogenic backgrounds. Theor Appl Genet 121(8):1601–1611

Qiu YF, Guo JP, Jing SL, Zhu LL, He GC (2012) Development and characterization of japonica rice lines carrying the brown planthopper-resistance genes *BPH12* and Bph6. Theoret Appl Genet 124(3):485–494

Qiu YF, Guo JP, Jing SL, Zhu LL, He GC (2014) Fine mapping of the rice brown planthopper resistance gene bph7 and characterization of its resistance in the 93-11 background. Euphytica 198(3):369–379

Qiu L, Fan J, Zhang B, Liu L, Wang X, Lei C, Lin Y, Ma W (2017) RNA interference knockdown of aminopeptidase N genes decrease the susceptibility of Chilo suppressalis larvae to Cry1Ab/Cry1Ac and Cry1Caexpressing transgenic rice. J Invertebr Pathol 145:9–12. https://doi.org/10.1016/j.jip.2017.03.001. Epub 2017 Mar 6. PMID: 28274717

Rahman MT, Khalequzzaman M, Khan MAR (2004) Assessment of infestation and yield loss by stem borers on variety of rice. J Asia Pac Entomol 7:89–95. https://doi.org/10.1016/s1226-8615(08)60203-4

Rahman ML, Jiang WZ, Chu SH, Qiao YL, Ham TH, Woo MO, Lee J, Khanam MS, Chin JH, Jeung JU, Brar DS, Jena KK, Koh HJ (2009) High-resolution mapping of two rice brown planthopper resistance genes, Bph20(t) and Bph21(t), originating from *Oryza minuta*. Theor Appl Genet 119(7):1237–1246

Ram T, Deen R, Gautam SK, Ramesh K, Rao YK, Brar DS (2010) Identification of new genes for brown planthopper resistance in rice introgressed from *Oryza glaberrima* and *O. minuta*. Rice Genet Newsl 25:67–68

Ramachandran R, Khan ZR (1991) Feeding site selection of first instar larvae of *Cnaphalocrocis medinalis* on susceptible and resistant rice plants. Entomol Exp Appl 60:43–49

Ramesh K, Padmavathi G, Dean R, Pandey MK, Lakshmi VJ, Bentur JS (2014) White-backed planthopper *Sogatella furcifera* (Horvath) (Homoptera: Delphacidae) resistance in rice variety Sinna Sivappu. Euphytica 200:139–148

Ranganathan S, Suvarchala V, Rajesh YBRD, Prasad MS, Padmakumari AP, Voleti SR (2006) Effect of silicon sources on its deposition, chlorophyll content, and disease and pest resistance in rice. Biol Plant 50(4):713–716

Rao KV, Rathore KS, Hodges TK et al (1998) Expression of snowdrop lectin (GNA) in transgenic rice plants confers resistance to rice brown planthopper. Plant J 15:469–477

Rao YC, Dong GJ, Zeng DL, Hua J, Zeng LJ, Gao ZY, Zhang GH, Guo LB, Qian Q (2010) Genetic analysis of leaffolder resistance in rice. J Genet Genomics 37:325–331

Reddy VA, Siddiq EA, Rao P et al (1997) Genetics of resistance to rice gall midge (*Orseolia oryzae*). Indian J Genet 57:361–372

Ren JS, Gao FY, Wu XT, Lu XJ, Zeng LH, Lv JQ, Su XW, Luo H, Ren GJ (2016) Bph32, a novel gene encoding an unknown SCR domain-containing protein, confers resistance against the brown planthopper in rice. Sci Rep 6:37645

Renganayaki K, Fritz AK, Sadasivam S, Pammi S, Harrington SE, McCouch SR, Kumar SM, Reddy AS (2002) Mapping and progress toward map-based cloning of brown planthopper biotype-4 resistance gene introgressed from *Oryza officinalis* into cultivated rice, *O. sativa*. Crop Sci 42(6):2112–2117

Renuka P, Madhav MS, Padmakumari AP, Barbadikar KM, Mangrauthia SK, Rao KVS, Marla SS, VR B (2017) RNA-seq of rice yellow stem borer *Scirpophaga incertulas* reveals molecular insights during four larval developmental stages. G3: Genes Genomes Genet 7. https://doi.org/10.1534/g3.117.043737

Reymond P, Farmer EE (1998) Jasmonate and salicylate as global signals for defense gene expression. Curr Opin Plant Biol 1:404–411

Reynolds OL, Padula MP, Zeng R, Gurr GM (2016) Silicon: potential to promote direct and indirect effects on plant defense against arthropod pests in agriculture. Front Plant Sci 7:744. https://doi.org/10.3389/fpls.2016.00744

Rezaul Karim ANM, Pathak MD (1982) New genes for resistance to green leafhopper, *Nephotettix virescens* (distant) in rice *Oryza sativa* L. Crop Prot 1:483–490

Rubia EG, Heong KL, Zaluki M, Gonzales B, Norton GA (1996) Mechanisms of compensation of rice plants to yellow stem borer *Scirpophaga incertulas* (Walker) injury. Crop Prot 15:335–340

Rubia-Sanchez E, Suzuki Y, Miyamoto K, Watanabe T (1999) The potential for compensation of the effects of the brown planthopper *Nilaparvata lugens* Stål (Homoptera: Delphacidae) feeding on rice. Crop Prot 18:39–45

Saha P, Majumder P, Dutta I, Ray T, Roy SC, Das S (2006) Transgenic rice expressing *Allium sativum* leaf lectin with enhanced resistance against sap-sucking insect pests. Planta 223:1329–1343. https://doi.org/10.1007/s00425-005-0182-z

Saidi A, Hajibarat Z (2019) Characterization of cis-elements in hormonal stress-responsive genes in *Oryza sativa*. Asia Pac J Mol Biol Biotechnol 27:95–102

Saka N, Tsuji T, Toyama T, Yano M, Izawa T, Sasaki T (2006) Development of cleaved amplified polymorphic sequence (CAPS) markers linked to a green rice leafhopper resistance gene, Grh3 (t). Plant Breed 125:140–143

Sama VSAK, Himabindu K, Naik BS, Sundaram RM, Viraktamath BC, Bentur JS (2012) Mapping and MAS breeding of an allelic gene to the Gm8 for resistance to Asian rice gall midge. Euphytica 187:393–400

Sama VSAK, Rawat N, Sundaram RM, Himabindu K, Naik BS, Viraktamath BC, Bentur JS (2014) A putative candidate for the recessive gall midge resistance gene *gm3* in rice identified and validated. Theor Appl Genet 127:113–124. https://doi.org/10.1007/s00122-013-2205-7

Sarao PS, Sahi GK, Neelam K, Mangat GS et al (2016) Donors for resistance to brown planthopper *Nilaparvata lugens* (Stål) from wild rice species. Rice Sci 23(4):219–224

Sardesai N, Kumar A, Rajyashri KR, Nair S, Mohan M (2002) Identification and mapping of an AFLP marker linked to Gm7, a gall midge resistance gene and its conversion to a SCAR marker for its utility in marker aided selection in rice. Theor Appl Genet 105:691–698

Savary S, Horgan F, Willocquet L, Heong KL (2012) A review of principles for sustainable pest management in rice. Crop Prot 32:54–63

Sebastian LS, Ikeda R, Huang N, Imbe T, Coffman WR, McCouch SR (1996) Molecular mapping of resistance to rice spherical virus and green leafhopper. Phytopathology 86:25–30

Selvi A, Shanmugasundaram P, Mohan Kumar S, Raja JAJ (2002) Molecular markers for yellow stem borer *Scirpophaga incertulas* (Walker) resistance in rice. Euphytica 124:371–377

Shao E, Lin L, Chen C et al (2016) Loop replacements with gut-binding peptides in Cry1Ab domain II enhanced toxicity against the brown planthopper, *Nilaparvata lugens* (Stål). Sci Rep 6:20106. https://doi.org/10.1038/srep20106

Sharoni AM, Nuruzzaman M, Satoh K, Moumeni A, Attia K, Venuprasad R, Serraj R, Kumar A, Leung H, Islam AKMR et al (2011) Comparative transcriptome analysis of AP2/EREBP gene family under normal and hormone treatments, and under two drought stresses in NILs setup by Aday Selection and IR64. Mol Gen Genomics 287:1–19

Shen W, Zhang X, Liu J, Tao K, Li C, Xiao S, Zhang W, Li JF (2022) Plant elicitor peptide signalling confers rice resistance to piercing-sucking insect herbivores and pathogens. Plant Biotechnol J 20(5):991–1005. https://doi.org/10.1111/pbi.13781

Shrivastava MN, Kumar A, Bhandarkar S et al (2003) A new gene for resistance in rice to Asian gall midge (*Orseolia oryzae* Wood-Mason) biotype 1 population at Raipur, India. Euphytica 130: 143–145

Sidhu GS, Khush GS, Madrano FG (1979) A dominant gene in rice for resistance to whitebacked planthopper and its relationship to other plant characteristics. Euphytica 28:227–232

Sidhu N, Bansal UK, Shukla KK, Saini RG (2005) Genetics of resistance to whitebacked planthopper in five rice stocks. SABRAO J Breed Genet 37(1):1–11

Sinha DK, Atray I, Bentur JS, Nair S (2012a) Expression of nucleoside diphosphate kinase (OoNDPK) is enhanced in rice gall midge feeding on susceptible rice hosts and its overexpression leads to salt tolerance in *E. coli*. Insect Mol Biol 21(6):593–603

Sinha DK, Nagaraju J, Tomar A, Bentur JS, Nair S (2012b) Pyrosequencing-based transcriptome analysis of the Asian rice gall midge reveals differential response during compatible and incompatible interaction. Int J Mol Sci 13:13079–13103

Sinha DK, Gupta A, Padmakumari AP, Bentur JS, Nair S (2022) Infestation of rice by gall midge influences density and diversity of *Pseudomonas* and *Wolbachia* in the host plant microbiome. Curr Genomics 23(2). https://doi.org/10.2174/1389202923666220401101604

Siwi BH, Khush GS (1977a) New genes for resistance to the green leafhopper in rice. Crop Sci 17 (1):17–20. https://doi.org/10.2135/cropsci1977.0011183X001700010006x

Soundararajan RP, Jeyaprakash P (2019) Rapid screening of rice genotypes for resistance to brown planthopper, *Nilaparvata lugens* (Stal). Electron J Plant Breed (EJPB) 10(1):76–82. https://doi.org/10.5958/0975-928X.2019.00009.7

Srivastava A, Kamini TV (2020) Screening of rice genotypes against rice hispa *Dicladispa armigera* Olivier. Indian J Entomol 82(4):678–680. https://doi.org/10.5958/0974-8172.2020.00172.8

Srivastava MN, Kumar A, Shrivastava SK et al (1993) A new gene for resistance to gall midge in rice variety Abhaya. Rice Genet Newsl 10:79–80

Su CC, Zhai HQ, Wang CM, Sun LH, Wan JM (2006) SSR mapping of brown planthopper resistance gene Bph9 in Kaharamana, an Indica rice (*Oryza sativa* L.). Acta Genet Sin 33(3): 262–268

Sun LH, Su CC, Wang CM, Zhai HQ, Wan JM (2005) Mapping of a major resistance gene to the brown planthopper in the rice cultivar RathuHeenati. Breed Sci 55(4):391–396

Sun LH, Liu Y, Jiang L, Su CC, Wang CM, Zhai HR, Wan JM (2007) Identification of quantitative trait loci associated with resistance to brown planthopper in the indica rice cultivar Col. 5 Thailand. Hereditas 144(2):48–52

Sundaram RM (2007) Fine-mapping of rice-gall midge resistance genes *Gm1* and *Gm2* and validation of the linked markers, Dissertation. University of Hyderabad, Hyderabad

Sundaram RM, Balachandran SM, Madhav MS, Viraktamath BC (2013) Biotechnological options for rice improvement. In: Shetty PK, Hegde MR, Mahadevappa M (eds) Innovation in rice production. National Institute of Advanced Studies, Bangalore, pp 167–202

Tamura K, Fukuta Y, Hirae M, Oya S, Ashikawa I, Yagi T (1999) Mapping of the Grh1 locus for green rice leafhopper resistance in rice using RFLP markers. Breed Sci 49:11–14

Tan G, Ren X, Weng Q, Zhu L, He G (2002) Two whitebacked planthopper resistance genes shared the same loci with brown planthopper resistance genes. In: Abtracts international rice congress. CAAS, Beijing

Thanysiriwat T, Pattwatang P, Angeles ER (2009) New biotypes of brown planthopper in Thailand. In: Proceedings of the rice and temperate cereal crops annual conference, Bangkok, Thailand, pp 386–389

Thapa RB (2008) Evaluation of selected wild rice genotypes to the yellow stem borer, *Scirpophaga incertulas* (Walker) at cellular level. J Inst Agric Anim Sci 29:79–83

Tong XH, Qi JF, Zhu XD, Mao BZ, Zeng LJ, Wang BH, Li Q, Zhou GX, Xu XJ, Lou YG, He ZH (2012) The rice hydroperoxide lyase OsHPL3 functions in defense responses by modulating the oxylipin pathway. Plant J 71(5):763–775

Trivellone V, Bougeard S, Giavi S, Krebs P, Balseiro D et al (2017) Factors shaping community assemblages and species co-occurrence of different trophic levels. Ecol Evol 7:4745–4754

Varshney RK (2014) R Terauchi SR, McCouch (2014) Harvesting the promising fruits of genomics: applying genome sequencing technologies to crop breeding. PLoS Biol 12:e1001883

Varshney RK (2016) Exciting journey of 10 years from genomes to fields and markets: Some success stories of genomics-assisted breeding in chickpea, pigeonpea and groundnut. Plant Sci 242:98–107. https://doi.org/10.1016/j.plantsci.2015.09.009

Velusamy R, Heinrichs EA (1986) Tolerance in crop plants to insect pests. Insect Sci Appl 7:689–696

Velusamy R, Heinrichs E, Medrano F (1986) Greenhouse techniques to identify field resistance to the brown planthopper, *Nilaparvata lugens* (Stål) (Homoptera: Delphacidae), in rice cultivars. Crop Prot 5(5):328–333. https://doi.org/10.1016/0261-2194(86)90112-2

Venkanna V, Hari Y, Rukminidevi K, Satish Chandra B et al (2018) Markers assisted selection for pyramiding of gallmidge resistance genes in Kavya, a popular rice variety. Int J Curr Microbiol App Sci 7(4):745–753. https://doi.org/10.20546/ijcmas.2018.704.083

Verma SK, Pathak PK, Singh BN, Lal MN (1979) Indian biotypes of the brown planthopper. Int Rice Res Newsl 4:6

Vijayalakshmi P, Amudhan S, Himabindu K, Cheralu C, Bentur JS (2006) A new biotype of the Asian rice gall midge *Orseolia oryzae* (Diptera: Cecidomyiidae) characterized from the Warangal population in Andhra Pradesh, India. Int J Trop Insect Sci 26:207–211

Voleti SR, Padmakumari AP, Raju VS, Babu SM, Ranganathan S (2008) Effect of silicon solubilizers on silica transportation, induced pest and disease resistance in rice (*Oryza sativa* L.). Crop Prot 27:1398–1402

Walling LL (2008) Avoiding effective defenses: strategies employed by phloem-feeding insects. Plant Physiol 146:859–866

Walling LL, Thompson GA (2013) Behavioral and molecular-genetic basis of resistance against phloem-feeding insects. In: Thompson GA, van Bel AJE (eds) Phloem: molecular cell biology, systemic communication biotic interactions. Wiley-Blackwell, Oxford, pp 328–351

Wang Y, Cao LM, Zhang YX, Cao CX, Liu F, Huang FK, Qiu YF, Li RB, Luo XJ (2015) Map-based cloning and characterization of BPH29, a B3 domain-containing recessive gene conferring brown planthopper resistance in rice. J Exp Bot 66(19):6035–6045

Wang L, Tang N, Gao X et al (2017a) Genome sequence of a rice pest, the white-backed planthopper (*Sogatella furcifera*). Gigascience 6:1–9

Wang Y, Jiang W, Liu H et al (2017b) Marker-assisted pyramiding of Bph6 and Bph9 into elite restorer line 9311 and development of functional marker for Bph9. Rice 10:51

Wang HY, Shi SJ, Guo Q, Nie LY, Du B, Chen RZ, Zhu LL, He GC (2018) High-resolution mapping of a gene conferring strong antibiosis to brown planthopper and developing resistant near-isogenic lines in 9311 background. Mol Breed 38(8):107

Wu CF, Khush GS (1985) A new dominant gene for resistance to whitebacked planthopper in rice. Crop Sci 25:505–509

Wu H, Liu YQ, He J, Liu YL, Jiang L, Liu LH, Wang CM, Cheng XN, Wan JM (2014) Fine mapping of brown planthopper (*Nilaparvata lugens* Stål) resistance gene *Bph28(t)* in rice (*Oryza sativa* L.). Mol Breed 33:909–918

Xu J, Qi-Xiang W, Jin-Cai W (2010) Resistance of cultivated rice varieties to Cnaphalocrocis medinalis (Lepidoptera: Pyralidae). J Econ Entomol 103:1166–1171

Xue J, Bao YY, Li BL, Cheng YB, Peng ZY, Liu H, Xu HJ, Zhu ZR, Lou YG, Cheng JA, Zhang CX (2010) Transcriptome analysis of the brown planthopper *Nilaparvata lugens*. PLoS One 5(12):e14233. https://doi.org/10.1371/journal.pone.0014233

Xue J, Zhou X, Zhang CX et al (2014) Genomes of the rice pest brown planthopper and its endosymbionts reveal complex complementary contributions for host adaptation. Genome Biol 15:521

Yamasaki M, Yoshimura A, Yasui H (2003) Genetic basis of ovicidal response to whitebacked planthopper (*Sogatella furcifera* Horvath) in rice (*Oryza sativa* L.). Mol Breed 12:133–143

Yang Y, Xu J, Leng Y, Xiong G, Hu J, Zhang G, Huang L, Wang L, Guo L, Li J, Chen F, Qian Q, Zeng D (2014) Quantitative trait loci identification, fine mapping and gene expression profiling for ovicidal response to whitebacked planthopper (Sogatella furcifera Horváth) in rice (Oryza sativa L.). BMC Plant Biol 14:145

Yang L, Li P, Li F, Ali S, Sun X, Hou M (2018) Silicon amendment to rice plants contributes to reduced feeding in a phloem-sucking insect through modulation of callose deposition. Ecol Evol 8:631–637. https://doi.org/10.1002/ece3.3653

Yang M, Cheng L, Yan LH, Shu W, Wang XY, Qiu YF (2019) Mapping and characterization of a quantitative trait locus resistance to the brown planthopper in the rice variety IR64. Hereditas 156(1):22

Yang M, Lin JB, Cheng L, Zhou HL, Chen S, Liu F, Li RB, Qiu YF (2020) Identification of a novel planthopper resistance gene from wild rice (*Oryza rufipogon* Griff.). Crop J 8(6):1057–1070

Yao C, Du L, Liu Q, Hu X, Ye W, Turlings TCJ, Li Y (2023) Stemborer-induced rice plant volatiles boost direct and indirect resistance in neighbouring plants. New Phytol. https://doi.org/10.1111/nph.18548

Yoshihara T, Sogawa K, Pathak MD, Juliano BO, Sakamura S (1979) Soluble silicic acid as a sucking inhibitory substance in rice against the brown planthopper (Delphacidae, Homoptera). Entomol Exp Appl 26:314–322

Yoshimura S, Komatsu M, Kaku K, Hori M, Ogawa T, Muramoto K, Kazama T, Ito Y, Toriyama K (2012) Production of transgenic rice plants expressing *Dioscorea batatas* tuber lectin 1 to confer resistance against brown planthopper. Plant Biotechnol 29:501–504. https://doi.org/10.5511/plantbiotechnology.12.0726b

Zhang KJ, Zhu WC, Rong X et al (2013) The complete mitochondrial genomes of two rice planthoppers, *Nilaparvata lugens* and *Laodelphax striatellus*: conserved genomere arrangement in Delphacidae and discovery of new characteristics of atp8 and tRNA genes. BMC Genomics 14:417

Zhang KJ, Zhu WC, Rong X et al (2014) The complete mitochondrial genome sequence of *Sogatella furcifera* (Horváth) and a comparative mitogenomic analysis of three predominant rice planthoppers. Gene 533:100–109

Zhang YX, Qin G, Ma QQ, Wei MY, Yang XH, Ma ZF, Liang HF, Liu C, Li ZJ, Liu F, Huang DH, Li RB (2020) Identification of a major resistance locus Bph35 to brown planthopper in rice (*Oryza sativa* L.). Rice Sci 27(3):237–245

Zhao X, Xu H, He K, Shi Z, Chen X, Ye X, Mei Y, Yang Y, Li M, Gao L, Xu L, Xiao H, Liu Y, Lu Z, Li F (2021) A chromosome-level genome assembly of rice leaffolder, *Cnaphalocrocis medinalis*. Mol Ecol Resour 21(2):561–572. https://doi.org/10.1111/1755-0998.13274

Zhu J, Jiang F, Wang X et al (2017) Genome sequence of the small brown planthopper, *Laodelphax striatellus*. Gigascience 6:1–12

Host Plant Resistance to Insect Pests in Wheat

5

Anamika Sharma, Venugopal Mendu, and Gadi V. P. Reddy

Abstract

Host plant resistance offers an excellent solution to pest problems, which reduces pesticide usage and environmental pollution. Host plant resistance to insect pests in wheat has enabled the management of major insect pests including *Mayetiola destructor*, *Cephus cinctus*, *Diuraphis noxia*, *Schizaphis graminum*, and *Rhopalosiphum padi.* The major sources of genetic diversity for pest resistance in wheat have been landraces cultivars of wheat and wild relatives. Several resistance genes have been identified and are incorporated into cultivated wheat (especially in *Triticum aestivum*). Nevertheless, scanty information is available about resistance to other economically important pests such as *Sitodiplosis mosellana* and *Oulema melanopus.* A coherent program to incorporate resistant varieties in the integrated pest management (IPM) of wheat pests is needed to better protect the crop and improve crop yields.

Keywords

Breeding · Transgenic wheat · Insect resistance · Hessian fly · Aphids · Wheat stem sawfly

A. Sharma (✉)
Center for Biological Control, College of Agriculture and Food Sciences, Florida A&M University, Tallahassee, FL, USA
e-mail: anamika.sharma@famu.edu

V. Mendu
Department of Plant Sciences and Plant Pathology, Montana State University, Bozeman, MT, USA

G. V. P. Reddy
USDA-ARS Southern Insect Management Research Unit, Stoneville, MS, USA

S. Kumar, M. Furlong (eds.), *Plant Resistance to Insects in Major Field Crops*,
https://doi.org/10.1007/978-981-99-7520-4_5

5.1 Introduction

Wheat (*Triticum* spp.) (Poaceae) is one of the widely consumed cereal grains in the world and is a staple food for over 30% of the world's population (FAO 2014). In terms of production, wheat is the second most-produced cereal after maize (FAO 2014). Wheat production increased in 2022 (FAO 2022) and is predicted to continue to expand due to its use in human and animal feeding purposes. Consumption has also increased due to the higher use of wheat in animal feed in China, the European Union, and the United States of America (FAO 2022), and demand is projected to rise at a rate of 1.6% annually until 2050 (FAO 2022). China remains the largest wheat producer in 2022 followed by the European Union, India, Russia, and the USA. The significant wheat exporters are Russia, Canada, Australia, and the USA, and the major importers are Egypt, Brazil, and Mexico.

Wheat is considered an environmentally friendly crop. In terms of climate change, emissions from cereals production are much lower than in the cattle and meat industries. Wheat emissions per hectare are about 20% of those of rice production, making wheat a desirable cereal crop (FAO 2022). There are several species of wheat, but those most widely cultivated are *Triticum aestivum* (common wheat, a hexaploid species, 95% of total wheat production), *T. durum* (durum or pasta wheat, a tetraploid species), and *T. dicoccum* (emmer wheat, a tetraploid species). Other less commonly cultivated species of wheat include *T. spelta* (spelt, a hexaploid species), *T. compactum* (club wheat, a hexaploid species), and *T. monococcum* (einkorn, a diploid species) (Moudrý et al. 2011). Species of wheat have different characteristics and uses and are adapted to various growing conditions and environments.

Both abiotic and biotic factors affect wheat yield and quality. Major abiotic stressors are drought, heat stress, cold stress, salinity, soil acidity, nutrient deficiency, water logging, and frost. Biotic stresses include pest species of fungi, bacteria, viruses, nematodes, weeds, and insects. Moreover, weather-related disorders, as well as chemical, genetic, and physiological problems also affect wheat production (Bockus et al. 2010). These stresses can act alone or in combination, leading to significant reductions in wheat yield and quality. Developing new wheat varieties more tolerant to stresses is a major goal of wheat breeding programs. Public and private organizations are working on developing insect-resistant wheat varieties, including the International Maize and Wheat Improvement Center (CIMMYT), the International Rice Research Institute (IRRI), the National Research Center for Wheat (NRCW), the United States Department of Agriculture (USDA), Agriculture and Agri-Food Canada (AAFC), and the China National Wheat Improvement Center (CNWIC).

Globally, insect losses in wheat are 5–35%, depending on the region and year (Dhaliwal et al. 2010). There are several research review articles available that comprise the basic details of insects, their biology, and insect resistance efforts. Some of the major articles in recent years are Luo et al. (2023), Sarthi et al. (2022), Arif et al. (2022), Mondal et al. (2016), Berzonsky et al. (2003). In this chapter, we are attempting to compile the basic mechanisms of producing resistant varieties, breeding methods, and the status of the current efforts.

5.2 Sources of Wheat Resistance to Arthropod Pests and Utilization in Breeding Programs

Plants naturally exhibit different levels of resistance to arthropods, which are categorized as *tolerance* (the ability of a plant to maintain its yield or quality under pest pressure) (Onstad and Knolhoff 2014), *antixenosis* (non-preference by a pest for a plant due to the plant's specific traits such as trichomes) (Gebretsadik et al. 2022; Achhami et al. 2020), and *antibiosis* (plant traits that interfere with the arthropod's physiological processes, such as plant chemistry) (Cao et al. 2015; El-Wakeil et al. 2010). Typically, antibiosis traits allow plants to actively resist predators and parasites but come with a plant fitness cost. These resistance traits evolve when the benefit is greater than the cost and may be used to produce resistant varieties. Such resistance traits create a strong selection pressure on the pest evolution to overcome the resistance. Plants then are selected to evolve different resistance traits, creating a cycle of resistance deployment followed by pest or pathogen adaptation (Smith 2005; Garcia et al. 2021; Tadesse et al. 2022a). Genetic diversity for pest resistance from wild species can be discovered by screening the germplasm collections. Increasing genetic diversity by incorporating genes from wild varieties could help improve the resistant traits of wheat (Moudrý et al. 2011). Finding potential sources of resistance to pests of interest proceeds through a series of steps as outlined here below.

5.2.1 Sources of Resistance

In developing new resistant varieties of a crop, the first step is to identify potential sources of resistance to the targeted stress. The level of resistance associated with sources must then be quantified, and the identified genetic factors used to modify crop genetic diversity (ideally enhancing the total number of genetic resistance characteristics), leading ultimately to crop improvement. New sources of genetic resistance are periodically re-collected over time because of the difficulties involved in carrying out systematic collections of new accessions and the decline in the condition of previously collected germplasm due to the deterioration during storage (Smith 2005). Wheat genetic resources are held at institutes such as GRIN (Germplasm Resources Information Network) maintained by the USDA-ARS, which has more than 400,000 wheat accessions from around the world, with information on the accessions' genetic and geographic origins, morphological and agronomic traits, and molecular data; IWIS (International Wheat Information System), which is an online platform maintained by CIMMYT that has information on wheat varieties, germplasm, and wheat genetic resources worldwide); WGRC (the Wheat Genetic Resource Center) of Kansas State University, USA, which has information on more than 26,000 wheat accessions from around the world); NCBI (the National Center for Biotechnology Information of the USA's National Library of Medicine), has information on wheat genome sequences and molecular markers for wheat germplasm characterization (Tadesse et al. 2019). However, only about 10% of the available wheat genetic resources have been used by breeders in their research programs. In part, this lack of use may be because some gene bank accessions are obsolete, and secondly, the germplasm collections of many accessions are poorly characterized or the available data might not be readily accessible. Such accessions cannot be easily matched with the interest of breeders. Finally, for some breeding objectives, there may be enough genetic diversity available in the best-studied, best-performing breeding lines and varieties (Tadesse et al. 2022a).

5.2.2 Screening Germplasm Collections and Identification of Resistant Varieties

Germplasm collections are repositories of genetic resources from a diverse collection of seeds from plants (landraces, wild relatives, cultivars, etc.) that can be used to identify potential sources of resistance. These collections can be screened for resistance to arthropod pests using various techniques, such as field observations, artificial infestations, and laboratory bioassays (Gebretsadik et al. 2022; Nasrollahi et al. 2019; Li et al. 2013; Luo et al. 2023). Varieties found in the above resources that appear to be resistant to a given arthropod pest can then be used as potential sources of resistance in breeding programs. Such candidate varieties can be identified through literature searches, expert consultations, and field observations (Xu et al. 2020).

5.2.3 Crossbreeding with Related Species and Genetic Mapping

When desired resistance traits are not located in wheat varieties as described, another approach is to crossbreed the crop with related species, such as wild wheat relatives, to introduce new genes and traits into the crop, which may confer resistance to arthropod pests. Advances in molecular biology, molecular genetics, and genomics allow scientists to map the genes that control resistance in wheat to specific locations on the genome. This information can be used to identify the source of observed resistance and to develop molecular markers that can be used to select the resistance traits in breeding programs to develop new wheat varieties with improved resistance to arthropod pests (Smith and Clement 2012; Smith 2005). Different types of markers can potentially be used for gene mapping, including RFLP, STS, RAPD, AFLP, SSRs, and SCAR (Smith 2005) for molecular breeding for rapid introgression of resistance traits.

5.3 Factors Influencing the Expression of Resistance in Wheat to Arthropods

The expression of resistance in wheat to arthropods is influenced by several factors, including genetic issues, environmental conditions, plant physiology, and the pest's traits.

5.3.1 Genetic Factors

Those that influence the level of resistance in wheat to a given pest can include (1) whether resistance is due to a single gene or by multiple genes, (2) the amount of gene expression, (3) whether the desired resistance is due to tolerance, non-preference, or antibiosis, and (4) the genetic background of the plant.

5.3.2 Environmental Factors

These include variables such as temperature, moisture, and light that can affect the expression of resistance in wheat to arthropods. For example, high temperature and low humidity may reduce the level of resistance in wheat to some pests (Tang et al. 2018), while moderate temperatures and adequate moisture can enhance the expression of resistance (Zhu et al. 2010).

5.3.3 Physiological Factors

These including the plant's physiological status can influence the expression of resistance in wheat to arthropods. For example, the level of resistance may be affected by the stage of plant growth, as well as by stress such as drought or nutrient deficiency (Zhang et al. 2017).

5.3.4 Arthropod Factors

Arthropod factors are those where an insect's biology or behavior influences the level of resistance in wheat. For example, the susceptibility of some arthropods to certain types of resistance may vary by pest species (Luo et al. 2023), and the resistance level may be affected by factors such as feeding behavior, mobility, and population dynamics (Smith and Clement 2012; Smith 2005).

5.4 Techniques Used to Measure Pest Resistance

The choice of technique to measure pest resistance in wheat depends on the type of resistance, the pest of interest, and the resources available for testing. A combination of techniques can be used to provide a more comprehensive assessment. Assessments often start with laboratory bioassays that expose plant varieties to the pest under controlled laboratory conditions. The insect responses are then measured, including variables such as insect oviposition responses, the number of insects that survive, and the developmental stage of the pest (Wu et al. 2003). Simultaneously, the effects on the plants are measured, such as the level and types of damage caused by the pest (Smith and Clement 2012). Bioassays of various designs can be used to test for various types of resistance, including antibiosis, antixenosis, and tolerance (Gebretsadik et al. 2022; Arif et al. 2022; Cao et al. 2015). Field observations can also be used to measure resistance based on monitoring the behavior and performance of arthropod pests on wheat plants under field conditions (Achhami et al. 2020; Xu et al. 2020; Gebretsadik et al. 2022). This type of information can provide information about the level of infestation, the timing of pest outbreaks, and the level of damage caused by the pests on different plant varieties. Both laboratory and field experiments involve visual rating scales to assess the level of damage caused by arthropod pests on wheat plants (Tadesse et al. 2022a, b). Molecular markers are then used to identify the genes that control resistance detected in plants and later track the presence or absence of these genes in different breeding lines. Another technique, the electrophysiological technique, measures the electrical activity of plant tissues in response to insect feeding (Arif et al. 2022). These techniques can provide a sensitive and rapid way to detect changes in plant physiology in response to insect feeding, which can be used to identify resistant varieties (Smith 2005; Smith and Clement 2012).

5.5 Inheritance of Arthropod Resistance in Wheat

The inheritance of arthropod resistance in wheat plants is a complex genetic process that can be influenced by various factors. The genetic basis of arthropod resistance in wheat plants is typically polygenic (controlled by multiple genes). The genes that control arthropod resistance in wheat plants can be inherited from both parents in breeding, and the inheritance pattern can be influenced by both the type of resistance and the genetic background of the plant. Some types of resistance, however, are controlled by single genes, such as the Hessian fly resistance gene (H6), which confers resistance to *Mayetiola destructor*. Inheritance of single-gene resistance follows Mendelian laws of inheritance, where resistance is associated with the dominant allele and susceptibility is associated with the recessive allele. Monogenic resistance is easier to introgress into varieties compared to polygenic resistance. Other types of resistance, such as antibiosis or tolerance, are controlled by multiple genes, and the inheritance pattern is more complex. In these cases, the expression of resistance is influenced by the interaction of multiple genes and environmental and physiological factors. Plant breeders use a variety of techniques, such as genetic mapping, marker-assisted selection, and genome editing, to identify and manipulate the genes that control arthropod resistance in wheat plants (Smith 2005; Smith and Clement 2012). These techniques can help to improve the effectiveness and durability of resistance in wheat plants (Smith 2005; Tadesse et al. 2022a).

5.6 Constitutive Versus Induced Host Resistance

Plants have evolved various resistance mechanisms to defend themselves from pests. Deploying resistance molecules is energy-intensive, hence some mechanisms are induced only in the presence of pests while others are constitutively active in the plants. Wheat resistance genes to arthropod pests can be classified as constitutive or induced. Constitutive resistance genes are always present in the plant and provide a baseline level of resistance. These genes can be inherited from naturally resistant parent plants or artificially introduced by breeders through genetic modification. For example, some wheat varieties may have constitutive resistance genes that produce toxic compounds or other defenses that deter or kill arthropod pests. Breeders apply techniques such as cloning, genetic engineering, or biotechnology to enhance constitutive resistance. In contrast, induced resistance genes are activated in response to arthropod infestations or other environmental cues. These genes are triggered by the presence of arthropod pests, damage to plant tissues, or exposure to other stress factors. Induced resistance genes can result in changes to plant physiology, such as enhanced production of defense compounds, or changes in plant morphology, such as increased trichome density or altered leaf shape. Induced resistance is accomplished by the induction of allelochemicals, elicitor proteins, or defense gene expression in arthropods-induced resistance plants. Some varieties of wheat also exhibit traits that enhance their tolerance to pests, for example, varieties that are tolerant to feeding byaphids (*Diuraphis noxia*; Aphididae), Hessian fly (*Mayetiola destructor*), or wheat stem sawfly (*Cephus cinctus*) (Onstad and Knolhoff 2014; Tadesse et al. 2022b).

5.7 Transgenic Approach for Arthropod Resistance

Transgenic resistance to pest arthropods involves the genetic modification of plants by introducing genes from other organisms (bacteria or other plants) that produce toxic compounds or other defense molecules harmful to pest arthropods. Currently, there are no transgenic varieties of wheat that are commercially approved for cultivation (Abbas 2018). Transgenic wheat has been developed and tested for insect resistance. Host-Induced Gene Silencing (HIGS)-mediated silencing of the Aphid gene resulted in reduced fecundity (Zhang et al. 2022). HIGS technology involves the expression of small RNAs in plants that target insect or pathogen genes (Koch and Wassenegger 2021; Qi et al. 2019). The small RNAs enter the insect and silence the host genes resulting in killing or weakening the insect's survival. Transgenic wheat expressing SmDSR33 RNA interference (RNAi) construct caused reduced expression of SmDSR33 in aphids resulting in resistance against aphids. Expression of barley trypsin inhibitor CMe (BTI-CMe) in wheat (Altpeter et al. 1999) resulted in a significant reduction in grain moth (*Sitotroga cerealella*). Expression of the Pinellia pedatisecta lectin gene in wheat showed enhanced resistance against wheat aphids (Duan et al. 2018). Successful insect resistance using transgenic and HIGS technologies suggests that it is possible to use transgenic technologies for arthropod insect resistance. In addition, potentially useful modifications include the incorporation of genes of *Bacillus thuringiensis* and genes that code for plant-derived insecticidal proteins such as lectins or protease inhibitors. A plant-derived insecticidal protein called cowpea trypsin inhibitor (CpTI and CpTI-Phyto) has been developed to enhance resistance to aphids and other sap-sucking insects. *Triticum aestivum*, the most widely planted wheat species, has a complicated hexaploid genome, and, therefore, its successful breeding and genetic manipulation will require a fundamental understanding of the resistance mechanisms using functional genomics.

5.8 Arthropod Biotypes and Development of Resistant Varieties

Insect biotypes evolve rapidly due to high fecundity and environmental/host pressure. Arthropod biotypes can affect the development of resistant varieties of wheat. Arthropod biotypes are subpopulations of arthropod species adapted to feed on specific host plants or overcome specific resistance mechanisms in their host plants (Smith 2005). In the context of wheat, different biotypes of arthropod pests may be able to overcome different types of resistance mechanisms in wheat plants. For example, specific biotypes of the *M. destructor* (Hessian fly) are able to overcome the resistance conferred by the wheat gene H13, which was previously considered to provide broad-spectrum resistance to the pest. This loss of efficacy led to developing new wheat varieties with resistance based on multiple genes (e.g., H5, H13, and H22 genes) to provide a more durable resistance (Tadesse et al. 2022b). Identifying arthropod biotypes in advance of crop resistance failure can help researchers anticipate such problems and prepare alternative resistant varieties. Visual or molecular tools can identify biotypes. Due to the low cost of sequencing technologies and the availability of reference genomes, in future, biotype identification by genome sequencing will help in the rapid identification of biotypes.

5.9 Historical Review of Plant Resistance to Wheat Pests

The first reports of wheat resistance to pest insects date back to the 1920s and 1930s, when researchers in the United States identified wheat varieties resistant to Hessian fly (*M. destructor*). In the 1970s, in response to problems with greenbug aphids, *Schizaphis graminum* (Aphidae), a major pest of wheat in the southern United States, a new resistant variety, "Greenbug Resistant 1,"was developed by USDA scientists (Royer et al. 2015). In the 1970s and 1980s, the use of resistant wheat varieties for pest management increased significantly in response to outbreaks of new and virulent insect pests (Armstrong and McNew 1976; Porter and Webster 1976). Since the 1980s, wheat breeding programs worldwide have developed insect-resistant varieties using traditional and molecular breeding methods based on naturally occurring resistance genes (Smith and Harris 1989; Porter and Webster 1982). More recently, molecular tools have been used to identify and introgress resistance genes from related species into desirable wheat varieties (Smith 2005; Smith and Clement 2012). Currently, there are several commercially available insect-resistant wheat varieties targeting the Hessian fly (*M. destructor*), green bug (*S. graminum*), and wheat stem sawfly (Tan et al. 2017; McCauley 2020; Peirce et al. 2022; Onstad and Knolhoff 2014). However, many of the varieties with resistance to Hessianfly or green bug are not resistant to important wheat diseases such as stem rust, leaf rust, and stripe rust and biotypes of Russian wheat aphid (*Diuraphis noxia*) (Zukoff et al. 2023). It is an ongoing challenge for researchers and breeders to identify new sources of resistance and develop effective strategies for managing insect pests in wheat production. There are many insect pests of wheat for which no known resistant variety is available (for example, wheat stem maggot (*Meromyza americana*) (Diptera: Chloropidae). In some cases, resistance has not yet been identified in wheat or its wild relatives. In other cases, resistant varieties may exist but have not yet been widely adopted by farmers due to various factors such as lack of availability or suitability to local growing conditions. It has been estimated that a new release of wheat varieties is needed every 5–10 years for a given pest just to overcome the development of resistant pest populations (Zhao et al. 2015).

5.10 Major Arthropod Pests of Wheat

Insect species or groups that often damage wheat include aphids (Homoptera: Aphididae), armyworms and cutworms (Lepidoptera: Noctuidae), leaf beetles (Coleoptera: Chrysomelidae), grasshoppers (Orthoptera: Acrididae), Hessian fly (Diptera: Cecidomyiidae), thrips (Thysanoptera: Thripidae), wheat curl mite (Acari: Eriophyidae), wheat jointworm (Hymenoptera: Eurytomidae), wheat stem maggot (Diptera: Chloropidae), wheat stem sawfly (Hymenoptera: Cephidae), wheat strawworm (Hymenoptera: Eurytomidae), white grubs (Coleoptera: Scarabaeidae), wireworms (Coleoptera: Elateridae), false wireworms (Coleoptera: Tenebrionidae), and stored-grain insects. Of these, pest management through host plant resistance has been considered for the following species.

5.10.1 Hessian Fly, *Mayetiola destructor* (Diptera: Cecidomyiidae)

This fly is an important pest of wheat in North Africa, North America, southern Europe, northern Kazakhstan, northwestern China, and New Zealand, attacking wheat, barley, and rye (Kamran et al. 2013). First instar larvae induce galls that function as feeding sites for all instars (Zhao et al. 2015). Third instars pupate inside the larval exuvia (the puparium, ashiny, protective case also known as flaxseed). The third instar enters a facultative diapause in which it overwinters inside wheat stubble or stems of volunteer wheat. Depending on environmental conditions, there are 2–5 generations/year (Tadesse et al. 2022b). Salivary secretions of this insect have a very high proportion of transcripts coding for Secreted Salivary Gland Proteins (SSGPs). Genome sequencing has identified many families of genes that collectively encode nearly 2000 putative active substances (effectors) in salivary secretion (Zhao et al. 2015; Aljbory et al. 2020). Plant resistance (*R*) genes encode proteins that elicit effector-triggered immunity (ETI) when they encountered SSGPs of Hessian fly (Zhao et al. 2015). So far, at least 34 genes conferring resistance to Hessian fly have been identified in wheat (numbered H1–H34) (Tan et al. 2017, 2018; Sardesai et al. 2005; Li et al. 2013). For example, the gene Hfr-2 expresses mannose-binding lectins in the lead sheath that have anti-insect properties and serve as storage proteins. Storage proteins that accumulate in the phloem sap in response to feeding by Hessian fly larvae includeWci-1 mRNAs and Hfr-1 (defender gene). These genes are used in resistant wheat varieties. The Hfr-1 gene is specifically active against Hessian fly and can protect crops from severe attack (Subramanyam et al. 2006).

Antibiosis as a basis for Hessian fly resistance includes the presence of elevated levels of silica in leaf sheaths and the production of free amino acids, organic acids, and sugars in plants. Novel jacalin-like lectin genes from wheat respond significantly to the infestation of Hessian fly larvae and could be used effectively in future breeding programs (Kamran et al. 2013). In areas where resistant varieties have been grown for several years, losses to Hessian fly have been reduced to $<1\%$.

5.10.2 Aphids

Several aphid species, including greenbug (*Schizaphis graminum*), English grain aphid (*Sitobion avenae*), bird cherry-oat aphid (*Rhopalosiphum padi*), corn leaf aphid (*Rhopalosiphum maidis*), rose-grain aphid (*Metopolophium dirhodum*), grain aphid (*Sitobion miscanthi*), and Russian wheat aphid (*Diuraphis noxia*), are known to feed on different species of wheat. *Rhopalosiphum padi* and *D. noxia* are considered to be the most damaging of aphids to wheat (Kamran et al. 2013).

5.10.2.1 Russian Wheat Aphid, *Diuraphis noxia* (Hemiptera: Aphididae)

This aphid sporadically causes significant yield losses to wheat and is found in South Africa, the western United States, central, and southern Europe, and the Middle East (Berzonsky et al. 2003). Damage appears as longitudinal chlorotic streaking and leaf rolling, which reduces photosynthesis. Leaf rolls protect aphids from contact with insecticides and natural enemies. Its primary hosts are wheat, barley, triticale, rye, and oat, while alternate hosts are cool season (crested) and wheat grasses (*Agropyron* spp.) (Kamran et al. 2013). Yield losses of 20–90% have been reported in different parts of the world (Archer et al. 1998). Host plant resistance has been the most effective and economic control method, through induction of antixenosis, antibiosis, or tolerance against this pest. Several biotypes of Russian wheat aphid have been recognized. About 15 different Dn (*Diuraphis noxia*) resistant genes have been identified in various wheat cultivars (Kisten et al. 2020). Dn4 has been the gene most extensively used in breeding resistant cultivars. However, multiple genes are usually required for resistance to different biotypes of *D. noxia*. Within the same breeding line, certain biotypes require two genes for resistance, while others only required one resistance gene (Kisten et al. 2020). Rye and common progenitors of wheat (*Triticum tauschii*) have served as sources of a number of resistance genes. The inclusion of several resistance genes can slow the development of resistant aphid biotypes. Although several resistant wheat varieties are available, Russian wheat aphids continue to develop resistance toward them. Currently, this pest is damaging to all commercial wheat varieties in western Kansas, USA (Zukoff et al. 2023).

5.10.2.2 Greenbug, *Schizaphis graminum* (Hemiptera: Aphididae)

Greenbug is distributed in Asia, southern Europe, Africa, and North and South America. It feeds on many genera of Poaceae, including *Agropyron*, *Avena*, *Bromus*, *Dactylis*, *Eleusine*, *Festuca*, *Hordeum*, *Lolium*, *Oryza*, *Panicum*, *Poa*, *Sorghum*, *Triticum*, and *Zea*. This pest transmits the Barley Yellow Dwarf Virus (BYDV), especially the *Schizaphis graminum* strain (SGV). Feeding causes chlorosis and necrosis at the feeding sites (Porter and Webster 2000). In 1970, the first successful resistant wheat variety was developed (Royer et al. 2015). Several biotypes (C, E, I, and K) have been identified, of which biotype I is predominant and most severe (Onstad and Knolhoff 2014). Both dominant and recessive resistance genes for this pest are known in wheat. Multiple quantitative trait loci for greenbug resistance in different genetic resistance sources have been located for use against greenbug biotypes C, E, I, and K. Gb6 is the most effective gene, conferring resistance against biotypes B, C, E, G, and I. It was recovered from a wheat-rye translocation germplasm (Crespo-Herrera et al. 2019a). Gene combinations should conform broad spectrum and long-lasting resistance against greenbug in wheat. Sequential use of resistant genes, along with monitoring of prevalent greenbug biotypes, could be helpful (Porter and Webster 2000; Tan et al. 2017).

5.10.2.3 Bird Cherry-Oat Aphid, *Rhopalosiphum padi* (Hemiptera: Aphididae)

This aphid is distributed worldwide (Elek et al. 2009). It transmits Barley Yellow Dwarf (BYD) and can overwinter on plants outside the *Poaceae*. This pest shows high biological plasticity, in both its holocyclic and anholocyclic life cycles, which causes contrary results in terms of host plant resistance to both types of life cycles. *Rhopalosiphum padi* can reduce yield by 31–62%, especially when damage is combined with BYDV infection. Plant traits or mechanisms that induce aphid nymphal mortality, increase aphid developmental time at the wheat seedling stage, or reduce the aphid birth rate at wheat flowering are the most effective resistance mechanisms to help manage this pest (Kamran et al. 2013). Also, plant traits that prevent the bird cherry-oat aphid from inoculating wheat phloem with BYDV also reduce the development of the winged females, limiting dispersal of BYDV to other plants (Kamran et al. 2013). At least four different Dn genes have been identified in wheat to help manage bird cherry-oat aphid.

5.10.3 Sunnpest, *Eurygaster integriceps* (Hemiptera: Scutelleridae)

This true bug is the most important pest of wheat and barley in western and central Asia, eastern Europe, and North Africa (Nasrollahi et al. 2019). This univoltine pest feeds on various parts of its host cereal plants, including the leaves, stems, and kernels, creating various amounts of damage. Losses can reach 100% yield reduction under severe infestations (Nasrollahi et al. 2019). Prolyl endoproteases injected into the grain during the pest's feeding can severely harm the quality of the resulting flour by degrading gluten proteins (Tadesse et al. 2022a). Only a few resistance sources have so far been identified for *E. integriceps* in wheat or its wild relatives (El-Bouhssini et al. 2013). Identification and deployment of additional resistance genes could prevent the development of new biotypes of the test (Nasrollahi et al. 2019).

5.10.4 Cereal Leaf Beetle, *Oulema melanopus* (Coleoptera: Chrysomelidae)

This leaf-feeding beetle is distributed in Europe, North America, Africa, and Asia. Reduction in yield can be from 23% to 55% (Kher et al. 2011; Herbert et al. 2007). Adult beetles overwinter in protected areas such as wind rows, crop stubble, and tree bark crevices (Buntin et al. 2004). Host plant resistance, including trichomes (pubescence) on leaf surfaces, is an important resistant trait, useful in managing this pest. However, few efforts have been made to develop resistant varieties due to a general lack of resistance sources and a negative correlation between known resistance traits and crop yield (Buntin et al. 2004). Putative quantitative trait loci (QTL), such as Ppd-D1 and Ppd-D1a, have been identified as potential sources of resistance to cereal leaf beetles. Resistance might be increased by classical phenotypic selection in fields with natural infestation. Alternatively, genomic selection might be a productive avenue; it is, however, more expensive and probably only worth pursuing if marker profiles become available (Würschum et al. 2020).

5.10.5 Wheat Stem Sawflies, *Cephus* spp. (Hymenoptera: Cephidae)

Several sawfly species in this genus affect wheat in North America, Europe, North Africa, and Asia. *Cephus cinctus* is a major pest in Europe, North America, and Asia. *C. pygmaeus* is common in Europe, North Africa, and West Asia (Morrill and Weiss 2007). Host plants of this insect include wheat and other cereal crops including barley, rye, and triticale (Shanower and Hoelmer 2004). Larval feeding damages the insides of stems, reducing the nutrient transfer capability of the plant and weakening the stems. Major losses occur if stems are girdled and topple to the ground before harvest. Larvae pass through four or five instars. There is only one generation per year. During severe infestations, there can be 35% yield reductions (Shanower and Hoelmer 2004). Using varieties with resistant genotypes (plants having solid stems) minimizes this damage. Several studies have identified multiple QTLs (quantitative trait loci) associated with resistance to wheat stem sawfly (*C. cinctus*) in wheat. Resistance due to this trait appears to be controlled by multiple genes. A solid stem has been the only well-characterized wheat trait used in resistant varieties. Solid stems are due to undifferentiated parenchyma cells that create a solid pith. In solid stem genotypes, genes involved with cell wall modification and degradation and in the regulation of programmed cell death are suppressed (Nilsen et al. 2017). However, stems become less solid as the plant matures. A solid stem inhibits egg hatching and serves as a mechanical barrier to the larva's movement, and early drying of the pith causes larval desiccation and death. Solid-stemmed genotypes also reduce female body weights, sizes, and fecundity, sometimes delay adult emergence, and affect sex ratio. Under high wheat stem fly infestations, solid-stemmed genotypes can increase yield compared to hollow-stemmed susceptible genotypes (Peirce et al. 2022).

5.10.6 Orange Wheat Blossom Midge, *Sitodiplosis mosellana* (Diptera: Cecidomyiidae)

This midge is a major pest in North America, Europe, Asia, and Africa. It is a small (~3 mm long), mosquito-like, orange fly. There is a single generation each year. Adult emergence coincides with flowering, and the first two larval instars feed on the developing seeds, reducing yield and quality. Larger seeds, if harvested, show undesirable changes in germination, protein content, and dough strength (Arif et al. 2022). Host plant resistance, including genotypes that produce antitoxic substances, can minimize wheat blossom midge infestation rates. Sm1 is the only described resistance gene and is the foundation for managing orange blossom midge (Kassa et al. 2016). These genotypes alter oviposition rates in the field and reduce the egg densities resulting in a smaller midge population.

5.11 Incorporation of Resistant Wheat Varieties into IPM Programs

Integrated pest management programs for wheat include chemical pesticides, cultural controls, biological control, and resistant varieties. Major insect pests such as Hessian fly and aphids use volunteer wheat as a host before attacking new wheat stands. Cultural controls, such as eradication of volunteer wheat or alternate hosts, crop rotation, tillage, and change in planting dates, help suppress densities of these pests (Kamran et al. 2013; Zukoff et al. 2023). Conservation and release of natural enemies (parasitoids, predators, and entomopathogenic fungi) are possible means for biological control of wheat pests (Kamran et al. 2013). Using resistant wheat varieties as part of the IPM program can benefit farmers economically. However, resistant wheat varieties can negatively impact the management of other pest arthropods, such as arthropods that depend on the targeted pests and the overall biological community (Shelton et al. 2002). Over time, repeated use of resistant wheat varieties can stimulate the development of pest populations that are able to overcome host plant resistance. This can make the resistant varieties less effective over time and require additional management methods. Fewer management strategies (due to resistance) can make the wheat crop more vulnerable to sudden pest outbreaks. The use of resistant varieties may or may not have unintended impacts on non-target arthropods, including natural enemies. For example, when antibiosis is based on the expression of toxins, then host plant resistance can have negative impact on natural enemies (Van Emden 2017). Nevertheless, majority of the time parasitism in aphid–wheat–parasitoid interaction was enhanced on resistant plants (Zanganeh et al. 2015; Cai et al. 2009). When selecting for resistance to arthropods, breeders may need to make trade-offs with other desirable traits, such as yield potential, drought tolerance, or disease resistance (Peirce et al. 2022).

On the other hand, transgenic arthropod resistance has the potential to improve the control of arthropod pests in wheat crops, reducing the need for chemical pesticides and promoting more sustainable agricultural practices without impacts on yield. However, the development and use of transgenic crops are subject to a complex and lengthy process for approval (Abbas 2018). Previously, genetically modified wheat varieties with traits such as herbicide tolerance, disease resistance, and insect resistance have been developed. However, due to concerns about consumer acceptance and market rejection, these varieties were not commercialized. In addition, some countries have restrictions or bans on cultivating genetically modified crops (Domingo 2016). Further, information on GM wheat's long-term health effects, including mutagenicity, teratogenicity, and carcinogenicity tests, is needed (Abbas 2018).

5.12 Conclusion

Wheat host plant resistance plays a key role in arthropod pest management by reducing pest damage and the use of chemical insecticides. The use of resistant wheat varieties can lead to significant reductions in pest populations and damage to crops, which in turn can lead to higher yields and greater profitability for farmers (Shiferaw et al. 2013; El-Bouhssini et al. 2021). Developing new resistant varieties is, however, a long, complex process, and a new variety can become ineffective in as little as 7–10 years (Crespo-Herrera et al. 2019b). It is also essential to develop and deploy resistant varieties suited for precise growing regions. Wild wheat species or landraces of *Aegilops* and *Triticum* species are potential sources of host plant resistance to both biotic and abiotic stressors (Crespo-Herrera et al. 2019b). Using the genetic diversity in landraces and wild wheat species should also lead to the discovery of new traits for biochemical responses in wheat. Further, the introgression of traits from close crossable wheat relatives is an excellent option. It is possible to cross wheat, barley, and oats in different combinations to transfer genes across these genera (Fedak and Armstrong 1980). Current advances in wheat functional genomics, metabolomics, and genome editing could provide methods for the identification and rapid introgression of desirable traits. The authors believe that developing new varieties is an effective strategy to manage wheat pests; however, better incorporation of resistant varieties into IPM programs is needed. Combining resistant traits for biotic and abiotic factors, including insect pests, diseases, and drought, in one variety could enable better management and help avoid the failure of a variety that is effective against one, but not another, type of stress.

References

Abbas MST (2018) Genetically engineered (modified) crops (*Bacillus thuringiensis* crops) and the world controversy on their safety. Egypt J Biol Pest Control 28:52. https://doi.org/10.1186/s41938-018-0051-2

Achhami BB, Reddy GVP, Sherman JD, Peterson RKD, Weaver DK (2020) Antixenosis, antibiosis, and potential yield compensatory response in barley cultivars exposed to wheat stem sawfly (Hymenoptera: Cephidae) under field conditions. J Insect Sci 20:9. https://doi.org/10.1093/jisesa/ieaa091

Aljbory Z, Aikins MJ, Park Y, Reeck GR, Chen M (2020) Differential localization of Hessian fly candidate effectors in resistant and susceptible wheat plants. Plant Direct 4:e00246. https://doi.org/10.1002/pld3.246

Altpeter F, Diaz I, McAuslane H, Gaddour K, Carbonero P, Vasil IK (1999) Increased insect resistance in transgenic wheat stably expressing trypsin inhibitor CMe. Mol Breeding 5:53–63

Archer TL, Peairs FB, Pike KS, Johnson GD, Kroening M (1998) Economic injury levels for the Russian wheat aphid (Homoptera: Aphididae) on winter wheat in several climate zones. J Econ Entomol 91:741–747

Arif MA, Waheed MQ, Lohwasser U, Shokat S, Alqudah AM, Volmar C, Borner A (2022) Genetic insight into the insect resistance in bread wheat exploiting the untapped natural diversity. Front Genet 13. https://doi.org/10.3389/fgene.2022.828905

Armstrong JS, McNew RW (1976) Resistance in wheat to the greenbug, *Schizaphis graminum* (Rondani) (Homoptera: Aphididae): mechanism and inheritance. Crop Sci 16:828–830

Berzonsky WA, Ding H, Haley SD, Harris MO, Lamb RJ et al (2003) Breeding wheat for resistance to insects. Plant Breeding Rev 22:221–296. https://doi.org/10.1002/9780470650202.ch5

Bockus WW, Bowden RL, Hunger RM, Morrill WL, Murray TD, Smiley RW (2010) Compendium of wheat diseases and pests, 3rd edn. The American Phytopathological Society, Minnesota

Buntin GD, Flanders KL, Slaughterand RW, Delamar ZD (2004) Damage loss assessment and control of the cereal leaf beetle (Coleoptera: Chrysomelidae) in winter wheat. J Econ Entomol 97:374–382

Cai QN, Ma XM, Zhao X, Cao YZ, Yang XQ (2009) Effects of host plant resistance on insect pests and its parasitoid: a case study of wheat-aphid–parasitoid system. Biol Control 49:134–138

Cao HH, Pan MZ, Liu HR, Wang SH, Liu TX (2015) Antibiosis and tolerance but not antixenosis to the grain aphid, *Sitobion avenae* (Hemiptera: Aphididae), are essential mechanisms of resistance in a wheat cultivar. Bull Entomol Res 105:448–455

Crespo-Herrera L, Singh RP, Reynolds M, Huerta-Espino J (2019a) Genetics of greenbug resistance in synthetic hexaploid wheat derived germplasm. Front Plant Sci 10:782. https://doi.org/10.3389/fpls.2019.00782

Crespo-Herrera LA, Singh RP, Sabraoui A, El-Bouhssini M (2019b) Resistance to insect pests in wheat—rye and *Aegilops speltoides* Tausch translocation and substitution lines. Euphytica 215:123

Dhaliwal GS, Jindal V, Dhawan AK (2010) Insect pest problems and crop losses: changing trends. Indian J Ecol 37:1–7

Domingo JL (2016) Safety assessment of GM plants: an updated review of the scientific literature. Food Chem Toxicol 95:12–18

Duan X, Hou Q, Liu G, Pang X, Niu Z, Wang X, Zhang Y, Li B, Liang R (2018) Expression of Pinellia pedatisecta lectin gene in transgenic wheat enhances resistance to wheat aphids. Molecules 23:748

El-Bouhssini M, Ogbonnaya FC, Chen M, Lhaloui S, Rihawi F, Dabbous A (2013) Sources of resistance in primary synthetic hexaploid wheat (*Triticum aestivum* L.) to insect pests: Hessian fly, Russian wheat aphid and Sunn pest in the Fertile Crescent. Genet Resour Crop Evol 60:621–627

El-Bouhssini M, Amri A, Lhaloui S (2021) Plant resistance to cereal and food legume insect pests in North Africa, West and Central Asia: challenges and achievements. Curr Opin Insect Sci 45:35–41

Elek H, Werner P, Smart L, Gordon-Weeks R, Nádasy M, Pickett J (2009) Aphid resistance in wheat varieties. Commun Agric Appl Biol Sci 74:233–241

El-Wakeil NE, Volkmar C, Sallam AA (2010) Jasmonic acid induces resistance to economically important insect pests in winter wheat. Pest Manag Sci 66:549–554

FAO (2014) Food and agricultural organization of United Nations, Food and Agriculture Organization, Statistics Division (FAOSTAT). FAOSTAT

FAO (2022) Food and agricultural organization of United Nations FAO's Statistical Yearbook for 2022. https://www.fao.org/newsroom/detail/fao-s-statistical-yearbook-for-2022-goeslive/en#:~:text=Sugarcane%20is%20the%20world's%20largest,factor%20of%203.7%20since%202000. Accessed Mar 2023

Fedak G, Armstrong KC (1980) Production of trigeneric (barley × wheat) × rye hybrids. Theor Appl Genet 56:221–224

Garcia M, Martinez M, Isabel D, Santamaria ME (2021) The price of the induced defense against pests: a meta-analysis. Front Plant Sci 11. https://doi.org/10.3389/fpls.2020.615122

Gebretsadik KG, Zhang Y, Chen J (2022) Screening and evaluation for antixenosis resistance in wheat accessions and varieties to grain aphid, *Sitobion miscanthi* (Takahashi) (Hemiptera: Aphididae). Plants 11:1094. https://doi.org/10.3390/plants11081094

Herbert DA, Van Duyn JW Jr, Bryan MD, Karren JB (2007) Cereal leaf beetle. In: Buntin GD, Pike KS, Weiss MJ, Webster JA (eds) Handbook of small grain insects. Entomological Society of America, Lanham, MD, p 120

Kamran A, Asif M, Hussain SB, Ahmad M, Hirani A (2013) Major insects of wheat: biology and mitigation strategies. Crop Prod. Accessed on 17 Feb 2023. https://doi.org/10.5772/55799

Kassa MT, Haas S, Schliephake E, Lewis C, You FM, Pozniak CJ, Krämer I et al (2016) A saturated SNP linkage map for the orange wheat blossom midge resistance gene Sm1. Theor Appl Genet 129:1507–1517

Kher SV, Dosdall LM, Cárcamo HA (2011) The cereal leaf beetle: biology, distribution and prospects for control. Prairie Soils Crop 4:32–41

Kisten L, Tolmay VL, Mathew I, Sydenham SL, Venter E (2020) Genome-wide association analysis of Russian wheat aphid (*Diuraphis noxia*) resistance in Dn4 derived wheat lines evaluated in South Africa. PLoS One 28(15):e0244455

Koch A, Wassenegger M (2021) Host-induced gene silencing—mechanisms and applications. New Phytol 231:54–59

Li C, Chen M, Chao S, Yu J, Bai G (2013) Identification of a novel gene, *H34*, in wheat using recombinant inbred lines and single nucleotide polymorphism markers. Theor Appl Genet 126: 2065–2071

Luo K, He D, Guo J, Li G, Li B, Chen X (2023) Molecular advances in breeding for durable resistance against pests and diseases in wheat: opportunities and challenges. Agronomy 13:628. https://doi.org/10.3390/agronomy13030628

McCauley D (2020) A Pith(y) solution to a growing problem. CSA News 65:6–11

Mondal S, Rutkoski JE, Velu G, Singh PK, Guzmán C, Bhavani S, Lan C, He X, Singh RP (2016) Harnessing diversity in wheat to enhance grain yield, climate resilience, disease and insect pest resistance and nutrition through conventional and modern breeding approaches. Front Plant Sci 7. https://doi.org/10.3389/fpls.2016.00991

Morrill W, Weiss M (2007) Wheat stem sawfly. In: Buntin G, Pike K, Weiss M, Webster J (eds) Handbook of small grain insects. Entomological Society of America, p 120

Moudrý J, Konvalina P, Stehno Z, Capouchová I, Moudrý J Jr (2011) Ancient wheat species can extend biodiversity of cultivated crops. Sci Res Essays 6:4273–4280

Nasrollahi S, Badakhshan H, Sadeghi A (2019) Analyzing Sunn pest resistance in bread wheat genotypes using phenotypic characteristics and molecular markers. Physiol Mol Biol Plants 25: 765–778

Nilsen KT, N'Diaye A, MacLachlan PR, Clarke JM, Ruan Y, Cuthbert RD, Knox RE, Wiebe K, Cory AT, Walkowiak S et al (2017) High density mapping and haplotype analysis of the major stem-solidness locus SSt1 in durum and common wheat. PLoS One 12:e0175285

Onstad DW, Knolhoff L (2014) Arthropod resistance to crops. In: Onstad DW (ed) Insect resistance management. Academic Press, Oxford, pp 293–326

Peirce ES, Cockrell DM, Mason E, Haley S, Peairs F, Nachappa P (2022) Solid stems and beyond: challenges and future directions of resistance to wheat stem sawfly (Hymenoptera: Cephidae). J Integr Pest Manag 13:29. https://doi.org/10.1093/jipm/pmac023

Porter DR, Webster JA (1976) Resistance in wheat to the Hessian fly. Crop Sci 16:227–230

Porter DR, Webster JA (1982) Mechanisms of Hessian fly resistance in wheat. Entomol Exp Appl 31:335–343

Porter DR, Webster JA (2000) Russian wheat aphid-induced protein alterations in spring wheat. Euphytica 111:199–203

Qi T, Guo J, Peng H, Liu P, Kang Z, Guo J (2019) Host-induced gene silencing: a powerful strategy to control diseases of wheat and barley. Int J Mol Sci 20:1

Royer TA, Pendleton BB, Elliott NC, Giles KL (2015) Greenbug (Hemiptera: Aphididae) biology, ecology, and management in wheat and sorghum. J Integr Pest Manag 6:19. https://doi.org/10.1093/jipm/pmv018

Sardesai N, Nemacheck JA, Subramanyan S, Williams CE (2005) Identification and mapping of *H32*, a new wheat gene conferring resistance to Hessian fly. Theor Appl Genet 111:1167–1173

Sarthi G, Singh AK, Sharma V (2022) Genetic basis of arthropod resistance in cereals. In: Plant defense against insect pests. Springer, Cham, pp 173–207

Shanower TG, Hoelmer KA (2004) Biological control of wheat stem sawflies: past and future. J Agric Urban Entomol 21:197–221

Shelton AM, Zhao JZ, Roush RT (2002) Economic, ecological, food safety, and social consequences of the deployment of *Bt* transgenic plants. Annu Rev Entomol 47:845–881

Shiferaw B, Smale M, Braun HJ, Duveiller E, Reynolds M, Muricho G (2013) Crops that feed the world 10. Past successes and future challenges to the role played by wheat in global food security. Food Security 5:291–317

Smith CM (2005) Plant resistance to arthropods-molecular and conventional approaches. Springer, Dordrecht, p 423

Smith CM, Clement SL (2012) Molecular bases of plant resistance to arthropods. Annu Rev Entomol 57:309–328

Smith CM, Harris MK (1989) Registration of KSU29, KSU30, and KSU31 wheat germplasms resistant to greenbug and Russian wheat aphid. Crop Sci 29:248–248

Subramanyam S, Sardesai N, Puthoff DP, Meyer JM, Nemacheck JA, Gonzalo M, Williams CE (2006) Expression of two wheat defense-response genes, Hfr-1 and Wci-1, under biotic and abiotic stresses. Plant Sci 170:90–103

Tadesse W, Sanchez-Garcia M, Tawkaz S, El-Hanafi S, Skaf P, El-Baouchi A, Eddakir K, El-Shamaa K, Thabet S, Gizaw AS, Baum M (2019) Wheat breeding handbook at ICARDA. https://hdl.handle.net/20.500.11766/10723

Tadesse W, Harris M, Crespo-Herrera LA, Mori BA, Kehel Z, El-Bouhssini M (2022a) Insect resistance. In: Reynolds MP, Braun H (eds) Wheat improvement-food security in a changing climate-CIMMYT. Springer, pp 361–378

Tadesse W, El-Fakhouri K, Imseg I, Rachdad FE, El-Gataa Z, El-Bouhssini M (2022b) Wheat breeding for hessian fly resistance at ICARDA. Crop J 10:1528–1535

Tan CT, Yu H, Yang Y, Xu X, Chen M, Rudd JC, Xue Q, Ibrahim AMH, Garza L, Wang S, Sorrells ME, Liu S (2017) Development and validation of KASP markers for the greenbug resistance gene Gb7 and the Hessian fly resistance gene H32 in wheat. Theor Appl Genet 130:1867–1884

Tan MK, El-Bouhssini M, Wildman O, Tadesse W, Chambers G, Luo S, Emebiri L (2018) Development of SNP assays for hessian fly response genes, Hfr-1 and Hfr-2, for marker-assisted selection in wheat breeding. BMC Genet 19:50

Tang G, Liu X, Chen GH, Witworth RJ, Chen MS (2018) Increasing temperature reduces wheat resistance mediated by major resistance genes to *Mayetiola destructor* (Diptera: Cecidomyiidae). J Econ Entomol 111:1433–1438

Van Emden HF (2017) Host-plant resistance. In: van Emden HF, Harrington R (eds) Aphids as crop pests, 2nd edn. CABI, Boston, MA, pp 515–532

Wu H, Sparks C, Amoah B, Jones HD (2003) Factors influencing successful Agrobacterium-mediated genetic transformation of wheat. Plant Cell Rep 21:659–668

Würschum T, Leiser WL, Langer SM, Tucker MR, Miedaner T (2020) Genetic architecture of cereal leaf beetle resistance in wheat. Plants 9:1117. https://doi.org/10.3390/plants9091117

Xu X, Li G, Carver BF, Puterka G (2020) Identification of wheat germplasm resistant to major Russian wheat aphid biotypes in the United States. Crop Sci 60:1428–1435. https://doi.org/10.1002/csc2.20041

Zanganeh L, Madadi H, Allahyari H (2015) Demographic parameters of *Diuraphis noxia* (Hemiptera: Aphididae) and *Hippodamia variegata* (Coleoptera: Coccinellidae) recorded in the context of *D. noxia* infesting resistant and susceptible cultivars of wheat. Eur J Entomol 112: 453–459

Zhang M, Wang H, Yi Y, Ding J, Zhu M, Li C, Guo W, Feng C, Zhu X (2017) Effect of nitrogen levels and nitrogen ratios on lodging resistance and yield potential of winter wheat (*Triticum aestivum* L.). PLoS One 12(11):e0187543

Zhang J, Li H, Zhong X, Tian J, Segers A, Xia L, Francis F (2022) Silencing an aphid-specific gene SmDSR33 for aphid control through plant-mediated RNAi in wheat. Front Plant Sci 13: 1100394

Zhao C, Escalante LN, Chen H, Benatti TR, Qu J, Chellapilla S (2015) A massive expansion of effector genes underlies gall-formation in the wheat pest *Mayetiola destructor*. Curr Biol 25: 613–620

Zhu Y, Qian W, Hua J (2010) Temperature modulates plant defense responses through NB-LRR proteins. PLoS Pathog 6:e1000844

Zukoff A, McCornack BP, Whitworth RJ (2023) Wheat insect pest management 2023. Kansas State University

Host Plant Resistance to Insect Pests in Maize

6

P. Lakshmi Soujanya, Chikkappa G. Karjagi, S. B. Suby, K. R. Yathish, and J. C. Sekhar

Abstract

Maize, a potential global crop is attacked by a wide array of insect pests. Among them, spotted stem borer, pink stem borer, fall armyworm, and shoot fly are the major insect pests affecting maize productivity. Significant yield losses and environmental concerns necessitated the need to develop potential and safe management strategies against insect pests. Maize genotypes with improved insect pest resistance help in minimizing yield losses. Adoption of host plant resistance-based approaches not only sustains maize production in long run but is also environmentally friendly. Exploration of wild relatives in breeding programs increases the genetic diversity and sources of resistance to key insect pests. Conventional breeding in combination with molecular techniques speeds up the progress in developing insect pest-resistant genotypes. Understanding plant–insect interactions and elucidation of the molecular regulation of signaling networks is important for formulating novel pest control strategies. Various phytohormones such as jasmonic acid (JA) and salicylic acid (SA) and their derivatives, elicitors present in the saliva/oral secretions/regurgitant/frass of herbivores play an important role in modulating plant defenses. In this chapter, host plant resistance-based approaches including screening techniques, identification, and utilization of resistant sources, mechanisms and factors associated with plant resistance, induced defenses, the role of the microbiome in insect pest management, novel breeding strategies, namely quantitative trait loci (QTL) mapping, genetic engineering, genome editing to develop insect-resistant maize genotypes are discussed.

P. L. Soujanya (✉) · K. R. Yathish · J. C. Sekhar
Winter Nursery Centre, ICAR-Indian Institute of Maize Research, Hyderabad, India

C. G. Karjagi · S. B. Suby
Unit Office, ICAR-Indian Institute of Maize Research, Pusa Campus, New Delhi, India

S. Kumar, M. Furlong (eds.), *Plant Resistance to Insects in Major Field Crops*,
https://doi.org/10.1007/978-981-99-7520-4_6

Keywords

Maize · Stem borers · Fall armyworm · Shoot fly · Screening techniques · Host plant resistance · Wild relatives · Structural traits · Cell wall components · Genome editing

6.1 Introduction

Maize (*Zea mays* L.) is an international crop popularly known as the "Queen of Cereals" because of its high genetic yield potential. It has different types like field corn, sweet corn, and popcorn apart from different variants of field corn like high oil corn, quality protein maize, and baby corn. The total maize produced in the world is used for various purposes like poultry and animal feed, starch, food, biofuel, and raw material in several other allied industries. It is being grown in diverse environmental conditions across >170 countries in the world. Globally, 1210 million MT of maize is produced from 205 million ha with an average productivity of 5.8 t/ha). In India, maize is the third most important cereal after rice and wheat constituting 10% of the total food grain production of the country. Currently, India produces 31.51 million MT from 9.2 million ha with an average productivity of 3.1 t/ha, which accounts for nearly 4% of the global maize area and 2% of global production (DES 2021). However, the maize productivity in India is around half of the world average. The major yield-limiting factors that affect maize productivity in India are various biotic and abiotic stresses. The emerging challenges like changing climate and depletion of natural resources put additional pressure on the production and productivity of maize. On the contrary, the demand for maize is also increasing gradually across the globe and more specifically, in developing countries including India. In this context, appropriate interventions to reduce the losses caused by major biotic stresses like insect pests play a crucial role.

In general, maize is damaged by more than 100 species of insect pests under field and storage conditions. However, few insect pests are considered the major ones across different geographical regions of the world; for example, spotted stem borer (SSB), *Chilo partellus* (Swinhoe), pink stem borer (PSB), *Sesamia inferens* Walker, Fall armyworm (FAW), *Spodoptera frugiperda* (J.E. Smith), Shoot fly, *Atherigona* spp., and *Helicoverpa armigera* are considered as the most economically important, which infest maize in one or the other growing seasons. The yield losses due to SSB (26–80%), PSB (25.7–78.9%), FAW (73%), and shoot fly (20–21.28%) in different agro-climatic regions have been reported (Moyal 1998; Panwar 2005; Rao 1983; Murua et al. 2006; Pathak et al. 1971).

Several approaches have been followed to reduce the losses caused due to insect pests in maize, and the most widely used method is the use of chemical insecticides. However, the use of chemicals on a sustainable basis, in the long run, is not a feasible option due to its inherent problems like increased cost of cultivation and adverse effects on humans, non-target organisms especially pollinators, and the environment. Further, continuous use of chemical insecticides may lead to the development of resistance in the insect pests against insecticides leading to the ineffectiveness of chemical control in long run. Apart from the above problems, the use of chemical insecticides increases the cost of cultivation and reduces the farmers' profitability. Therefore, novel, environmentally friendly, economical crop protection strategies are necessary to mitigate the negative impact of major insect pests in maize agro-ecosystems on a sustainable basis. In this context, the adoption of a host plant resistance approach can play an important role in effective insect pest management in maize.

6.2 Screening Techniques for Insect Resistance

Identification of resistant sources against insect pests is the first step before employing resistance sources for the management of insect pests. Evaluation of a large number of germplasm against different insect pests under artificial infested conditions in different locations over the years is necessary to identify reliable sources of resistant genotypes. The screening techniques have been developed for each of the major insect pests of maize. Screening maize germplasm for stem borer under artificial infestation is being done by releasing 10 neonate larvae into the whorl of a 10–12-day-old maize plant. Observation on leaf injury rating (LIR) is recorded at 30–35 days after infestation using a 1–9 scale. Based on LIR, the germplasm is classified into resistant (1.0–3.0), moderately resistant (3.1–6.0), and susceptible (6.1–9.0) (Sarup et al. 1978; Reddy et al. 2003). A detailed description of recording LIR by following a 1–9 scale for spotted stem borer and pink stem borer is given in Tables 6.1 and 6.2, respectively.

Similarly, the screening technique for FAW is also based on leaf damage rating (LDR) on a 1–9 scale, which was developed based on the number of shot holes/

Table 6.1 Leaf injury rating scale for *Chilo partellus* (Sarup et al. 1978)

Rating	Description
1.	Apparently healthy plant
2.	Plant with parallel, oval or oblong holes, slightly bigger than pin-sized (2–3 mm) on 1–2 leaves
3.	Plant with more elongated holes (4–5 mm or match stick head-sized) or shot holes on 3–4 leaves
4.	Plant with injury (oval holes, shot holes and slits of 1–4 cm) in about 1/3 of total number of leaves
5.	Plants with about 50% leaf damage
6.	Plants with a variety of leaf injuries in about two-thirds of the total number of leaves (ragged appearance) or one or two holes or slits at the base of the stem (>10 cm streaks are observed)
7.	Plants with every type of leaf injury and almost all the leaves damaged (ragged or crimpled appearance), with tassel stalk boring or circular dark ring at the base of stem
8.	Plants with stunted growth in which all the leaves are damaged
9.	Plants with dead heart

Table 6.2 Leaf injury rating scale for *Sesamia inferens* (Reddy et al. 2003)

Rating	Description
1.	Apparently healthy plant
2.	Plant with parallel, oval or oblong holes, slightly bigger than pin-sized (2–3 mm) on 1–2 leaves
3.	Plant with more elongated holes (4–5 mm or match stick head-sized) or shot holes on 1–2 leaves
4.	Plant with injury (oval holes, shot holes and slits of 1–4 cm) in about 1/3 of total number of leaves and midrib damage on 1–2 leaves
5.	Plants with about 50% leaf damage, oblong holes, shot holes, slits, and streaks of 5–10 cm and midrib damage on leaves
6.	Plants with a variety of leaf injuries to about two-thirds of the total number of leaves (ragged appearance) or one or two holes or slits at the base of the stem (>10 cm streaks are observed)
7.	Plants with every type of leaf injury and almost all the leaves damaged (ragged or crimpled appearance), with tassel stalk boring or circular dark ring at the base of stem
8.	Plants with stunted growth in which all the leaves are damaged
9.	Plants with dead heart

Table 6.3 Leaf Damage Rating (LDR) scale to categorize maize germplasm for resistance to FAW (Lakshmi Soujanya et al. 2022)

Rating	Description/symptoms
1.	Healthy plant/no damage/visible symptoms
2.	Few short /pin size holes/scraping on few leaves (1–2)
3.	Short/pin size holes/scraping on several leaves (3–4)
4.	Short/pin size holes/scraping on several leaves (5–6) and a few long elongated lesions (1–3 Nos) up to 2.0 cm length present on whorl and or adjacent fully opened leaves
5.	Several holes with elongated lesions (4–5 Nos) up to 4.0 cm length and uniform/ irregular shaped holes present on whorl and or adjacent fully opened leaves
6.	Several leaves with elongated lesions (6–7 Nos) up to 6.0 cm length and uniform/ irregular shaped holes present on whorl and adjacent fully opened leaves
7.	Several long lesions (>7 Nos) up to 10 cm length and uniform/ irregular shaped holes common on one-half of the leaves present on whorl and adjacent fully opened leaves
8.	Several long lesions >10 cm length and uniform/ irregular shaped holes common on one half to two-thirds of leaves present on whorl and adjacent fully opened leaves
9.	Complete defoliation of whorl of the plant

pinholes and lesion length to categorize maize genotypes based on their reaction to FAW infestation (Lakshmi Soujanya et al. 2022). In the case of FAW, 15–20 neonate larvae will be released manually into the whorl of each maize plant at the V_5 phenological stage as part of the artificial infestation. The degree of leaf feeding damage is visually rated thrice, i.e., on the 7th, 14th, and 28th day after infestation. Based on LDR, the genotypes are classified into resistant (1–4), moderately resistant (4.1–6.0), and susceptible (6.1–9.0). A detailed description of observation recorded on LDR for FAW in the 1–9 scale is given in Table 6.3.

Table 6.4 Modified rating scale of Sharma et al. (1992) for shoot fly in maize

S. no.	Damage symptoms/description	Response
1.	<10% plants with dead hearts	Resistant
2.	>10–20% plants with dead hearts	Moderately resistant
3.	>20–30% plants with dead hearts	Moderately susceptible
4.	>30–50% plants with dead hearts	Susceptible
5.	>50% plants with dead hearts	Highly susceptible

The screening of maize genotypes against shoot fly species is done under natural conditions but in hot spot locations only, because the occurrence of shoot flies is nearly uniform and also high in hot spot locations during a particular window of the spring season. In general, artificial infestation with the shoot fly is not followed while screening maize germplasm for shoot fly because of difficulties in rearing shoot fly in the laboratory. On the contrary, shoot fly occurrences are regular and also high during a particular window of the spring season in hot spot locations of north-west India including Punjab, Haryana, Delhi, western Uttar Pradesh, and Uttarakhand. Therefore, screening is done under natural conditions at hot spot locations. However, spreading of fish meal in and around the screening area (fish meal technique) is followed to attract a large number of shoot flies toward maize genotypes while screening to simulate artificial infested conditions. Observations are recorded on the number of plants with eggs, plants with dead hearts, the total number of eggs, and the total number of plants with dead hearts at 14 and 21 days after seedling emergence. Based on the percentage of dead hearts, the germplasm is grouped into resistant (<10), moderately resistant (>10–20), moderately susceptible (>20–30), susceptible (>30–50), and highly susceptible (>50) (Sharma et al. 1992) (Table 6.4).

6.3 Identification and Utilization of Resistant Sources including Wild Relatives

Repeated screening of maize germplasm with different and diverse genetic backgrounds for major insect pests, namely stem borers, FAW, and shoot fly has led to the identification of resistant/moderately resistant genotypes. In some cases, the probable reasons for plant resistance were also reported. Several promising germplasms, namely Antigua group 1 and Antigua group 2, CML 139, CML 67, PFSRS 2, AEBYC 534-1-1, P 390AM/CMLC4F230-B-2, AEBCYC534-3-1, CML 384X176F3-100-9, P 63C2-BBB-17B have been identified as resistant or moderately resistant sources to spotted stem borer (Chatterji et al. 1966; Panwar and Sarup 1980; Kumar et al. 2005).

Similarly, the following genotypes, namely CML 421, CAO 3141, CAO 3120, CAO 0106, WNZPBTL 9 (3.2), WNZPBTL 8 (3.5), CML 338 (3.6), WNZ EXOTIC POOL DC2 (3.1), CML 424 (3.2), WNZPBTL9-1 (3.4), BGS 86, CM111/*Zeadiploperennis*/CM111, CML 141, CML 33#-4 (2.4), DML 1432 (3.0), EC

619101 (2.5) are resistant to PSB (Sekhar et al. 2008, 2016a, b; Soujanya et al. 2019).

FAW was one of the serious insect pests in the maize-growing ecologies across different countries in the world. Previously, several efforts have been made elsewhere to identify the resistant sources against FAW. The previous efforts made by several researchers/breeders in maize have led to the identification and development of numerous FAW-resistant lines. Some of the FAW-resistant genotypes developed were Mp 496, Mp 703, Mp 704, Mp 705, Mp 706, Mp 707, Mp 701, Mp 702, Mp 713, Mp 714, Mp 716, Mp 708 (Williams et al. 1990); CML 333, CML 336, MP 708 (Ni et al. 2008); Mp 708, FAW 7061 (Ni et al. 2011); UR 11003:S0302, CUBA 164-1; DK 7 (Ni et al. 2014); CML 338, CKSBL 10008, CKIR 04002, CKIR 04005 (Prasanna 2019). It was interesting to note that most of the FAW-resistant maize lines were derived from Caribbean maize germplasm and Tuxpeño landrace accessions from Mexico (Mihm 1997). The introgression of insect-resistant maize populations and inbred lines in the regular breeding programs at CIMMYT and many African countries has led to the development of elite maize germplasm against lepidopteran insect pests (Murenga et al. 2015; Tefera et al. 2016a, b; Matova et al. 2020; Kasoma et al. 2020a).

In India, FAW being the recent invasive insect pest, several efforts have been made to identify the resistant sources against FAW based on leaf damage rating (LDR) under artificial infestation conditions during 2019–2022. Some of the promising germplasms with high resistance or moderate resistance to FAW are DMRE 63, DML163 1, CML 71, CML 141, CML 337, and CML 346 (Lakshmi Soujanya et al. 2022). The potential lines identified for resistance to PSB and FAW are being used in breeding programs at ICAR-IIMR and implementing extensive efforts to develop insect-resistant maize genotypes. Based on per cent dead hearts, five resistant germplasm lines against shoot fly have been identified, namely CML 420 (8.3%), ACC 263214 (9.1%), WINPOP 8 (9.1%) AEB(Y) (10.0%), and CML 49 (10%) (AICRP Annual Report IIMR 2015).

Efforts have been made to evaluate some of the most widely cultivated landraces to identify resistant sources. Meckenstock et al. (1991) reported that landraces of the Maicillo group (San Bernardo III, Pina61, Hilate179, and Lerdo104), close alternative hosts of FAW (like sorghum) from Honduras, exhibited antibiosis to FAW, which might be due to increased selection pressure on the landraces as a result of intercropped maize. Similarly, Zapalote Chico 2451F (ZC2451F), derived from Zapalote Chico landrace, ZM 4236 and ZM 7114 collected from Zambia are reported as improved sources of resistance to FAW (Nuessly et al. 2007; Widstrom et al. 2003; Kasoma et al. 2020b). Waiss et al. (1979) identified that part of the Zapalote Chico silk's resistance is due to the presence of maysin, which confers an antibiosis-based mechanism of resistance when FAW larvae fed on silk content (Byrne et al. 1996). Tamiru et al. (2011) observed that maize landraces emit herbivore-induced plant volatiles against egg deposition of SSB that attracted parasitic wasps, which was absent in commercial maize hybrids. Similarly, Luciano et al. (2018) observed reduced oviposition, delayed larval, and larva to an adult developmental period when FAW larvae fed on landrace Perola. In another study, abnormal growth, resulting in lower pupal weight (1.05-fold lower) and longer larvae-pupae development (1.12-fold), was observed when FAW fed on maize landrace Tuxpeno (De La Rosa-Cancino et al. 2016).

Apart from elite breeding material and landraces, several efforts have been made to evaluate wild relatives of maize against different insect pests to identify the resistance sources. Ramirez (1997) reported that wild relatives of maize, namely *Z. mays* ssp. mexicana, *Z. mays* ssp. diploperennis, and *Z. mays* ssp. perennis are resistant to Asian corn borer. Similarly, *T. dactyloides* exhibited resistance to corn rootworms via non-preferences and/or antibiosis (Branson 1971; Moellenbeck et al. 1995; Eubanks 2001). In a similar study, Mammadov et al. (2018) reported that morphological traits such as leaf toughness and leaf trichome density in *Z. mays* ssp. parviglumis are responsible for resistance to FAW. Further, Lakshmi Soujanya et al. (2022) also confirmed the finding that the wild relative maize, *Zea mays* ssp. parviglumis was found to be resistant to FAW. The in-depth analysis of resistance in some of the studies involving wild species has led to an understanding of the mechanism of resistance. For example, the higher expression of herbivore resistance genes, wound inducible protein (wip1), a maize protease inhibitor (mpi), and pathogenesis-related protein (PR-1) in *Z. mays* ssp. parviglumis impart resistance to FAW (Szczepaniec et al. 2012). In *Z. mays* ssp. diploperrennis, the chemical composition of leaves such as apimaysin and 3′-methoxymaysin in leaves or silks (Gueldner et al. 1992), caffeoylquinic acids, and other luteolin derivates (Farias-Rivera et al. 2003), wip1, PR-1, and chitinase gene, maysin, and chlorogenic acid (Szczepaniec et al. 2012) contribute resistance to FAW. In certain cases, the intercrosses between elite lines and wild relatives have led to the development and identification of resistance against insect pests. For example, a cross between maize × *Z. mays* ssp. mexicana showed some degree of resistance against European corn borer (*Ostrinia nubilalis*) (Pasztor and Borsos 1990).

6.4 Mechanisms of Resistance

The response of host plants upon infestation by insect pests is complex. However, the resistance reaction or response by the host plant to insect pests is categorized into three types such as antixenosis, antibiosis, and tolerance. Antixenosis resistance is defined as the non-preference of the pest for a resistant plant and influences the behavioral traits of a pest (Painter 1951; Smith 2005). Antibiosis resistance influences the biology of the pest to diminish its population and subsequent damage, resulting in higher mortality or reduced longevity and reproduction of the insect. Tolerance is a resistance where a plant can resist or recover from damage caused by the pest population (Smith 2005). Three types of resistance mechanisms are specific to target insect pests. Thus, the development of insect-resistant genotypes provides a stable and cumulative effect on the pests' population and has no harmful effect on the non-target organisms and environment.

Several researchers observed oviposition behavior (indicative of antixenosis mechanism of resistance) of stem borers and FAW in field conditions by growing resistant and susceptible genotypes side by side or in construction cages under greenhouse conditions and observed significant differences in oviposition on resistant and susceptible genotypes (Ampofo 1985; Dubey and Sarup 1982). Oviposition for non-preference by SSB was shown in genotypes WNZPBTL 2, CM 500, WNZPBTL 6, AEB(Y)C5F38-1 under multi-choice conditions (Cholla et al. 2018a). The differential ovipositional behavior of SSB might be due to variations in the humidity stimuli in the vicinity of the plants. Due to contact-perceivable characters such as surface waxes, and trichomes, the lowest number of eggs were laid by the SSB females on the resistant maize genotypes compared to distance-perceivable characters (hygro, visual and olfactory stimuli) (Kumar and Saxena 1985). Divekar et al. (2019) recorded fewer eggs on resistant genotypes WP 21, E 63, and HKI 193-1 by PSB than on susceptible genotypes. Further, a significantly higher number of PSB eggs were deposited on the first leaf sheath followed by the second leaf sheath, basal leaf sheath, and then the third leaf sheath in susceptible ones. Minimum oviposition by FAW females was observed on the tolerant maize genotype CKH191221 compared to susceptible ones, which might be due to components of host plant quality such as carbon, nitrogen, and defensive metabolites, which directly affect the fecundity (Awmack and Leather 2002; Anyanda et al. 2022).

Antibiosis mechanisms of resistance express mainly through the plant's biochemical components such as the phenols, tannins, amino acids, and fiber content, which have adverse effects on the biological attributes of insect pests. Antibiosis is a component of resistance to stem borers and FAW. It affects survival, larval weight, pupal weight, and prolongation of larval and pupal periods. Antibiosis results in extended larval and pupal periods and a reduced percentage of pupation thus directly reflecting the resistance to SSB in maize (Arabjafari and Jalali 2007). The biological parameters of SSB have also been studied in an artificial diet through the impregnation of dry leaf powders of resistant and susceptible maize genotypes (Kumar 1993). Sekhon and Sajjan (1990) reported that the antibiosis mechanism of resistance depends upon plant age; it was observed that antibiosis became operative in a 15-day-old maize plant, and it increases with plant age. The mechanisms of antibiosis differ between genotypes. For example, AEB(Y) C5 55-1 (32 days), PFSR S3 (31 days), and HKIPC4B (31 days) genotypes showed antibiosis against PSB in terms of the prolonged larval period, whereas genotype E 30 showed antibiosis in terms of least larval weight (20.53 mg) (Kaur et al. 2016). Similarly, low larval viability, small larval weight, short adult longevity, and low emergence rates were found to impart an antibiosis form of resistance against FAW in maize when FAW was fed on resistant maize genotypes (Lima et al. 2006; de Paiva et al. 2016; Anyanda et al. 2022).

Tolerance refers to host plant's or cultivar's ability to yield well despite pest infestations that severely harm and lower the yield of susceptible host plants (Painter 1951). Genotypes CML 338 and Mp708 were classified as tolerant to FAW as uninfested and damaged plants did not exhibit any variations in photosynthetic rate and light response curves (Ni et al. 2008).

6.5 Factors Associated with Insect Resistance

Host plants respond to herbivore attacks by activating their native but complex defense mechanisms and in the process several changes either at structural and/or at biochemical and/or molecular levels.

6.5.1 Structural Traits

Among direct defense, plants can modify their morphology by increasing trichome densities that create a physical barrier for insects (Tian et al. 2012). Plant structural traits are the first line of defense against insect pests, which play a predominant role in plant resistance. It was reported that trichomes, surface wax, leaf thickness and toughness, and silica content result in avoidance behavior in insects (Gatehouse 2002; Schoonhoven et al. 2007). The role of trichomes in inhibiting oviposition by spotted stem borer was reported as a mechanism of antixenosis (Ampofo 1985; Kumar and Saxena 1985). The role of pith puncture resistance (PPR), rind puncture resistance (RPR), rind thickness, length of the meristematic area (LMA), and pith parenchyma inner lumen thickness (PPIT) was identified as resistance-imparting structures against PSB (Santiago et al. 2003). Similarly, Suby et al. (2020) reported the choice of second above ground internode by stem boring larvae in V6-10 stage maize and its pith and rind penetration resistance as the predictors of plant tolerance and antixenosis to SSB. Cholla et al. (2018b) reported that leaf damage as indicated by leaf injury rating (LIR) and stem tunneling are good indicators of resistance to SSB. The role of cuticular lipids in imparting resistance against FAW was also been reported. For example, FAW larvae fed on leaves free from cuticular lipids weighed more and developed faster as compared to larvae fed on leaves with cuticular lipids (Yang et al. 1993a). Further, it was also reported that FAW neonate larvae traveled longer distances on upper leaves, which have a smooth appearance, compared to lower leaves, which contain a dense array of wax crystals (Yang et al. 1993b). The other morphological traits attributed to FAW resistance include a thicker cell wall complex of the epidermal layer (Davis et al. 1995), a very tight husk cover, and kernel hardness that minimizes insect feeding. Whereas in shoot fly, seedling vigor, less leaf area, less number of leaves, and a thin stem are reported to impart resistance (Goyal et al. 2020).

6.5.2 Biochemical Constituents

Biochemical compounds in the host plants form the second level of defense mechanisms. An array of biochemicals such as nutritional compounds and secondary metabolites present in the different parts of plants (tissues/cells) either individually or in conjunction with other compounds contribute to imparting resistance to key insect pests of maize. The role of different biochemical compounds in plant defenses is discussed below.

6.5.2.1 Nutritional Factors

The crop plants are the primary source of energy for insect pests. The suitability of a plant or preference of crop plants by insect pests determines its status as the host plant. In most cases, insect pests derive their energy or depend on food from several types of crop plants. However, the preference of host plants by insect pests varies between crop plants and/or between different varieties of the same crop plant, which again depends on the nutritional status of host plants. For example, the suitability of host plants is influenced by some of the nutritional factors, namely nitrogen, phosphorous, total sugars, and amino acid content. Higher nitrogen (Haile and Hofsvang 2001) and low phosphorous contents (Sharma and Chatterji 1971) are attributed to the intensity of insect pest infestation. However, insect physiology and its overall biology are also responsible for differential preference to different genotypes. Since carbohydrates form the universal source of energy, simple sugars form the most preferred source of energy. It was reported that the stem borer–resistant maize genotypes contain lower reducing and total sugars because the reducing sugars and total sugars are found to be positively related to infestation by a stem borer. Similarly, higher protein content was found to be responsible for susceptibility to SSB in maize (Kabre and Ghorpade 1999). Thus, a positive correlation has been observed between the per cent SSB larvae completing the life cycle and major biochemical compounds (protein, total sugars, reducing sugars) and nutrient elements (nitrogen and phosphorus). Rao and Panwar (2002) reported that SSB-resistant genotypes possessed higher lignin content, which contributes toward the strengthening of stem tissues. On the contrary, the higher concentration of amino acids, namely aspartic acid and tyrosine, imparts resistance to FAW (Hedin et al. 1990). Goyal et al. (2020) reported that total soluble sugars (TSS), proteins, and amino acids influence dead heart incidence caused by shoot fly in maize. A positive correlation was observed between dead heart incidence and total soluble sugars (TSS), proteins, and amino acids in maize genotypes.

6.5.2.2 Secondary Metabolites

Total phenols and tannins and activities of defensive enzymes, namely phenylalanine ammonia-lyase (PAL) and polyphenol oxidase (PPO) are associated with resistance to shoot fly in maize genotypes. A similar negative correlation was found between potassium and total phenol content and SSB infestation (Lokesh and Mehla 2017). Phenolic compounds and hydroxamic acids are the most common groups of secondary metabolites identified as constitutive chemical compounds that are involved in resistance to stem borers (Ortego et al. 1998; Barros-Rios et al. 2015). DIMBOA, the major hydroxamic acid, and its degradation products, MBOA and HMBOA, were negatively associated with European corn borer (Barry et al. 1994); pink stem borer (Gutierrez et al. 1988); corn root worm (Xie et al. 1990). Wiseman et al. (1992) observed that maysin imparts resistance to FAW and corn earworm *Helicoverpa zea*. Genotypes CML 333 (with moderate silk maysin), CML 336 (with low silk maysin), and CML 338 (with high silk maysin) were identified as resistant to FAW, whereas CML 335 without silk maysin was found susceptible

(Ni et al. 2008). Santiago et al. (2006) reported that phenyl propanoids in pith tissues, *p*-coumaric acid, and ferulic acid impart resistance to *Sesamia nonagrioides*.

6.5.2.3 Cell Wall–Bound Components

Plant cell wall acts as a physical barrier against insect attack on the cell from the outside, it protects all the internal organelles and cytoplasm. The components of the cell wall play a major role in imparting resistance to insect pests (Read and Stokes 2006). Santiago et al. (2016) identified that the cell wall components, namely fiber and hydroxycinnamates impart resistance to feeding by several stem borers in maize. The fiber content in the cell wall disturbs feeding insect pests both from the nutritional and physical point of view. Tissue toughness is one of the important traits that regulate insect damage in plants (Raupp 1985). High content of lignin could make plants less palatable to herbivores (Zheng et al. 2021). The high levels of fiber content in the cell wall increase the bulk density of the diet, hence larvae are unable to ingest sufficient amounts of nutrients and water. Besides, tissues with high fiber content are tougher and more resistant to the cutting or chewing action of the mandibles. Therefore, maize genotypes with elevated levels of cell wall fiber and lignin (Beeghly et al. 1997) or fortification of the epidermal cell wall (Bergvinson et al. 1995) contribute resistance to the European corn borer, *Ostrinia nubilalis* in maize. Several researchers reported that variation in cell wall–bound phenolic constituents like *p*-CA and FA in resistant and susceptible maize genotypes contributes contrasting levels of resistance to stem borers (Bergvinson et al. 1997; Santiago et al. 2013; Lakshmi Soujanya et al. 2020). Lakshmi Soujanya et al. (2020) reported that the cell wall–bound *p*-CA and FA contribute resistance to PSB infesting maize. Similarly, Gundappa et al. (2013) identified through laboratory bioassays that *p*-CA was detrimental to SSB.

6.5.2.4 Induced Defenses

The presence of elicitors in the saliva/oral secretions/regurgitates/frass of insects or herbivores helps the plant sensors to perceive the physical and chemical signals (Wu and Baldwin 2009). As a result, the induced defense signaling process in the host plant is activated, leading to the production of defense chemical compounds through chemical changes in plants. The induced defense signaling processes triggered by the insect infestation serve as a powerful chemical weapon against insect pests. Plant response to insect pests is genotype-dependent and varies from one pest to another. It was reported that several plant defense mechanisms such as genes encoding jasmonic acid biosynthetic enzymes, proteinase inhibitors, defense-related transcription factors, and proteins involved in cell-wall reorganization activated due to Mediterranean corn borer attack (Rodríguez et al. 2012). Similarly, several studies have found that the accumulation of Benzoxazinoids increases due to insect attack. Malook et al. (2021) reported the accumulation of higher levels of benzoxazinoids in Chinese maize inbred line Xi 502 upon FAW attack. The first gene in the benzoxazinoid biosynthesis pathway, i.e., the transcript levels of *Bx1* increased after insect attack (Erb et al. 2009; Tzin et al. 2015). Further, insect feeding led to significant increases in DIMBOA and HDMBOA-Glc (the methylated form of

DIMBOA-Glc) and decreased levels of DIMBOA-Glc (Glauser et al. 2011). Tamayo et al. (2000) observed maize proteinase inhibitor accumulation in tissues adjacent to the wound site when fed by *Spodoptera littoralis* on maize leaves. Further, the level of inhibitor accumulation was higher in leaves chewed by larvae than in leaves that had been damaged mechanically.

Sometimes, the induced defense response of host plants upon insect attack provides resistance to multiple insects. For example, defense responses in the Mp708 genotype negatively impacted both European corn borer and Western corn rootworm larval weights providing evidence for changes in root physiology in response to ECB feeding on shoot tissues. In the genotype Mp708, resistance to insect pests is mediated by the defense proteins including Maize Insect Resistance1-Cysteine Protease (Mir1-CP) (Pechan et al. 2000, 2002); protease inhibitors, and peroxidases, which could hamper caterpillar growth (Castano-Duque and Luthe 2018). Lopez et al. (2007) reported an accumulation of Mir1-CP in the roots due to FAW infestation in the whorl regions of maize, which also provides enhanced resistance to root-feeding herbivores (Gill et al. 2011). Thus, the Mp708 genotype provides resistance to both above and below-ground feeding caterpillars (Gill et al. 2011; Castano-Duque et al. 2017). Significant downregulation of genes in the root tissues following short-term ECB feeding was observed (Pingault et al. 2021).

Mason et al. (2022) reported that the expression of plant resistance to insects varies with the fertilization regime. The study reported the increasing pattern in gene and protein expression with herbivory, but it varied between fertilization levels. For example, low fertilization reduced induced defenses in the resistant maize genotype. However, the soluble protein concentrations did not change with fertilization levels but were higher in the susceptible ones. Wang et al. (2022) reported that FAW attack and JA application enhanced the maize defense when supplemented with 156.6 mg/kg of N, which led to the decreased contents of amino acids and soluble sugars. Insect attack also led to increased levels of phenolics in the plants supplemented with 156.6 mg/kg of N and, thus, had negative effects on insects. Further, significantly higher volatiles of acrylic acid, indoles, and myristic aldehyde might be the reason for the reduced selection of maize supplemented with 156.6 mg/kg of N by FAW.

6.5.2.4.1 Elicitors and Signaling Regulation of Insect Resistance

Induced defense responses are complex; generally, they are initiated by wounding; the elicitors derived from insect-feeding herbivore-associated molecular patterns (HAMPS) or oviposition activate a complex regulatory network that mediates the biosynthesis of secondary metabolites. Among them, saliva, frass, oral secretions, and regurgitation of chewing insects play a major role in insect–plant interactions (Basu et al. 2018). To identify these insect HAMPS, plants utilize surface immune receptors and further induce a multitude of downstream defenses to impair insect development and/or colonization on host plants (Zogli et al. 2020). Some of the reported signaling molecules in different insects and the subsequent downstream processes are given below. In European corn borer (ECB), indole-3-acetic acid (IAA) present in its OS (Dafoe et al. 2013), saliva (Louis et al. 2013), frass (Ray et al. 2016) modulates maize defense responses. ECB saliva-induced defense-related transcripts including lipoxygenase (LOX) and 12-phytodienoic acid (OPR) genes are involved in the biosynthesis of jasmonic acid (Louis et al. 2013). The oral secretion of *Mythimna separata* feeding on maize contains more than ten kinds of fatty acid amino conjugates (FACs), the most abundant one is the hydroxylated FAC volicitin (Qi et al. 2016). The elevated levels of the hormones JA and its derivative JA-Ile were observed when the oral secretion of *Mythimna separata* was applied to maize wounds. Louis et al. (2015) reported that feeding by corn leaf aphids induces the rapid accumulation of *mir1* transcripts in the resistant maize genotype, Mp708, which is regulated by the ethylene signaling pathway. Furthermore, Mp708 provided elevated levels of antibiosis and antixenosis-mediated resistance to aphids compared to B73 and Tx601 maize susceptible lines. Block et al. (2019) observed elevated production of the defense hormone salicylic acid and an increase in FAW resistance to the combinatorial stress of flooding and infestation with FAW. In the same study, it was also observed the remodeling of phenylpropanoid pathways led to increased production of the C-glycosyl flavones (maysins) and the herbivore-induced volatile phenolics, benzyl acetate, and phenethyl acetate.

6.5.2.4.2 Phytohormones

Phytohormones play a key role in modulating maize defense against chewing insects. Among them, JA, salicylic acid (SA), and ethylene (ET) are the critical players in providing resistance to chewing pests (Erb et al. 2012). ECB feeding on maize foliage resulted in an increased accumulation of phytoalexins such as benzoxazinoids and kauralexins (Dafoe et al. 2011). Zhang et al. (2015) reported that MeJA-induced proteins, including pathogenesis-related protein 1 (PR1) and thioredoxin M-type chloroplastic precursor (TRXM), as the recombinant proteins of PR1 and TRXM inhibited the development of Asian corn borer larvae and pupa.

Similarly, ZmLOX10 (maize lipoxygenase) mutant, involved in maize JA biosynthesis, showed strongly reduced wounding-induced JA levels and reduced volatile emissions. Further, attractiveness to larval parasitoid wasps was observed. These results confirm that JA mediates direct and indirect defense responses in maize. Ethylene is also an important phytohormone that regulates *Mythimna separata* resistance in maize (Qi et al. 2011). In Mp708, a maize insect resistance1 (mir1) gene was identified that encodes a cysteine (Cys) proteinase, named Mir1-Cys protease, which disrupts the peritrophic matrix and thus reduces FAW growth (Pechan et al. 2002). It was observed that blocking ETH synthesis or perception in Mp708 reduces Mir1-Cys protease accumulation, resulting in more susceptibility to FAW and corn leaf aphids (Harfouche et al. 2006; Louis et al. 2015).

6.5.2.4.3 Plant Volatiles

Herbivore-induced plant volatiles (HIPVs) defend plants with the help of natural enemies of insect pests. Volatiles are also released by secondary metabolites that attract the insect's natural enemies. Plant volatiles from resistance and susceptible genotypes were equally effective in eliciting oviposition by SSB (Kumar 1994). Alborn et al. (1997) reported that the application of synthetic volicitin to maize attracts the predators of insects. In some maize landraces, oviposition by SSB induces the release of volatiles, which attract egg parasitoid *Trichogramma* spp. and larval parasitoids (*Cotesia sesamiae*) (Tamiru et al. 2011). Veyrat et al. (2016) reported that indole, a volatile aromatic compound, decreases insect food consumption and the survival rate of *Spodoptera littoralis*. Further, applying indole to maize elevates herbivory-induced JA and JA-Ile contents. Methyl salycilate (MeSA) is another aromatic compound that is released by maize after sensing insect feeding/oviposition (Tamiru et al. 2011). Maize landrace Braz1006 releases eightfold of (*E*)-bcaryophyllene, owing to the higher TPS23 transcript levels, resulting in more attraction to egg parasitoid (*C. sesamiae*) of stem borer (SSB) (Tamiru et al. 2017). Similarly, maize landraces C-2101, B-3016, and H-2034 also release (*E*)-b-caryophyllene and other terpenes ((*E*)-4,8-dimethyl-1,3,7, nonatriene (DMNT) and (*E*,*E*)-4,8,12-Trimethyl-1,3,7,11-tridecatetraene (TMTT)) after sensing SSB oviposition (Tamiru et al. 2011). Apart from these, maize can perceive other insect-derived elicitors, such as caeliferin and inceptin, but differences between genotypes do exist in their responses to these elicitors.

6.5.2.4.4 Kauralexins

Kauralexins are a group of maize diterpenoid phytoalexins that accumulate in response to insect infestation and phytohormone treatment, which are highly attractive to female parasitic wasps (Fontana et al. 2011). These volatile blends of herbivore-damaged maize are mostly dominated by sesquiterpenes (Degen et al. 2004). The herbivore-induced sesquiterpenes of maize are mainly produced by two terpene synthases, TPS10 and TPS23, in which TPS10 forms (*E*)-α-bergamotene and (*E*)-β-farnesene along with 13 minor sesquiterpenes in herbivore-damaged leaves (Schnee et al. 2006). The herbivore-induced sesquiterpene synthase TPS23 forms mostly (*E*)-β-caryophyllene, which can also attract parasitic wasps to leaf-feeding herbivores (Kollner et al. 2008), as well as the entomopathogenic nematodes to the root-damaging western corn rootworm (Rasmann et al. 2005; Kollner et al. 2008). These elicitations generally induce expression of genes of phytoalexin biosynthesis through transcriptional or translational regulation. Schmelz et al. (2011) observed that the accumulation of phytoalexins resulted in significant increase in antifeedant activity in *O. nubilalis*. The expression of TPS10 (terpene synthase) is highly induced by herbivory (Kollner et al. 2013) and also by MeJA treatment. Maize ZmWRKY79, a common transcription factor, physically binds to the W-boxes or WLE cis-elements in the promoters of terpenoid phytoalexin biosynthetic genes (An2 and ZmTPS6) in maize, regulating their expression (Fu et al. 2018). Li et al. (2015) reported that the MeJA- and herbivore-inducible transcription factor EREB58 directly promoted the expression of *TPS10*. In vivo and in vitro assays indicated that *EREB58* promotes *TPS10* expression by directly binding to the GCC-box within the region from −300 to −200 of the *TPS10* promoter.

6.6 Role of the Microbiome in Plant Resistance to Insect Pests

All plants are associated with both beneficial and pathogenic microbes, which play the predominant role in altering the expression of traits. Insect pests depend upon plant-produced cues to locate their hosts that provide information about suitable plant characters (Bruce et al. 2005). The microbes that establish mutualistic interactions with plants indirectly influence the plant–insect interactions and impact host-plant selection by insect pests (Hassani et al. 2018). The beneficial plant-associated microbes include plant growth-promoting rhizobacteria (PGPR), nitrogen-fixing rhizobia, arbuscular mycorrhizal fungi (AMF), and foliar and root endophytes. Root colonization by rhizobacteria induces plant resistance against insect pests in numerous crops (Dean et al. 2009). These microbes interact with both above- and below-ground plant organs and live endophytically/ectophytically, depending on the species of a microbe and the specificity or type of interaction (Gibert et al. 2019). Beneficial rhizobacterium, *Azospirillum brasilense* influences the host plant selection by the South American rootworm, *Diabrotica speciosa*, which infests maize roots. It is observed that the *A. brasilense* inoculation triggers higher emissions of (*E*)-β-caryophyllene as compared to non-inoculated plants. Non-preference of *D. speciosa* was observed for inoculated plants due to sesquiterpenes emission, which is a well-known compound to mediate belowground insect–plant interactions (Santos et al. 2014).

Arbuscular mycorrhizal fungi (AMF) also improve plant nutrient uptake and trigger defense responses. As a result, the ability of a plant to resist insect attack increases. The production of secondary metabolites and defense gene upregulation in plants enhance due to AMF colonization. Song et al. (2011) reported that 2,4-dihydroxy-7-methoxy-1,4-benzoxazin-3-one production increased in maize seedlings by AMF colonization. The AMF colonization increases the induced jasmonic acid pathway plant defense responses and reduces the rate of survival and development of European corn borer in maize. AMF colonization also upregulates induced defense production in response to insect attack that reduces the herbivore's performance and/or improves plant fitness (Karban and Myers 1989). Murrell et al. (2019) measured the cascading consequences of cover crop treatment (canola/radish/rye/oats) on maize root AMF colonization, maize growth, and performance of European corn borer.

6.7 Breeding for Insect Pests Resistance

Host plant resistance (HPR) breeding is an important component of IPM. HPR is the inherent ability of the host plant to resist insect pests' attack, and it is largely determined by its genetic makeup. Understanding the genetic basis of resistance is important to develop resistant maize genotypes. Conventional breeding for resistance is a bit challenging due to the nature of the inheritance of resistance to insect pests. Further, several efforts toward long-term planning and strategy are required for the successful development of resistant lines. Recently, success in identifying quantitative trait loci (QTL) in maize using various molecular markers provides an alternative approach for speeding up conventional breeding programs. Currently, several molecular-marker-based breeding techniques including mapping of QTLs, development of transgenics, and genome editing are being employed. A brief note on each of the approaches is described below.

6.7.1 QTL Mapping

As insect pests inflict significant yield losses, it is necessary to know the genetic mechanisms that regulate the resistance traits. Therefore, several researchers have tried to identify the genomic regions controlling resistance to stem borers in maize using quantitative trait loci (QTL) mapping. The QTL mapping studies have been reported in major insect pests of maize, namely spotted stem borer (Munyiri and Mugo 2017); Mediterranean corn borer (Santiago et al. 2016; Jimenez-Galindo et al. 2017); European corn borer (Papst et al. 2004); southwestern corn borer (Brooks et al. 2007); sugarcane borer (Groh et al. 1998). Willcox et al. (2002) performed QTL mapping to identify genomic regions for resistance to southwestern corn borer using BC_1F_2 mapping population. The study identified three putative QTLs on chromosomes 7, 9, and 10 together accounting for 28% of the total phenotypic variance. QTL mapping for resistance to Mediterranean corn borer (MCB) using two different recombinant inbred lines (RILS) mapping populations, namely EP42 × A637 (Samayoa et al. 2014) and A637 × A509 (Jimenez-Galindo et al. 2017) has led to the identification of genomic regions determining resistance to MCB. The studies have found that the favorable QTL alleles were contributed by the A637 genotype. Jimenez-Galindo et al. (2017) have identified 12 QTLs for both yield (agronomic) and resistance to MCB using 285 SNP markers. QTLs identified are based on phenotypic data of 171 RILs, evaluated along with the two parental inbred lines A637 and A509 in a 13 × 14 single lattice design for MCB. The QTL located at bins 1.10 and 5.04 improved both yield and insect resistance simultaneously. On the contrary, QTLs located in the region 8.04–8.05 showed opposite effects for yield and resistance. Further, several QTLs located in the region 10.02–10.03 were identified for indexes that combine yield and resistance traits.

Munyiri and Mugo (2017) identified QTL associated with resistance to SSB and *Busseola fusca* in maize. The study identified two QTLs for reduced stem tunneling on chromosome 4, one each for *B. fusca* and SSB and one QTL for reduced stem borer exit holes on chromosome 5. The study used 203 $F_{2:3}$ individuals, derived from a cross between CML 442 (susceptible) and CKSBL10026 (resistant) and 152 SNPs for mapping the QTLs. Badji et al. (2018) meta-analyzed QTLs for resistance to stem borers in maize and cell wall constituents imparting resistance using the IBM2 2008 Neighbors as a reference map. The study generated 24 leaf injury rating (LIR), 42 stem injury rating (SIR), and insect resistance meta-QTL (MQTL) of a diverse genetic background. Some of the LIR MQTLs such as LIR4, LIR17, and LIR22 involved in imparting resistance to the European corn borer, sugarcane borer, and southwestern corn borer. Eleven out of the 42 SIR MQTLs related to resistance to

European corn borer and Mediterranean corn borer. Samayoa et al. (2019) tested the suitability of marker-assisted selection (MAS) for improving resistance to stem tunneling by *S. nongrioides* without adverse effects on yield. The results revealed that marker information can be utilized for simultaneous improvement of resistance and yield especially if genome-wide approaches are applied.

Brooks et al. (2007) investigated genomic regions encoding resistance to FAW in maize using family-based QTL analyses. The cross of A619 (susceptible parent) and Mp708 (resistant parent) was mapped using 91 simple sequence repeat (SSR) markers and 213 $F_{2:3}$ families. The study has found QTLs determining resistance to FAW on chromosomes 1, 2, 6, 7, and 9. Womack et al. (2020) identified two major QTLs in bins 4.06 and 9.03 for resistance to FAW in maize. The linkage map was constructed using 1276 single-nucleotide polymorphisms and simple sequence repeat molecular markers.

Vikal et al. (2020) identified genomic regions associated with shoot fly resistance. A total of 107 F_2 population derived from the cross CM 143 (resistant) and CM 144 (susceptible) was genotyped with 120 SSR markers. The major QTLs determining shoot fly resistance, namely *qDH9.1* (dead heart) and *qEC9.1* (oviposition) were co-localized on chromosome 9. These QTLs are syntenic to regions of chromosome 10 of sorghum, which also accounted for dead heart and oviposition suggesting that the same gene block may be responsible for shoot fly resistance in maize. Some of the candidate genes predicted within the QTL region are cysteine protease, subtilisin-chymotrypsin inhibitor, and cytochrome P_{450}. The candidate genes are involved in the synthesis of allelochemicals, receptor kinases, *glossy15*, and ubiquitin-proteasome degradation pathways.

6.7.2 Transgenics

Globally, transgenic maize expressing *Cry* gene(s) of *Bacillus thuringiensis* (Bt), responsible for the production of insecticidal proteins (delta-endotoxins), has been adopted to provide resistance to insect pests, reduce the losses due to insect pests, and also sustain the productivity (James 2015; Ashfaq et al. 2010). *Bt* toxins act on epithelial cells of the gut of insects and promote osmotic disruption of these cells, leading to the death of the caterpillar (Gill 1995). Currently, around 208 events comprising various insect-resistant genes in maize have been approved for cultivation depending on the prevalence of insect pests (Kumar et al. 2020). Sun et al. (2015) developed transgenic maize with resistance to Asian corn borer and glyphosate tolerance by fusion gene transformation. Linker peptide LP4/2A was used in the study to connect the *Bt* cry1Ah gene with the 2mG2-epsps gene and combined the widely used manA gene as a selective marker to construct one coordinated expression vector called p2EPUHLAGN. Liu et al. (2018) developed transgenic maize event ZD12-6 expressing a *Bacillus thuringiensis* (*Bt*) fusion protein Cry1Ab/Cry2Aj and a modified 5-enolpyruvylshikimate-3-phosphate synthase (EPSPS) protein G10. Insect bioassays revealed that the transgenic plants are highly resistant to Asian corn borer (*Ostrinia furnacalis*), cotton boll worm (*Helicoverpa armigera*), and armyworm (*Mythimna separata*). Li et al. (2022) evaluated the lethal effects of the transformation events of DBN9936 (Bt-Cry1Ab), DBN9936 × DBN9501 (Bt-Cry1Ab + Vip3A), Ruifeng 125 (Bt-Cry1Ab/Cry2Aj), and MIR162 (Bt-Vip3A) on major lepidopteran pests, namely *Spodoptera frugiperda*, *Helicoverpa armigera*, *Ostrinia furnacalis*, *Conogethes punctiferalis*, *Mythimna separata*, *Leucania loreyi*, and *Athetis lepigone*, using an artificial diet containing lyophilized *Bt* maize tissue at a concentration representing a 25-fold dilution of tissue. The study found that the mortalities of DBN9936 (*Bt*-Cry1Ab), DBN9936 × DBN9501 (*Bt*-Cry1Ab + Vip3A), Ruifeng 125 (*Bt*-Cry1Ab/Cry2Aj), and MIR162 (*Bt*-Vip3A) to *Spodoptera frugiperda*, *Helicoverpa armigera*, *Ostrinia furnacalis*, *Conogethes punctiferalis*, *Mythimna separata*, *Leucania loreyi*, and *Athetis lepigone*, were in the range of 53.80–100%, 62.98–100%, 57.09–100%, and 41.02–100%, respectively. Even though *Bt* maize provides an area-wide suppression of insect pests, the development of resistance is a major threat to the durability of *Bt* crops (Huang et al. 2011). Insect resistance management strategies are to be implemented wherever *Bt* crops are being grown to slow down/minimize the development of resistance in insect pests.

6.7.3 Genome Editing

Genome editing of plants for the management of insect pests has been less exploited compared to diseases in various crops. It is a new technology that can serve as a potential tool to tackle insect pests as well (Razzaq et al. 2019; Vats et al. 2019). This technique utilizes either a sequence-specific nuclease (SSN) to produce precise gene knockout and knock-in edits or synthetic oligonucleotides to introduce specific point mutations in the target DNA region (Songstad et al. 2017). Recently, Gui et al. (2020) reported that genome editing facilitated by (CRISPR)/CRISPR-associated nuclease protein (clustered regularly interspaced short palindromic repeats) is precise, and efficient, for the manipulation of target genes associated with insect resistance. Herbivores identify host plants based on the plant's own volatile, gustatory clues, and the oviposition site's visual appearance (Larsson et al. 2004). Beale et al. (2006) identified that changes in volatile blends deter insects from host plants. Bringing changes in the blends of plant volatiles through genome editing is one of the promising approaches in insect pest management. However, utmost care is to be taken so that the changes will not result in deleterious effects on natural enemies. Infestation by aphids results in the emission of a sesquiterpene hydrocarbon (*E*)-β-farnesene (Eβf), which retracts feeding by other host populations and attracts a parasitic wasp *Diaeretiella rapae*, which manages aphid. Apart from plant volatiles, the visual appearance of host plants also plays an important role in the recognition of insects. Change in leaf color acted as a deterrent to the insect pests, *Spodoptera litura*, and *Helicoverpa armigera*. Malone et al. (2009) reported that alteration in plant pigmentation has been found to modify insect host preferences, which were observed in transgenic red-leaf tobacco, which was developed by the modification of the anthocyanin pathway.

6.8 The Way Forward

The development of sustainable integrated pest management is essential for strengthening the nation's food security. Host plant resistance forms the backbone of IPM for improved maize production. However, understanding insect pests' biology, herbivore–plant interactions, and their impact on natural enemies, i.e., multi-trophic interactions are vital to deploy defensive traits in pest management. To overcome insect pest attacks, manipulation and higher expression of plant secondary metabolites biosynthesis pathway-related genes could be advantageous. Further, the elicitors of induced defense can be exploited for pest management in maize. The deployment of modern breeding techniques provides enormous potential to develop insect-resistant maize. Wild relatives are important sources of insect-resistant traits, and through improved molecular technologies such as marker-assisted breeding and interspecific hybridization techniques, insect-resistant traits can be introduced into cultivated maize. Emphasis is to be given to unravel molecular regulation of host plant defense for the improvement of insect-resistant traits in maize so that the target genes could be identified and deployed for conferring resistance to insect pests through genetic transformation. Genome editing through CRISPR/CAS 9 also offers new opportunities and long-term solutions for reducing yield losses by insect pests.

References

Alborn HT, Turlings TCJ, Jones TH, Stenhagen G, Loughrin JH, Tumlinson JH (1997) An elicitor of plant volatiles from beet armyworm oral secretion. Science 276:945–949

Ampofo JKO (1985) *Chilo partellus* (Swinhoe) oviposition on susceptible and resistant genotypes. Insect Sci Appl 6:323–330

Annual Report (2015–16) AICRP annual report, Indian Institute of Maize Research, New Delhi

Anyanda GN, Bruce AY, Makumbi D, Ahonsi M, Kahuthia-Gathu R, Namikoye SE, Beyene Y, Prasanna BM (2022) Reproductive potential of fall armyworm *Spodoptera frugiperda* (J.E. Smith) and effects of feeding on diverse maize genotypes under artificial infestation. Front Insect Sci 2:950815

Arabjafari HK, Jalali SK (2007) Identification and analysis of host plant resistance in leading maize genotypes against spotted stem borer, *Chilo partellus* (Swinhoe) (Lepidoptera: Pyralidae). Pak J Biol Sci 10:1885–1895

Ashfaq M, Ane MN, Zia K, Nasreen A, Hasan M (2010) The correlation of abiotic factors and physic-morphic charateristics of (*Bacillus thuringiensis*) Bt transgenic cotton with whitefly, *Bemisia tabaci* (Homoptera: Aleyrodidae) and jassid, *Amrasca devastans* (Homoptera: Jassidae) populations. Afr J Agric Res 5(22):3102–3107

Awmack CS, Leather SR (2002) Host plant quality and fecundity in herbivorous insects. Annu Rev Entomol 47:817–844

Badji A, Otim M, Machida L, Odong T, Kwemoi DB, Okii D, Agbahoungba S, Mwila N, Kumi F, Ibanda A, Mugo S, Kyamanywa S, Rubaihayo P (2018) Maize combined insect resistance genomic regions and their co-localization with cell wall constituents revealed by tissue-specific QTL meta-analyses. Front Plant Sci 9:895

Barros-Rios J, Santiago R, Jung HJG, Malvar RA (2015) Covalent cross-linking of cell-wall polysaccharides through esterified diferulates as a maize resistance mechanism against corn borers. J Agric Food Chem 63:2206–2214

Barry D, Alfaro D, Darrah LL (1994) Relation of European corn borer (Lepidoptera: Pyralidae) leaf-feeding resistance and DIMBOA content in maize. Environ Entomol 23:177–182

Basu S, Varsani S, Louis J (2018) Altering plant defenses: herbivore associated molecular patterns and effector arsenal of chewing herbivores. Mol Plant Microbe Interact 31:13–21. https://doi.org/10.1094/MPMI-07-17-0183-FI

Beale MH, Birkett MA, Bruce TJ, Chamberlain K, Field LM, Huttly AK, Martin JL, Parker R, Phillips AL, Pickett JA, Prosser IM (2006) Aphid alarm pheromone produced by transgenic plants affects aphid and parasitoid behaviour. Proc Natl Acad Sci U S A 103(27):10509–10513

Beeghly HH, Coors JG, Lee M (1997) Plant fibre composition and resistance to European corn borer in four maize populations. Maydica 42:297–303

Bergvinson DJ, Hamilton RI, Arnason JT (1995) Leaf profile of maize resistance factors to the European corn borer, *Ostrinia nubilalis*. J Chem Ecol 21:343–354

Bergvinson DJ, Arnason JT, Hamilton RI (1997) Phytochemical changes during recurrent selection for resistance to the European corn borer. Crop Sci 37:1567–1572

Block AK, Hunter CT, Sattler SE, Rering C, McDonald S, Basset GJ, Christensen SA (2019) Fighting on two fronts: elevated insect resistance in flooded maize. Plant Cell Environ 43:223. https://doi.org/10.1111/pce.13642

Branson TF (1971) Resistance in the grass tribe Maydeae to larvae of the Western corn rootworm. Ann Entomol Soc Am 64:861–863

Brooks TD, Bushman BS, Williams WP, McMullen MD, Buckley PM (2007) Genetic basis of resistance to fall armyworm (Lepidoptera: Noctuidae) and southwestern corn borer (Lepidoptera: Crambidae) leaf-feeding damage in maize. J Econ Entomol 100:1470–1475

Bruce TJA, Wadhams LJ, Woodcock CM (2005) Insect host location: a volatile situation. Trends Plant Sci 10:269–274

Byrne PF, Mcmullen MD, Snook ME, Musket TA, Theuri JM, Widstrom NW, Wiseman BR, Coe EH (1996) Quantitative trait loci and metabolic pathways: genetic control of the concentration of maysin, a corn earworm resistance factor, in maize silks. Proc Natl Acad Sci U S A 93:8820–8825

Castano-Duque L, Luthe DS (2018) Protein networks reveal organ-specific defense strategies in maize in response to an aboveground herbivore. Arthropod Plant Interact 12:147–175. https://doi.org/10.1007/s11829-017-9562-0

Castano-Duque L, Loades KW, Tooker JF, Brown KM, Paul Williams W, Luthe DS (2017) A maize inbred exhibits resistance against western corn root worm, *Diabrotica virgifera virgifera*. J Chem Ecol 43:1109–1123

Chatterji SM, Young WR, Sharma GC, Panwar VPV, Siddiqui KH (1966) Differential resistance of some maize composites, composite crosses, hybrids and varieties to *Chilo zonellus* (Swinhoe) (Lepidoptera: Pyralidae). Indian J Entomol 28:427–434

Cholla A, Kumar P, Subash C, Suby SB (2018a) Antixenosis in maize to the spotted stem borer *Chilo partellus* (Swinhoe). Indian J Entomol 80(3):1027–1034

Cholla A, Chander S, Kaur J, Suby SB, Kumar P (2018b) Improved method of screening maize germplasm for resistance against *Chilo partellus* (Swinhoe). Indian J Genet Plant Breed 78(4): 454–459

Dafoe NJ, Huffaker A, Vaughan MM, Duehl AJ, Teal PE, Schmelz EA (2011) Rapidly induced chemical defenses in maize stems and their effects on short-term growth of *Ostrinia nubilalis*. J Chem Ecol 37:984–991. https://doi.org/10.1007/s10886-011-0002-9

Dafoe NJ, Thomas JD, Shirk PD, Legaspi ME, Vaughan MM, Huffaker A et al (2013) European corn borer (*Ostrinia nubilalis*) induced responses enhance susceptibility in maize. PLoS One 8: e73394. https://doi.org/10.1371/journal.pone.0073394

Davis FM, Baker GT, Williams WP (1995) Anatomical characteristics of maize resistant to leaf feeding by southwestern corn borer (Lepidoptera: Pyralidae) and fall armyworm (Lepidoptera: Noctuidae). J Agric Entomol 12(1):55–65

De La Rosa-Cancino W, Rojas JC, Cruz-Lopez L, Castillo A, Malo EA (2016) Attraction, feeding preference, and performance of *Spodoptera frugiperda* larvae (Lepidoptera: Noctuidae) reared on two varieties of maize. Environ Entomol 45:384–389

de Paiva LA, Correa F, Silva CLT, Moura TL, Silva FC, Araujo MS et al (2016) Resistance of corn genotypes to fall armyworm *Spodoptera frugiperda* (Lepidoptera: Noctuidae). Afr J Biotechnol 15:1877–1882

Dean JM, Mescher MC, De Moraes CM (2009) Plant-rhizobia mutualism influences aphid abundance on soybean. Plant Soil 323:187–196

Degen T, Dillmann C, Marion-Poll F, Turlings TCJ (2004) High genetic variability of herbivore-induced volatile emission within a broad range of maize inbred lines. Plant Physiol 135:1928–1938

DES (2021) Directorate of Economics and Statistics, Second advance estimates of production of foodgrains for 2020–21. Ministry of Agriculture and Farmers Welfare, Department of Agriculture, Cooperation and Farmers Welfare

Divekar P, Kumar P, Suby SB (2019) Oviposition preference of pink stem borer, *Sesamia inferens* (Walker) in maize germplasm. J Entomol Zool Stud 7(3):1115–1119

Dubey SL, Sarup P (1982) Ovipositional responses of moths of *Chilo partellus* (Swinhoe) on different maize germplasm. J Entomol Res 6:1–9

Erb M, Flors V, Karlen D, De Lange E, Planchamp C, D'Alessandro M, Turlings TCJ, Ton J (2009) Signal signature of aboveground-induced resistance upon belowground herbivory in maize. Plant J 59:292–302

Erb M, Meldau S, Howe GA (2012) Role of phytohormones in insect specific plant reactions. Trends Plant Sci 17:250–259. https://doi.org/10.1016/j.tplants.2012.01.003

Eubanks MW (2001) The origin of maize: evidence for *Tripsacum* ancestry. In: Janick J (ed) Plant breeding reviews, vol 20. John Wiley & Sons, Inc, New York, pp 15–66

Farias-Rivera LA, Hernandez-Mendoza JL, Molina-Ochoa J, Pescador-Rubio A (2003) Effect of leaf extracts of teosinte, *Zea diploperennis* L, a Mexican maize variety, *criollo 'uruapeño'*, on the growth and survival of the fall armyworm (Lepidoptera: Noctuidae). Fla Entomol 86:239–243

Fontana A, Held M, Fantaye CA, Turlings TC, Degenhardt J, Gershenzon J (2011) Attractiveness of constitutive and herbivore-induced sesquiterpene blends of maize to the parasitic wasp *Cotesia marginiventris* (Cresson). J Chem Ecol 37:582–591

Fu J, Liu Q, Wang C et al (2018) ZmWRKY79 positively regulates maize phytoalexin biosynthetic gene expression and is involved in stress response. J Exp Bot 69:497–510

Gatehouse JA (2002) Plant resistance towards insect herbivores: a dynamic interaction. New Phytol 156:145–169

Gibert A, Tozer W, Westoby M (2019) Plant performance response to eight different types of symbiosis. New Phytol 222:526–542

Gill SS (1995) Mechanism of action of *Bacillus thuringiensis* toxins. Mem Inst Oswaldo Cruz 90(1):69–74

Gill TA, Sandoya G, Williams P, Luthe DS (2011) Belowground resistance to western corn rootworm in lepidopteran-resistant maize genotypes. J Econ Entomol 104:299–307. https://doi.org/10.1603/EC10117

Glauser G, Marti G, Villard N, Doyen GA, Wolfender JL, Turlings TCJ, Erb M (2011) Induction and detoxification of maize 1,4-benzoxazin-3-ones by insect herbivores. Plant J 68:901–911

Goyal S, Jindal J, Kaur K (2020) Physico-chemical bases of resistance in maize against shoot fly, *Atherigona naqvii* Steyskal during spring season in north India. Euphytica 216:154

Groh S, Gonzalez-de-Leon D, Khairallah MM, Jiang C, Bergvinson D, Bohn M et al (1998) QTL mapping in tropical maize: III. Genomic regions for resistance to *Diatraea* spp. and associated traits in two RIL populations. Crop Sci 38:1062–1072

Gueldner RC, Nook MES, Idstorm NWW, Wiseman BER (1992) TLC screen for maysin, chlorogenic acid and other possible resistance factors to the fall armyworm and the corn ear worm in *Zea mays*. J Agric Food Chem 40:1211–1213

Gui S, Taning CNT, Wei D, Smagghe G (2020) First report on CRISPR/Cas9-targeted mutagenesis in the Colorado potato beetle *Leptinotarsa decemlineata*. J Insect Physiol 121:104013

Gundappa B, Kumar P, Suby SB (2013) Antibiosis effect of phenolic acids (ferulic acid and p-coumaric acid) on maize spotted stem borer, *Chilo partellus* (Swinhoe) (Lepidoptera: Pyralidae). Indian J Entomol 75(3):247–250

Gutierrez C, Castarera P, Torres V (1988) Wound-induced changes in DIMBOA (2,4 dihydroxy-7-methoxy-2H-1,4 benzoxazin-3(4H)-one) concentration in maize plants caused by *Sesamia nonagrioides* (Lepidoptera: Noctuidae). Ann Appl Biol 113:441–454

Haile H, Hofsvang T (2001) Effect of sowing dates and fertilizer on the severity of stem borer (*Busseola fusca* Fuller, Lepidoptera: Noctuidae) on sorghum in Eritrea. Int J Pest Manag 47: 259–264

Harfouche AL, Shivaji R, Stocker R et al (2006) Ethylene signaling mediates a maize defense response to insect herbivory. Mol Plant Microbe Interact 19:189–199

Hassani MA, Duran P, Hacquard S (2018) Microbial interactions within the plant holobiont. Microbiome 6:58

Hedin PA, Williams WP, Davis FM, Buckley PM (1990) Roles of amino acids, protein, and fiber in leaf-feeding resistance of corn to the fall armyworm. J Chem Ecol 16(6):1977–1995

Huang FN, Andow DA, Buschman LL (2011) Success of the high-dose/refuge resistance management strategy after 15 years of Bt crop use in North America. Entomol Exp Appl 140(1):1–16

James C (2015) Global status of commercialized biotech/GM crops: ISAAA Brief No. 51. International Service for the Acquisition of Agri-biotech Applications (ISAAA), Ithaca, NY

Jimenez-Galindo JC, Ordas B, Butron A, Samayoa LF, Malvar RA (2017) QTL mapping for yield and resistance against Mediterranean corn borer in maize. Front Plant Sci 8:698

Kabre GB, Ghorpade SA (1999) Susceptibility to maize stem Borer, *Chilo partellus* (Swinhoe) in relation to sugars, proteins and free amino acids content of maize germplasm and hybrids. J Insect Sci 12(1):37–40

Karban R, Myers JJH (1989) Induced plant responses to herbivory. Annu Rev Ecol Syst 20:331–348

Kasoma C, Shimelis H, Laing M, Shayanowako AIT, Mathew I (2020a) Screening of inbred lines of tropical maize for resistance to fall armyworm, and for yield and yield-related traits. Crop Prot 136:105218

Kasoma C, Shimelis H, Laing MD, Shayanowako AIT, Mathew I (2020b) Revealing the genetic diversity of maize (*Zea mays* L.) populations by phenotypic traits and DArTseq markers for variable resistance to fall armyworm. Genet Resour Crop Evol 9:243–259

Kaur J, Kumar P, Suby SB, Singh J, Jha GH (2016) Measurement of antibiosis in maize genotypes against pink stem borer. Indian J Agric Sci 86(3):340–346

Kollner TG, Held M, Lenk C, Hiltpold I, Turlings TC, Gershenzon J, Degenhardt J (2008) A maize (E)-beta-caryophyllene synthase implicated in indirect defense responses against herbivores is not expressed in most American maize varieties. Plant Cell 20:482–494

Kollner TG, Lenk C, Schnee C, Kopke S, Lindemann P, Gershenzon J, Degenhardt J (2013) Localization of sesquiterpene formation and emission in maize leaves after herbivore damage. BMC Plant Biol 13:15

Kumar H (1993) Responses of *Chilo partellus* (Lepidoptera: Pyralidae) and *Busseola fusca* (Lepidoptera: Noctuidae) to hybrids of a susceptible and a resistant maize. J Econ Entomol 86:962–068

Kumar H (1994) Components of resistance in maize (*Zea mays* L) to first and second generation *Chilo partellus* (Swinhoe). Maydica 39:165–170

Kumar H, Saxena KN (1985) Ovipositional responses of *Chilo partellus* (Swinhoe) to certain susceptible and resistant maize genotypes. Insect Sci Appl 6:331–335

Kumar P, Sekhar JC, Chaudhary R (2005) Management of insect pests of maize in tropics. In: Zaidi PH, Singh NN (eds) Stresses on maize in tropics. Directorate of Maize Research, New Delhi, pp 298–323

Kumar K, Gambhir G, Dass A, Tripathi AK, Singh A, Jha AK, Yadava P, Choudhary M, Rakshit S (2020) Genetically modified crops: current status and future prospects. Planta 251(4):91

Lakshmi Soujanya PL, Sekhar JC, Ratnavathi CV, Shobha E, Karjagi CG, Suby SB, Sunil N, Rakshit S (2020) Role of soluble, cell wall bound phenolics, tannin and flavonoid contents in maize resistance to pink stem borer *Sesamia inferens* Walker. Maydica 65(1):12

Lakshmi Soujanya PL, Sekhar JC, Yathish KR, Karjagi CG, Rao KS, Suby SB, Jat SL, Kumar B, Kumar K, Vadessery J, Subaharan K, Patil J, Kalia VK, Dhandapani A, Rakshit S (2022) Leaf damage based phenotyping technique and its validation against fall armyworm, *Spodoptera frugiperda* (J. E. Smith) in maize. Frontiers. Plant Sci 13:906207

Larsson MC, Domingos AI, Jones WD, Chiappe ME, Amrein H, Bm VL (2004) *Or83b* encodes a broadly expressed odorant receptor essential for *Drosophila* olfaction. Neuron 43(5):703–714

Li SY, Wang H, Li FQ et al (2015) The maize transcription factor EREB58 mediates the jasmonate-induced production of sesquiterpene volatiles. Plant J 84:96e308

Li G, Ji T, Zhao S, Feng H, Wu K (2022) High-dose assessment of transgenic insect-resistant maize events against major lepidopteran pests in China. Plants 11:3125

Lima FWN, Ohashi OS, Souza FRS, Gomes FS (2006) Avaliação de acessos de milho para a resistência a *Spodoptera frugiperda* (Smith) (Lepidoptera: Noctuidae) em laboratório. Acta Amaz 32:147–150

Liu MM, Zhang XJ, Gao Y, Shen ZC, Lin CY (2018) Molecular characterization and efficacy evaluation of a transgenic corn event for insect resistance and glyphosate tolerance. J Zhejiang Univ Sci B 19(8):610–619

Lokesh, Mehla JC (2017) Study on morphological and biochemical features associated with maize (*Zea mays L.*) resistance against *Chilo partellus*. Insect Biochem Mol Biol 5(4):1011–1021

Lopez L, Camas A, Shivaji R, Ankala A, Williams P, Luthe D (2007) Mir1-CP, a novel defense cysteine protease accumulates in maize vascular tissues in response to herbivory. Planta 226: 517–527

Louis J, Peiffe M, Ray S, Luthe DS, Felton GW (2013) Host-specific salivary elicitor(s) of European corn borer induce defenses in tomato and maize. New Phytol 199:66–73. https://doi.org/10.1111/nph.12308

Louis J, Basu S, Varsani S, Castano-Duque L, Victoria Jiang W, Williams P, Felton GW, Luthe DS (2015) Ethylene contributes to maize insect resistance1-mediated maize defense against the phloem sap-sucking corn leaf aphid. Plant Physiol 169(1):313–324

Luciano N, Eduardo NC, Mirella MDB, Juno FSD, Zulene AR, Arlindo LB Jr (2018) Oviposition preference and antibiosis to *Spodoptera frugiperda* (Lepidoptera: Noctuidae) in Brazilian maize landraces. J Econ Entomol 20:1–9

Malone LA, Barraclough EI, Lin-Wang K, Stevenson DE, Allan AC (2009) Effects of red-leaved transgenic tobacco expressing a MYB transcription factor on two herbivorous insects, *Spodoptera litura* and *Helicoverpa armigera*. Entomol Exp Appl 133(2):117–127

Malook S, Xiao-Feng Liu X-F, Liu W, Qi J, Zhou S (2021) The race goes on a fall armyworm resistant maize inbred line influences insect oral secretion elicitation activity and nullifies herbivore suppression of plant defense. BioRxiv. https://doi.org/10.1101/2021.05.17.444416

Mammadov J, Buyyarapu R, Guttikonda SK, Parliament K, Abdurakhmonov IY, Kumpatla SP (2018) Wild ancestors of maize, rice, cotton, and soybean: treasure troves for tolerance to biotic and abiotic stresses. Front Plant Sci 9:886

Mason CJ, Ray S, Davidson-Lowe E, Ali J, Luthe DS, Felton G (2022) Plant nutrition influences resistant maize defense responses to the fall armyworm (*Spodoptera frugiperda*). Front Ecol Evol 10:844274. https://doi.org/10.3389/fevo.2022.844274

Matova PM, Kamutando CN, Magorokosho C, Kutywayo D, Gutsa F, Labuschagne M (2020) Fall armyworm invasion, control practices and resistance breeding in sub-Saharan Africa. Crop Sci 60:2951–2970

Meckenstock DH, Castro MT, Pitre HN, Gomez F (1991) Antibiosis to fall armyworm in honduran landrace sorghum. Environ Entomol 20:1259–1266

Mihm JA (ed) (1997) Insect resistant maize: recent advances and utilization; Proceedings of an international symposium held at CIMMYT, 27 Nov–3 Dec 1994. CIMMYT, Mexico DF

Moellenbeck DJ, Barry BD, Darrah LL (1995) *Tripsacum dactyloides* (Gramineae) seedlings for host plant resistance to the Western corn rootworm (Coleoptera: Chrysomelidae). J Econ Entomol 88:1801–1803

Moyal P (1998) Crop losses caused by maize stem borer (Lepidoptera: Pyralidae). J Econ Entomol 91:512–516

Munyiri SW, Mugo SN (2017) Quantitative trait loci for resistance to spotted and African maize stem borers (*Chilo partellus* and *Busseola fusca*) in a tropical maize (*Zea mays* L.) population. Afr J Biotechnol 16:1579–1589

Murenga M, Derera J, Mugo S, Tongoona P, Gichuru L (2015) Evaluation of tropical maize inbred lines for resistance to two stem borer species, *Busseola fusca* and *Chilo partellus*. J Plant Breed Crop Sci 8:23–33

Murrell EG, Ray S, Lemmon ME, Luthe DS, Kaye JP (2019) Cover crop species affect mycorrhizae-mediated nutrient uptake and pest resistance in maize. Renew Agric Food Syst 1–8:467. https://doi.org/10.1017/S1742170519000061

Murua G, Molina-Ochoa J, Coviella C (2006) Population dynamics of the fall armyworm, and its parasitoids in northwestern Argentina. Fla Entomol 89(2):175–182

Ni X, Da K, Buntin GD, Brown SL (2008) Physiological basis of fall armyworm (Lepidoptera: Noctuidae) resistance in seedlings of maize inbred lines with varying levels of silk maysin. Fla Entomol 91:537–545

Ni X, Chen Y, Hibbard BE, Wilson JP, Williams WP, Buntin GD, Ruberson JR, Li X (2011) Foliar resistance to fall armyworm in corn germplasm lines that confer resistance to root-and ear-feeding insects. Fla Entomol 94(4):971–981

Ni X, Xu W, Blanco MH, Williams WP (2014) Evaluation of fall armyworm resistance in maize germplasm lines using visual leaf injury rating and predator survey. Insect Sci 21(5):541–555

Nuessly GS, Scully BT, Hentz MG, Beiriger R, Snook ME, Widstrom NW (2007) Resistance to *Spodoptera frugiperda* (Lepidoptera: Noctuidae) and *Euxesta stigmatias* (Diptera: Ulidiidae) in sweet corn derived from exogenous and endogenous genetic systems. J Ecol Entomol 100: 1887–1895

Ortego F, Ruiz M, Castañera P (1998) Effect of DIMBOA on growth and digestive physiology of *Sesamia nonagrioides* (Lepidoptera: Noctuidae) larvae. J Insect Physiol 44:95–101

Painter RH (1951) Insect resistance in crop plants. Soil Sci 72(6):481. https://doi.org/10.1097/00010694-195112000-00015

Panwar VPS (2005) Management of maize stalk borer, *Chilo partellus* (Swinhoe) in maize. In: Zaidi PH, Singh NN (eds) Stresses on maize in the tropics. Directorate of Maize Research, New Delhi, pp 376–395

Panwar VPS, Sarup P (1980) Differential development of *Chilo partellus* (Swinhoe) in various maize varieties. J Entomol Res 4:28–33

Papst C, Bohn M, Utz HF, Melchinger AE, Klein D, Eder J (2004) QTL mapping for European corn borer resistance (*Ostrinia nubilalis* Hb.), agronomic and forage quality traits of testcross progenies in early-maturing European maize (*Zea mays* L.) germplasm. Theor Appl Genet 108:1545–1554

Pasztor K, Borsos O (1990) Inheritance and chemical composition in inbred maize (*Zea mays* L.) × teosinte (*Zea mays* ssp. *mexicana* (Schräder)) hybrids. Növénytermelés 39:193–213

Pathak PK, Sharma VK, Singh JM (1971) Effect of date of planting on the spring maize on the incidence of shoot fly, *Atherigona* spp. and loss in yield due to its attack. In: Annual report 1970-71 experiment station. UPAU, Pantnagar, pp 279–281

Pechan T, Ye L, Chang Y, Mitra A, Lin L, Davis FM et al (2000) A unique 33-kD cysteine proteinase accumulates in response to larval feeding in maize genotypes resistant to fall armyworm and other Lepidoptera. Plant Cell 12:1031–1041

Pechan T, Cohen A, Williams WP, Luthe DS (2002) Insect feeding mobilizes a unique plant defense protease that disrupts the peritrophic matrix of caterpillars. Proc Natl Acad Sci U S A 99:13319–13323. https://doi.org/10.1073/pnas.202224899

Pingault L, Varsani S, Palmer N, Ray S, Williams WP, Luthe DS, Ali JG, Sarath G, Louis J (2021) Transcriptomic and volatile signatures associated with maize defense against corn leaf aphid. BMC Plant Biol 21:138

Prasanna BM (2019) Host plant resistance to fall armyworm: status and prospects. WTO Meeting, Mar 19

Qi JF, Zhou GX, Yang LJ et al (2011) The chloroplast-localized phospholipases d alpha 4 and alpha 5 regulate herbivore-induced direct and indirect defenses in rice. Plant Physiol 157:1987–1999

Qi J, Sun G, Wang L et al (2016) Oral secretions from *Mythimna separata* insects specifically induce defense responses in maize as revealed by high-dimensional biological data. Plant Cell Environ 39:1749–1766

Ramirez DA (1997) Gene introgression in maize (*Zea mays* ssp *mays* L.). Philipp J Crop Sci 22:51–63

Rao AB (1983) Technique of scoring for resistance in maize stalk borer (*Sesamia inferens*). In: Techniques for scoring for resistance to the major insect pests of maize. AICMIP, IARI, New Delhi, pp 16–26

Rao CN, Panwar VPS (2002) Biochemical factors affecting resistance to *Chilo partellus* (Swinhoe) in maize. Ann Plant Sci 10(1):28–30

Rasmann S, Kollner TG, Degenhardt J, Hiltpold I, Toepfer S, Kuhlmann U, Gershenzon J, Turlings TC (2005) Recruitment of entomopathogenic nematodes by insect-damaged maize roots. Nature 434:732–737

Raupp MJ (1985) Effects of leaf toughness on mandibular wear of the leaf beetle, *Plagiodera versicolora*. Ecol Entomol 10:73–79

Ray S, Basu S, Rivera-Vega LJ, Acevedo FE, Louis J, Felton GW et al (2016) Lessons from the far end: Caterpillar FRASS-induced defenses in maize, rice, cabbage, and tomato. J Chem Ecol 42:1130–1141. https://doi.org/10.1007/s10886-016-0776-x

Razzaq A, Saleem F, Kanwal M, Mustafa G, Yousaf S, Imran Arshad HM, Hameed MK, Khan MS, Joyia FA (2019) Modern trends in plant genome editing: an inclusive review of the CRISPR/Cas9 toolbox. Int J Mol Sci 20(16):4045

Read J, Stokes A (2006) Plant biomechanics in an ecological context. Am J Bot 93:1546–1565

Reddy MLK, Ramesh Babu T, Venkatesh S (2003) A new rating scale for *Sesamia inferens* (Walker) (Lepidoptera: Noctuidae) damage to maize. Insect Sci Appl 23:293–299

Rodríguez VM, Santiago R, Malvar RA, Butrón A (2012) Inducible maize defense mechanisms against the corn borer *Sesamia nonagrioides*: a transcriptome and biochemical approach. Mol Plant Microbe Interact 25:61–68

Samayoa LF, Butrón A, Malvar RA (2014) QTL mapping for maize resistance and yield under infestation with *Sesamia nonagrioides*. Mol Breed 34:1331–1344

Samayoa LF, Butrón A, Malvar RA (2019) Usefulness of marker-assisted selection to improve maize for increased resistance to *Sesamia nonagrioides* attack with no detrimental effect on yield. Ann Appl Biol 174(2):219–222

Santiago R, Souto XC, Sotelo J, Butro A, Malvar RA (2003) Relationship between maize stem structural characteristics and resistance to pink stem borer (Lepidoptera: Noctuidae) attack. J Econ Entomol 96(5):1563–1570

Santiago R, Butron A, Arnason JT, Reid LM, Souto XC, Malvar RA (2006) Putative role of pith cell wall phenylpropanoids in *Sesamia nonagrioides* (Lepidoptera: Noctuidae) resistance. J Agric Food Chem 54:2274–2279

Santiago R, Barros-Rios J, Malvar RA (2013) Impact of cell wall composition on maize resistance to pests and diseases. Int J Mol Sci 14:6960–6980

Santiago R, Malvar RA, Barros-Rios J, Samayoa LF, Butrón A (2016) Hydroxycinnamate synthesis and association with Mediterranean corn borer resistance. J Agric Food Chem 64:539–551

Santos F, Penaflor MFGV, Paré PW, Sanches PA, Kamiya AC, Tonelli M et al (2014) A novel interaction between plant-beneficial rhizobacteria and roots: colonization induces corn resistance against the root herbivore *Diabrotica speciosa*. PLoS One 9(11):e113280

Sarup P, Panwar VPS, Marwaha KK, Siddiqui KH (1978) Management of maize pests with particular reference to the stalk borer, *Chilo partellus* (Swinhoe) under resource constraints. J Entomol Res 2(1):5–14

Schmelz EA, Kaplan F, Huffaker A et al (2011) Identity, regulation, and activity of inducible diterpenoid phytoalexins in maize. Proc Natl Acad Sci U S A 108:5455–5460

Schnee C, Kollner TG, Held M, Turlings TC, Gershenzon J, Degenhardt J (2006) The products of a single maize sesquiterpene synthase form a volatile defense signal that attracts natural enemies of maize herbivores. Proc Natl Acad Sci U S A 103:1129–1134

Schoonhoven LM, van Loon JJA, Dicke M (2007) Insect-plant biology, 2nd edn. Oxford University Press, New York, p 421

Sekhar JC, Kumar P, Rakshit S, Mehrajuddin AM, Dass S (2008) Differential response of CMLS and their hybrid combinations to pink borer, *Sesamia inferens* Walker. Ann Plant Prot Sci 16: 404–406

Sekhar JC, Lakshmi Soujanya P, Chikkappa GK, Sunil N, Kaul J, Singh KP, Kumar P (2016a) Response of different maize accessions to pink stem borer *Sesamia inferens* Walker (Lepidoptera: Noctuidae). Electron J Plant Breed 7. https://doi.org/10.5958/0975-928X.2016.00017.X

Sekhar JC, Soujanya PL, Chikkappa GK, Kumar P, Sunil N, Singh KP (2016b) Resistance to the spotted stem borer *Chilo partellus* (Swinhoe) and pink stem borer *Sesamia inferens* Walker in maize. Indian J Entomol 78:346–352

Sekhon SS, Sajjan SS (1990) Antibiosis in maize to maize borer *Chilo partellus* Swinhoe in relation to plant age. Indian J Entomol 52(4):579–582

Sharma VK, Chatterji SM (1971) Studies on some chemical constituents in relation to differential susceptibility of some maize germplasm to *Chilo zonellus* (Swinhoe). Indian J Entomol 33(4): 419–424

Sharma HC, Taneja SL, Leuschner K, Nwanze KF (1992) Techniques to screen sorghums for resistance to insect pests. Information Bulletin No. 32. International Crops Research Institute for the Semi-Arid Tropics (ICRISAT), Patancheru

Smith CM (ed) (2005) Plant resistance to arthropods: molecular and conventional approaches. Springer, Dordrecht

Song YY, Cao M, Xie LJ, Liang XT, Zeng RS, Su YJ, Huang JH, Wang RL, Luo SM (2011) Induction of DIMBOA accumulation and systemic defense responses as a mechanism of enhanced resistance of mycorrhizal corn (*Zea mays* L.) to sheath blight. Mycorrhiza 21:721–731

Songstad DD, Petolino JF, Voytas DF, Reichert NA (2017) Genome editing of plants. Crit Rev Plant Sci 36:1–23

Soujanya L, Sekhar JC, Chikkappa K, Suby SB, Sunil N, Yathish KR, Reddy MLK, Jindal J, Rakshit S (2019) Field screening of maize inbred lines for resistance to stem borers *Chilo partellus* (Swinhoe) and *Sesamia inferens* Walker. Maize J 8(1):8–14

Suby SB, Soujanya PL, Yadava P, Patil J, Subaharan K, Prasad GS, Babu KS, Jat SL, Yathish KR, Vadassery J, Kalia VK, Bakthavatsalam N, Shekhar JC, Rakshit S (2020) Invasion of fall armyworm (*Spodoptera frugiperda*) in India: nature, distribution, management and potential impact. Curr Sci 119(1):44–51

Sun H, Lang Z-H, Lu W, Zhang J, He K-L, Zhu L, Lin M, Huang D-F (2015) Developing transgenic maize (*Zea mays* L.) with insect resistance and glyphosate tolerance by fusion gene transformation. J Integr Agric 14(2):305–313

Szczepaniec A, Widney SE, Bernal JS, Eubanks MD (2012) Higher expression of induced defenses in teosintes (*Zea* spp.) is correlated with greater resistance to fall armyworm, *Spodoptera frugiperda*. Entomol Exp Appl 146:242–251

Tamayo MC, Rufat M, Bravo JM, Segundo BS (2000) Accumulation of a maize proteinase inhibitor in response to wounding and insect feeding, and characterization of its activity toward digestive proteinases of *Spodoptera littoralis* larvae. Planta 211(1):62–71

Tamiru A, Bruce TJA, Woodcock CM, Caulfield JC, Midega CAO, Ogol CKPO, Mayon P, Birkett MA, Pickett JA, Khan ZR (2011) Maize landraces recruit egg and larval parasitoids in response to egg deposition by a herbivore. Ecol Lett 14:1075–1083

Tamiru A, Bruce TJA, Richter A et al (2017) A maize landrace that emits defense volatiles in response to herbivore eggs possesses a strongly inducible terpene synthase gene. Ecol Evol 7: 2835–2845

Tefera T, Mugo S, Beyene Y (2016a) Developing and deploying insect resistant maize varieties to reduce pre-and post-harvest food losses in Africa. Food Security 8:211–216

Tefera T, Mugo S, Mwimali M et al (2016b) Resistance of *Bt*-maize (MON 810) against the stem borers *Busseola fusca* (Fuller) and *Chilo partellus* (Swinhoe) and its yield performance in Kenya. Crop Prot 89:202–208

Tian D, Tooker J, Peiffer M, Chung SH, Felton GW (2012) Role of trichomes in defense against herbivores: comparison of herbivore response to woolly and hairless trichome mutants in tomato (*Solanum lycopersicum*). Planta 236:1053–1066. https://doi.org/10.1007/s00425-012-1651-9

Tzin V, Lindsay PL, Christensen SA, Meihls LN, Blue LB, Jander G (2015) Genetic mapping shows intraspecific variation and transgressive segregation for caterpillar-induced aphid resistance in maize. Mol Ecol 24:5739–5750

Vats S, Kumawat S, Kumar V, Patil GB, Joshi T, Sonah H, Sharma TR, Deshmukh R (2019) Genome editing in plants: exploration of technological advancements and challenges. Cell 8(11):1386

Veyrat N, Robert CAM, Turlings TCJ, Erb M (2016) Herbivore intoxication as a potential primary function 574 of an inducible volatile plant signal. J Ecol 104:591–600

Vikal Y, Kaur A, Jindal J, Kaur K, Pathak D, Garg T, Singh A, Singh P, Yadav I (2020) Identification of genomic regions associated with shoot fly resistance in maize and their syntenic relationships in the sorghum genome. PLoS One 15:e0234335. https://doi.org/10.1371/.journal.pone.0234335

Waiss ACJ, Chan BG, Elliger CA, Wiseman BR, McMillian WW, Widstrom NW, Zuber MS, Keaster AJ (1979) Maysin, a flavone glycoside from corn silks with antibiotic activity toward corn earworm. J Econ Entomol 72:256–258

Wang WM, Wang X, Liao H, Feng Y, Guo Y, Shu Y, Wang J (2022) Effects of nitrogen supply on induced defense in maize (*Zea mays*) against fall armyworm (*Spodoptera frugiperda*). Int J Mol Sci 23:10457

Widstrom NW, Wiseman BR, Snook ME, Nuessly GS, Scully BT (2003) Registration of the maize population Zapalote Chico 2451F. Crop Sci 43:444–445

Willcox MC, Khairallah MM, Bergvinson D, Crossa J, Deutsch JA, Edmeades GO, Gonzalez-de-Leon D, Jiang C, Jewell DC, Mihm JA, Williams WP, Hoisington D (2002) Selection for resistance to southwestern corn borer using marker-assisted and conventional backcrossing. Crop Sci 42:1516–1528

Williams WP, Davis FM, Windham GL (1990) Registration of Mp708 germplasm line of maize. Crop Sci 30(3):757

Wiseman BR, Snook ME, Isenhour DJ, MihmN JA, Widstrom W (1992) Relationship between growth of corn earworm and fall armyworm larvae (Lepidoptera: Noctuidae) and maysin concentration in corn silks. J Econ Entomol 85:2473–2477

Womack ED, Williams WP, Smith JS, Warburton ML, Bhattramakki D (2020) Mapping quantitative trait loci for resistance to fall armyworm (Lepidoptera: Noctuidae) leaf-feeding damage in maize inbred Mp705. J Econ Entomol 113(2):956–963

Wu J, Baldwin IT (2009) Herbivory-induced signalling in plants: perception and action. Plant Cell Environ 32:1161–1174

Xie YS, Arnason JT, Philogene BJR, Lambert JDH, Atkinson J, Morand P (1990) Role of 2,4-dihydroxy-7-methoxy-1,4-benzoxazin-3-one (DIMBOA) in the resistance of maize to western corn rootworm, *Diabrotica virgifera virgifera* (LeConte) (Coleoptera: Chrysomelidae). Can Entomol 122:1177–1186

Yang G, Wiseman BR, Isenhour DJ, Espelie KE (1993a) Chemical and ultrastructural analysis of corn cuticular lipids and their effect on feeding by fall armyworm larvae. J Chem Ecol 19:2055–2074

Yang G, Espelie KE, Wiseman BR, Isenhour DJ (1993b) Effect of corn foliar cuticular lipids on the movement of fall armyworm (Lepidoptera: Noctuidae) neonate larvae. Fla Entomol 76:302–316

Zhang YT, Zhang YL, Chen SX, Yin GH, Yang ZZ, Lee S, Liu CG, Zhao DD, Ma YK, Song FQ, Bennett JW, Yang FS (2015) Proteomics of methyl jasmonate induced defense response in maize leaves against Asian corn borer. BMC Genomics 16(1):224

Zheng Y, Zhang X, Liu X, Qin N, Xu K, Zeng R, Liu J, Song Y (2021) Nitrogen supply alters rice defense against the striped stem borer *Chilo suppressalis*. Front Plant Sci 12:691292

Zogli P, Pingault L, Grover S, Louis J (2020) Ento(o)mics: the intersection of 'omic' approaches to decipher plant defense against sap-sucking insect pests. Curr Opin Plant Biol 56:153–161. https://doi.org/10.1016/j.pbi.06.002

Host Plant Resistance to Insects in Pulse Crops

7

Abou Togola, Patrick Obia Ongom, Saba Baba Mohammed, Christian Fatokun, Manuele Tamò, and Ousmane Boukar

Abstract

Pulses are important sources of protein, dietary fiber, vitamins, and minerals for human nutrition and are widely cultivated in many countries around the world. Crops belonging to pulse include mainly chickpeas, mung bean, lentils, pigeon peas, and cowpeas. Pulse production and storage are negatively affected by a variety of biotic and abiotic factors, such as insect pests, diseases, parasites, low soil fertility, heat, and drought. Among these, insect pests have the most significant impact on pulse productivity worldwide. The extent of insect pest damage is a major challenge for pulse farmers and can lead to significant quantitative and qualitative losses. While synthetic insecticides are commonly used to control insect pests in pulses, their harmful effects on humans, animals, and the environment, as well as the development of insecticide resistance in pests, have prompted research efforts to identify eco-friendly alternatives. One such alternative is host plant resistance, which involves developing pulses with genetic traits that make them resistant to insect pests. The chapter underlined the importance of host plant resistance as a sustainable approach to protecting pulse crops against insect pests. It reviews past and recent studies that have contributed to improving pulse crops and sustaining their production through host plant resistance measures. Genomic tools and resources could be further utilized to develop insect resistance in all types of pulses and accelerate the pace of research in this area.

A. Togola (✉) · P. O. Ongom · S. B. Mohammed · O. Boukar
International Institute of Tropical Agriculture (IITA), Kano, Nigeria
e-mail: a.togola@cgiar.org

C. Fatokun
International Institute of Tropical Agriculture (IITA), Ibadan, Nigeria

M. Tamò
International Institute of Tropical Agriculture (IITA), Cotonou, Benin

S. Kumar, M. Furlong (eds.), *Plant Resistance to Insects in Major Field Crops*,
https://doi.org/10.1007/978-981-99-7520-4_7

Keywords

Pulse crops · Vigna unguculata · Insect pests · Genetic traits · Eco-friendly alternatives · Crop improvement

7.1 Introduction

Pulses are important crops for human nutrition and food security in many countries around the world. They are a rich source of proteins, dietary fiber, complex carbohydrates, vitamins, and essential minerals (Patterson et al. 2009). However, pulses also contain enzyme inhibitors, lectins, oligosaccharides, polyphenols, phytates, and saponins that affect the digestibility and bioavailability of nutrients in humans and animals (Bressani 1993).

Despite their importance, pulse production and storage are highly challenged by many constraints leading to quantitative and qualitative losses. These losses are imputable to a complex of biotic and abiotic stress factors including insect pests, diseases, parasitic weeds, heat, low soil fertility, and drought. Above all, biotic stresses in general and insect pests in particular have the most negative impact on pulse productivity worldwide. The extent of insect pest diversity and the severity of their attacks in pulses vary from crop to another and also from one location to another. Average grain yield losses between 50 and 80% in untreated cowpea fields were reported by Singh and Allen (1979) that can go up to 90–100% under high insect infestation conditions, on susceptible varieties (Jackai and Daoust 1986; Singh and Singh 2014; Togola et al. 2017). Over 85 insect species attack cowpea (Singh et al. 1990) with about 20 of them having regular occurrence and being of economic importance in various cowpea production areas worldwide (Oyewale and Bamaiyi 2013). Legume pod borer (*Maruca vitrata* Fabricius), cowpea aphid (*Aphis craccivora* Koch), flower bud thrips (*Megalurothrips sjostedti* Trybom), pod sucking bugs (*Clavigralla tomentosicollis* Stål), and cowpea weevils (*Callosobruchus maculatus*) are the most widespread and damaging insect pests of cowpea (Hassan 2009; Tamò et al. 2019; Oyewale and Bamaiyi 2013; Togola et al. 2017, 2019, 2020).

Other pulses like chickpea are similarly damaged by some 60 insect species known to feed on this crop, of which cut worm *Agrotis* spp., beet armyworm *Spodoptera exigua,* leaf miner *Liriomyza cicerina*, aphid *Aphis craccivora*, pod borer *Helicoverpa armigera*, and bruchid *Callosobruchus chinensis* and *Callosobruchus maculatus* are the major pests worldwide (Sharma et al. 2007, 2014; Malo and Hore 2020). Also, *Liriomyza cicerina* (Rondani) is one of the devastating and dominant leaf miner species on the crop (Chrigui et al. 2020).

Among the devastating insect pests of lentil, there are legume pod borer *Helicoverpa armigera* (Hub.), spiny pod borer *Etiella zinckenella* (Treit.), aphids *Aphis craccivora* Koch and *Acyrthosiphon pisum* (Harris), pea leaf weevil *Sitona crinitus lineatus* and weevil *Bruchus lentis*, pod sucking bug *Clavigralla gibbosa* Spin (Clement et al. 1994; Erskine et al. 1994; Sharma and Crouch 2004; Malo and

Hore 2020). As for the pigeon pea, over 150 insect species damage the crop (Sharma 2016). According to Shanower et al. (1999), the most serious and primary insect pests of pigeon pea are those that attack reproductive structures, including buds, flowers, and pods. Nearly 30 species of Lepidoptera and a large number of Hemiptera feed on the reproductive structures of the crop. The legume pod borer *Helicoverpa armigera* (Hübner), the spotted pod borer *Maruca vitrata* Geyer, the pod fly *Melanagromyza obtusa* Malloch, the pod wasp *Tanaostigmodes cajaninae* La Salle, the spiny pod borer *Etiella zinckenella* Triet., the pod sucking bugs *Clavigralla* spp. and the bruchids *Callosobruchus chinensis* L. are the major pests (Sharma 2016). Similarly, pod borer and pod fly were reported as major insect pests affecting pigeon pea production (Malo and Hore 2020; Rajeswari and Jayamani 2020). A total of 64 insect species attacking mung bean in the field were reported by Lal (1985). However, the major insect pests are bean fly *Ophiomyia phaseoli* Tryon, stem fly *Melanagromyza sojae* Zehntner, thrips, *Megalurothrips* spp. and *Caliothrips indicus* Bagnall, aphids *Aphis craccivora* Koch, whitefly *Bemisia tabaci* Gennadius, pod borers *Helicoverpa armigera* (Hubner), spotted caterpillar *Maruca vitrata* Fabricius, pod sucking bugs *Clavigralla gibbosa* Spinola and bruchids *Callosobruchus maculatus* Fab., and *Callosobruchus chinensis* L. (Lal 1985; Yadav and Singh 2006; Gentry 2010; Swaminathan et al. 2012; Duraimurugan and Tyagi 2014; Mbeyagala et al. 2017; Laosatit et al. 2020; Pratap et al. 2020). According to Rana et al. (2016), insect pests alone can lead to a yield reduction of 20–55% on mung bean, depending upon the stage of the crop, the severity of the stress, and prevailing environmental factors.

Application of synthetic insecticides remains the most common measure used to control the insect pests in pulses. However, knowing the harmful hazards caused by chemicals to humans, animals, and environment and possible development of insecticide resistance in pests, many research efforts were made to seek eco-friendly alternative options such as host plant resistance. Varietal resistance appears to be the most cost-effective and environment friendly approach to mitigate pest damage in pulses. This chapter highlights the importance of the host plant resistance as sustainable protection measure of pulse crops against insect pests and reviews some achievements of past and recent studies to support pulse crops improvement and sustain their production.

7.2 Host Plant Resistance to Insects in Pulse Crops

Over the past decades, significant progress has been made in developing resistant varieties of pulse crops to insect pests. The research efforts of identifying resistant genotypes to insects went through the process of phenotypic screening where the genotypes were evaluated for their resistance to various insects of economic importance to pulses.

As a result, hundreds of insect tolerant/resistant pulse genotypes were identified, improved, or developed through field, laboratory and screen house screening, genetic improvement, and biotechnology applications to mitigate insect pests' effects while sustaining pulse production and value-chain.

7.2.1 Progress in Host Plant Resistance to Insect Pests of Cowpea

On cowpea (*Vigna unguiculata*), concerted efforts were undertaken to boost productivity through the development of resistant lines to insect pests and also the deployment of quantitative modern genetics and genomic tools (Boukar et al. 2016, 2019; Ongom et al. 2022). These are expected to accelerate the rate of genetic gain, allowing farmers to benefit from the full genetic potentials of the crop (Ongom et al. 2021).

In recent decades, three accessions of the International Institute of Tropical Agriculture (IITA) mini core collection (TVu6464, TVu1583, and TVu15445) were reported to be resistant to *Aphis craccivora,* all having a low sucrose content in the plant and a high kaempferol and quercetin content (Togola et al. 2020). Earlier studies found the cowpea wild relative, TVNu1158, as resistant to aphid in seedling stage (Boukar and Fatokun 2009; Souleymane et al. 2013; Boukar et al. 2020). Single dominant genes designated as *Rac1* and *Rac2* were identified to control aphid resistance in cowpea (Boateng 2015). Same efforts were made to identify sources of resistance to *Megalurothrips sjostedti* where Togola et al. (2019) identified four cowpea mini core accessions TVu8631, TVu16368, TVu8671, and TVu7325 as resistant to this pest. Other cowpea accessions reported to be resistant to *M. sjostedti* are Sanzisabinli (Abudulai et al. 2006; Alabi et al. 2011; Sobda et al. 2018), Moussa local, TVu1509, TVx3236 and Sewe (Alabi et al. 2004), and IT93K-452-1 (Dormatey et al. 2015).

The cowpea accessions IT06K-123-1, ALEGI*SECOW3B, IT86D-1038, WC35B, IT86D-1033, TOUMKALAM, KPLOBEROUGE, WC66*NE50, IT06K-270, IT84S-2246-4, WC36, and TVu1471 were identified to be resistant to *Callosobruchus maculatus* in Benin (Kpoviessi et al. 2019). Similarly, TVu13677, WC36, and WC66*5T were identified as resistant to *C. maculatus* (Kpoviessi et al. 2021). Dabire-Binso et al. (2010) identified IT86D-716 as resistant to *Clavigralla tomentosicollis* implying the presence of cyanogenic heterosides, flavonoids, tannins, and trypsin inhibitors in the pods.

Recent development of genomic resources will support the molecular breeding and facilitate the development of durable resistance to cowpea insect pests. To accelerate such breeding process, advanced biotechnology methods and tools are being explored. Gene pyramiding is being explored by IITA and associated national agricultural research systems (NARS) partners to develop high desired cultivars combining resistance genes to different insects of cowpea (Boukar and Fatokun 2009; Togola et al. 2017). Genetic engineering option had been undertaken for the development of Maruca resistant transgenic cowpea by inserting the Cry1Ab Bt gene (Popelka et al. 2006; Togola et al. 2017). The Bt gene has been transferred through backcrossing to some improved and released cowpea varieties in different countries.

7.2.2 Progress in Host Plant Resistance to Insect Pests of Lentil

Research efforts have made it possible to find sources of plant resistance to some important field and storage insect pests of lentil (*Lens culinaris* Medik) as reported by Clement et al. (1994). Field tests were conducted to successfully identify some lentil genotypes for resistance to attack by *Bruchus lentis* Froel, *Aphis craccivora* Koch, and *Sitona* sp. (Sedivy 1972; Chopra and Pajni 1987; Clement et al. 1994). However, host plant resistance to *Sitona* sp. has not yet received major focus as a control alternative (Erskine et al. 1994).

Genetic differences in resistance against aphid species such as *A. craccivora* and *Acyrthosiphon pisum* were reported but no strong resistant genotype was obtained. Although resistance against *Etiella zinckenella*, has been found, but no attempt has been made to breed for resistance (Erskine et al. 1994). Similarly genetic variations in response to seed weevils, *Bruchus lentis* have been found, but breeding for resistance has not been initiated (Erskine et al. 1994). Laserna-Ruiz et al. (2012) recorded lower infestation of seed by bruchids in *Lens culinaris* Medikus subsp. *culinaris*, *Lens culinaris* Medikus subsp. *orientalis* (Boiss.) Ponert, *Lens nigricans* (M. Bieb.) Godr., and *Lens lamottei* Cezfr., than the check.

Genotypes, ILL 9924, RL 83, ILL 10856, ILL 6458 and RL 67 were found to be less susceptible to *A. craccivora* and had higher grain yield (Neupane et al. 2020). Low aphid incidence was associated with green or yellowish green foliage and slightly pubescent leaves on lentil genotypes (Kumari et al. 2009; Neupane et al. 2020).

7.2.3 Progress in Host Plant Resistance to Insect Pests of Chickpea

In South Asia, progress was made to breed resistance or tolerance of chickpea to reduce losses caused by *H. armigera* in this crop (ICRISAT 1990). Since 1976 extensive germplasm collections have been screened at the International Crops Research Institute for Semi-Arid Tropics (ICRISAT), India, where more than 14,000 chickpea germplasm accessions were screened for resistance to *H. armigera* of which ICC 506EB, ICC 10667, ICC 10619, ICC 4935, ICC 10243, ICCV 95992, and ICC 10817 were found to be resistant to the pod borer (Lateef 1985; Sharma et al. 2014). Germplasm accessions such as ICC506, 10667, 10619, 6663, 10817, ICCX 730008-8, ICCX 730041-8, ICCX 730094-18, ICCX 730020-11-I, ICC 10870, and ICC 5264 have also been identified as borer-resistant (Lateef 1985; Ahmed et al. 1990). Similarly, Lateef and Sachan (1990) found good level of resistance of *H. armigera* in ICCV 7, ICC 6663, ICC 10817, ICCL 86102, CCL 86103, ICC 4935-E2793, PDE-2, and PDE-5 across the agroecological zones of India. Varieties such as ICCV 7, ICCV 10, and ICCL 86103 with moderate levels of resistance to *H. armigera* have been released for cultivation in India (Sharma et al. 2014). Accessions belonging to *Cicer bijugum* (ICC 17206, IG 70002, IG 70003, IG 70006, IG 70012, IG70016, and IG 70016), to *C. judaicum* (IG 69980, IG 70032, and IG 70033), to *C. pinnatifidum* (IG 69948), and to *C. reticulatum* (IG 70020, IG

72940, IG 72948 and IG 72949, and IG 72964) showed resistance to *H. armigera* as well (Sharma et al. 2014). Sarwar et al. (2009) found germplasm C-727 resistant to *H. armigera.* Salimath et al. (2008) found the least pod damage in chickpea genotypes ICCL 87317, ICC 12479, ICC506, ICC 86102, ICCV 95992, ICCV 96752, ICCL 87315, ICCL 87314, ICCL 87316 and ICC 12494, and ICCV 2. Most of these lines showed the highest content of lignin and cellulose in pod husk. In addition, the tolerant genotypes, viz. ICCL 87315, ICC 506, and ICC 12479 had higher number of trichomes and exhibited less percent pod damage to *H. armigera*. Ruttoh et al. (2013) found genotypes EC58318, ICCV10, ICC14831, EC583260, EC583264, and EC583250 resistant to *H. armigera* in Kenya.

Germplasm accessions ILWC 39, ILC 3800, ILC 5901, and ILC 7738 were identified as resistant to leaf miner damage (Chen et al. 2011; Sharma et al. 2014). Sharma et al. (2014) found that oxalic acid content in chickpea leaves was correlated with leaf miner resistance in this crop. Genotypes CPI 29973, CPI 29975, CPI 29976, NCS 960003, K 902, CM 72, CMN 122, and BG 372 have been reported to be resistant to *C. maculatus*. Apart from the cultigens, wild relatives of several grain legumes have shown high levels of resistance to bruchids (Lateef 1985). Lines showing resistance to bruchids usually have small seeds with a rough seed coat. Accessions belonging to *C. bijugum, C. pinnatifidum,* and *C. echinospermum* have shown resistance to the bruchid, *Callosobruchus chinensis* (Sharma et al. 2014).

Against *Liriomyza cicerina,* some genotypes with different degrees of resistance were available. Chrigui et al. (2020) found superior RILs with resistance to the leaf miner indicating that introgression of resistance to leaf miner from chickpea could be possible using interspecific crosses. The inheritance pattern of resistance to the leaf miner in RILs was shown to be quantitative. Succinic acid exhibited the highest direct effects on resistance to the leaf miner suggesting that a high level of this compound could be used as a potential biochemical selection criterion for resistance to leaf miner in chickpea.

Divija et al. (2020) found 5 germplasm accessions resistant to *C. chinensis* as well. These germplasms were ICC397375, ICC372351, ICC3552, ICC3089, and PG186. The resistance in ICC37235 was established to be an antibiosis implying high content of some secondary metabolites such as phenol, flavonoid, and trypsin (protease) inhibitor. As for germplasms ICC 397375 and ICC 3552, their resistance was associated with some seed characters (e.g., seed roughness, medium seed size) implying a non-preference mechanism (Divija et al. 2020). Also, a laboratory investigation conducted by Sathish et al. (2020) determined resistance in PI 599066 to *C. chinensis*. This genotype exhibited a complete resistance to *C. chinense* due to its small seed size and hairiness.

7.2.4 Progress in Host Plant Resistance to Insect Pests of Pigeon Pea

Host plant resistance to insects is one of the components of pest management in pigeon pea, and considerable progress has been made in developing techniques to screen for resistance to major insect pests (Sharma 2016). Genotypes with resistance to *H. armigera, M. vitrata, M. obtusa*, and *C. chinensis* have been identified, but the levels of resistance are low to moderate in the cultivated germplasm (Sharma 2016). Rathinam et al. (2020) reported that the wild relative of pigeon pea *Cajanus platycarpus* was resistant to the devastating pigeon pea pod borer *H. armigera.* Another wild relative of pigeon pea, *Cajanus scarabaeoides* was found to be resistant to the pod borer by Njaci et al. (2020). The mechanism of resistance was associated with an enhanced induction of phytohormone and calcium/calmodulin, plant volatiles, and secondary metabolites all showing an elevated pest-induced gene expression (Njaci et al. 2020). Kumar et al. (2020) found the pigeon pea genotypes LRG-208 and CRG 2015-007 to be the most resistant to *H. armigera.* Ngugi-Dawit et al. (2020) found good resistance in IBS 3471 to *H. armigera*. They found that the mechanisms of resistance in this genotype were both antibiosis and antixenosis by inhibiting the growth and development of *H. armigera*. Vanambathina et al. (2021) found a high level of resistance to *H. armigera* in 5 Australian Cajanus wild genotypes including *C. acutifolius* (AGG316925WCAJ1), *C. latisepalus* (AGG309208WCAJ1), *C. lanceolatus* (AGG300129WCAJ1), *C. pubescens* (AGG309206WCAJ1), and *C. reticulatus* var. *reticulates* (AGG300159WCAJ1). The resistance mechanism was driven by the total phenolic content in the wild genotypes of pigeon pea. The progress in transferring insect resistance into the improved pigeon pea varieties has been limited, and there is a need to introgress resistance genes from the wild relatives into the culigen and/or develop pigeon pea cultivars to confer resistance to pod borers (Sharma 2016). According to Naresh et al. (1993), extra-early and determinate types of pigeon pea genotypes are more susceptible to pod borer damage. Genotypes P54(b), ICPL 5EB-EB, Phule T 1, Prabhat, T 21, Phule T 3 and 7411, DL-78-1, ICPL 155, TAT 9 and TAT 10, ICPL 1, H 79-6, UPAS 120, GP Nos. 17, 20, 24, 33, 30, 40, 43 and 45, Bahar, ICPL 94, ICPL 154 and ICPL 85059, ICPL 332, PPE 45-2 (ICP 1964), MA 2 and ICPL 84060, ICPL 6, PPE 45-2, ICP 1903, MA 1, ICPL 187-1, ICPL 288, T 21, ICP 909, ICPL 86040, MAZ, ICPL 2, TA 10, ICPL 1, Pant A1, ICP 7345-1-5, BDN 7, DA 2, ICP 4070, ICP 3615, BSMR 1, ICP 10531, ICPL 201, ICP 109BB, (AUT 82-1) ICPX 77303, ICPL 87089, Bahar, ICPL 87088, ICP 7946-E, and ICP 9889 have been reported to be relatively resistant to *H. armigera* (Chauhan and Dahiya 1987; Kushwaha and Malik 1988; Reed and Lateef 1990; Gupta et al. 1990; Naresh et al. 1993; Sharma 2016). Also, short-duration genotypes ICP 7, ICP 13011, ICPB 2089, ICPL 187-1, ENT 11, and ICPL 98008 showed moderate levels of resistance to pod borer damage (Sharma 2016).

7.2.5 Progress in Host Plant Resistance to Insect Pests of Mung Bean

Various mung bean insect-resistant varieties have been identified through a series of screening methods. Against *Bemisia tabaci*, several varieties were identified. These resistant sources included ML 803, ML 839, PDM 91-249, and PBM 5 (Yadav and

Dahiya 2000); NM 92 (Khattak et al. 2004); ML 1265 and ML 1229 (Kooner and Cheema 2007); TMB 36 and RMG 1004 (Singh and Singh 2014); ML 1774 and ML 1779 (Cheema et al. 2015). Other mung bean resistant lines to *B. tabaci* comprise ML 1, ML 6, ML 7, P 290, P 292, P 131, P 293, P 325, P 364, and 11,148 (Kooner et al. 1997). Against bean blossom or flower thrips (*Megalurothrips distalis* Karny), some resistant sources were found including cultivars Co 3, Co 4, and Co 5 (Lal 1987); genotypes SML 77, UPM 82-4 and Pusa 107, MGG 347, SML 832 (Malik 1990; Laosatit et al. 2020), genotypes PIMS 2, PIMS 3, CO 3, ML 5 and ML 337 (Chhabra 2001); NM-92 (Khattak et al. 2004) and MH 3153 (Nadeem et al. 2014).

Against stem fly, *Ophiomyia* sp., sources of resistance comprise cultivars, viz. V2396, V3495 and V4281, G05253, G05776, G02005, and G02472 (Abate 1990; Talekar 1990). Against, *Aphis craccivora* Koch, the resistant genotypes identified include JRUM 1, JRUM 11, JRUM 33, DPI 703, LAM 14-2, UPM 83-6 and UPM 83-10, Pusa 115, PDM 116 and ML 353 (Sahoo and Hota 1991). Against *Helicoverpa armigera* (Hub.) resistant cultivars include J1, LM 11, P526 and Co3 (Lal 1987), ML 337, ML 423 and ML 428 (Chhabra et al. 1988). Against *Maruca vitrata,* resistant sources include LGG 505, ML 267, LGG 502, LGG 407, LGG 460 and LGG 485 (Swarnalatha 2007); MGG 364, MGG 365 and MGG 363 (Choragudi et al. 2012). A wild mung bean accession TC1966 (*V. radiata* var. sublobata Roxb.) was identified as a potential source of resistance to *Callosobruchus maculatus* (Fab.) and *C. chinensis* (Fujii et al. 1989; Lambrides and Imrie 2000). Similarly, accessions (V2802 and V2709) were confirmed to possess complete resistance to *C. chinensis* and *C. maculatus* (Somta et al. 2007). Also, bruchid resistant accessions comprise V2709, V2802, TC1966, ACC41, VC1973A, Jangan Mung (Laosatit et al. 2020). The first resistance source identified in wild mung bean accession was TC1966 that exhibited complete resistance to both *C. chinensis* and *C. maculatus* (Fujii et al. 1989). Other wild mung bean accessions such as ACC23 and ACC41 were reported to be resistant to *C. chinensis* and *C. maculatus* (Lambrides and Imrie 2000). Cultivated mung bean accessions V1128, V2709, V2802, and V2817 were resistant to *C. chinensis* and *C. maculatus* (Somta et al. 2007). Two breeding lines V-02-802 × DGGV-7 and V-02-802 × DGGV-2 from F_4 generation were reported to be resistant to *C. maculatus* (Majhi et al. 2020). Hema et al. (2022) found three inter sub-specific lines, viz. GGISC 124, GGISC 150, and GGISC 140 resistant to *C. chinensis*. Variety Pant Moong-1 was found to be resistant to *B. tabaci* by Khaliq et al. (2017). Three lines of Green Gram (*Vigna radiata*), namely GGISC-2, GGISC-21, and GGISC-49 were found resistant to *Callosobruchus chinensis* (Harshitha et al. 2022) in India. The mechanism of resistance in these lines was antibiosis affecting some biological parameters of the beetle (e.g., delayed adult emergence, uneven adult emergence, prolonged developmental period, small and malformed adults).

The recent release of a reference genome sequence of mung bean and current advanced sequencing technology has enabled fast and efficient DNA marker development; and fine-mapping and identification of candidate gene(s) for the biotic resistance in mung bean have been made possible (Laosatit et al. 2020). The resistance to *C. chinensis* in the wild mung bean TC1966 is controlled by a single dominant gene locus, designated Br, possibly in combination with minor gene(s), and the resistance is dependent on genotypes of the seeds (Kitamura et al. 1988; Fujii et al. 1989). Similarly, Miyagi et al. (2004) found that the resistance to *C. chinensis* in wild mung bean ACC41 was controlled by a single dominant locus. Secondary metabolites, storage proteins, and enzyme inhibitors in legume seeds are major biochemical compounds causing resistance (antibiosis) to bruchid (Laosatit et al. 2020). The cyclopeptide alkaloids (vignaticacids A and B) were isolated from BC20F4 isogenic lines carrying the Br gene from TC1966 (Sugawara et al. 1996).

7.3 Conclusion

Pulse crops play an important role in human and livestock nutrition worldwide but their production remains challenged by quantitative and qualitative losses due to insect pests. In this chapter, major insect pests of pulse crops were described and sustainable management focusing on host plant resistance was highlighted. Especially, past and recent achievements in plant breeding for genetic improvement as well as screening efforts to identify sources of resistance among existing germplasms were reviewed. Although advanced efforts and more research synergy are still needed, existing achievements are likely to mitigate insect pests' effects while sustaining pulse production and value-chain when adequately explored. Genomic tools and resources could still be exploited in all the pulses to accelerate development of insect resistance.

References

Abate T (1990) Studies on genetic, cultural and insecticidal controls against bean fly, *Ophiomyia phaseoli* (Tryon) (Diptera: Agromyzidae) in Ethiopia (Dissertation). Simon Fraser University, p 220

Abudulai M, Salifu AB, Haruna M (2006) Screening of cowpea for resistance to the flower bud thrips, *Megalurothrips, sjostedti* Trybom (Thysanoptera: Thripidae). J Appl Sci 6(7):1621–1624

Ahmed K, LaP SS, Morris H, Khallque F, Malik B (1990) Insect pest problems and recent approaches to solving them on chickpea in South Asia. In: Chickpea in the nineties: proceedings of the second international workshop on chickpea improvement. ICRISAT Center, Hyderabad

Alabi OY, Odebiyi JA, Tamo M (2004) Effect of host plant resistance in some cowpea (*Vigna unguiculata* {L.} Walp.) cultivars on growth and developmental parameters of the flower bud thrips, *Megalurothrips sjostedti* (Trybom). Crop Prot 23(2):83–88

Alabi OY, Odebiyi JA, Tamo M, Omoloye AA (2011) The roles of plant secondary metabolites from cowpea floral structures in resistance to the flower bud thrips. J Agric Sci Technol 1:262–269

Boateng AB (2015) Genetic studies of aphids (*Aphis craccivora* Koch) resistance in cowpea. Kwame Nkrumah University of Science and Technology, p 67

Boukar O, Fatokun C (2009) Strategies in cowpea breeding. New approaches to plant breeding of orphan crops in Africa. Rome, FAO, pp 69–92

Boukar O, Fatokun CA, Huynh BL, Roberts PA, Close TJ (2016) Genomic tools in cowpea breeding programs: status and perspectives. Front Plant Sci 7:757

Boukar O, Belko N, Chamarthi S, Togola A, Batieno J, Owusu E, Fatokun C (2019) Cowpea (*Vigna unguiculata*): genetics, genomics and breeding. Plant Breed 138(4):415–424

Boukar O, Abberton M, Oyatomi O, Togola A, Tripathi L, Fatokun C (2020) Introgression breeding in cowpea [*Vigna unguiculata* (L.) Walp.]. Front Plant Sci 11:567425

Bressani R (1993) Grain quality of common legumes. Food Rev Int 9:217

Chauhan R, Dahiya B (1987) Damage due to pod borer and pod fly in early maturing genotypes of pigeonpea in Haryana. Indian J Plant Prot 15:5–9

Cheema HK, Singh RS, Taggar GK (2015) Screening of some advanced mung bean genotypes against sucking insect–pests. In: Abstracts of the brain storming meeting on promotion of pulses in Indo-Gangetic Plains of India. Punjab Agricultural University, Ludhiana, pp 63–64

Chen W, Sharma HC, Muehlbauer F (2011) Chickpea and lentil crop protection compendium. American Phytopathological Society, St Paul

Chhabra KS (2001) Advances in pest management in grain legume crops—*Vigna* spp. Indian J Entomol 47(1):103–105

Chhabra KS, Kooner BS, Sharma AK, Saxena AK (1988) Sources of resistance in mungbean (*Vigna radiata*) to insect pests and mungbean yellow mosaic virus. In: Shanmugasundaram S, McLean BT (eds) Proceedings of the second international mungbean symposium. The World Vegetable Center (AVRDC), Shanhua, pp 308–314

Chopra N, Pajni HR (1987) Resistance of different lentil varieties to the attack of *Bruchuslentis* [*Lens culinaris*]. Lens Newsl 14:23–26

Choragudi SR, Eswari KB, Sudarshanam A (2012) Reaction of greengram (*Vigna radiata* L.) OVT entries against major insect pests in rabi season. Andhra Agric J 59(1):87–92

Chrigui N, Sari D, Sari H, Eker T, Cengiz MF, Ikten C, Toker C (2020) Introgression of resistance to leaf miner (*Liriomyza cicerina* Rondani) from *Cicer reticulatum* Ladiz. to *C. arietinum* L. and relationships between potential biochemical selection criteria. Agronomy 11(1):57

Clement NS, El-Din SW, Lateef SS (1994) Research achievements in plant resistance to insect pests of cool season food legumes. Euphytica 73:41–50

Dabire-Binso CL, Ba NM, Sanon A, Drabo I, Bi KF (2010) Resistance mechanism to the pod-sucking bug *Clavigralla tomentosicollis* (Hemiptera: Coreidae) in the cowpea IT86D-716 variety. Int J Trop Insect Sci 30(4):192–199

Divija SD, Agnihotri M, Reddy MS (2020) Biophysical and biochemical basis of host plant resistance in chickpea germplasm against *Callosobruchus chinensis* (L.). J Entomol Zool Stud 8(5):769–774

Dormatey R, Atokple IDK, Ishiyaku MF (2015) Genetics of thrips resistance in cowpea. Int J Agric Sci Res 2:2348–3997

Duraimurugan P, Tyagi K (2014) Pest spectra, succession and its yield losses in mungbean and urdbean under changing climatic scenario. Legum Res 37(2):212–222

Erskine W, Tufail M, Russell A, Tyagi MC, Rahman MM, Saxena MC (1994) Current and future strategies in breeding lentil for resistance to biotic and abiotic stresses. Euphytica 73:127–135

Fujii K, Ishimoto M, Kitamura K (1989) Patterns of resistance to bean weevils (Bruchidae) in *Vigna radiata–mungo–sublobata* complex inform the breeding of new resistant variety. Appl Entomol Zool 24:126–132

Gentry J (2010) Mungbean management guide, 2nd edn. Queensland Government, Brisbane, p 24

Gupta SC, Jain KC, Saxena KB, Lateef SS (1990) ICPL 85059—a short-duration dwarf pigeonpea line. Int Pigeonpea Newsl 12:6–9

Harshitha GP, Jayamani P, Manimegalai S, Muthuswamy A (2022) Host plant resistance for Bruchids in pre-breeding lines of green gram [*Vigna radiata* (L.) Wilczek]. Legume Res Int J 1:5

Hassan S (2009) Effect of variety and intercropping on two major cowpea [*Vigna unguiculata* (L.) Walp] field pests in Mubi, Adamawa State, Nigeria. J Hortic For 1(2):014–016

Hema T, Jayamani P, Gnanamalar RP, Rajeswari E, Vishnupriya R (2022) Host plant resistance and analysis of chemical compounds responsible for Bruchid resistance in greengram *Vigna radiata* (L.) Wilczek. Legume Res Int J 1:6

ICRISAT (1990) ICRISAT annual report 1989 (Patancheru, India). ICRISAT, Hyderabad, pp 93–94

Jackai LEN, Daoust RA (1986) Insect pests of cowpeas. Annu Rev Entomol 31:95–119

Khaliq N, Koul V, Shankar U, Ganai SA, Sharma S, Norboo T (2017) Screening of mungbean (*Vigna radiata* (L.) Wilczek) varieties against whitefly (*Bemisia tabaci* Genn.) and mungbean yellow mosaic virus (MYMV). Int J Curr Microbiol App Sci 6:129–132

Khattak MK, Shafqat A, Chisti JI (2004) Varietal resistance of mungbean (*Vigna radiata* L.) against whitefly (*Bemisia tabaci* Genn.), jassid (*Amrasca devastans* Dist.), and thrips (*Thrips tabaci* Lind.). Pak Entomol 26:9–12

Kitamura K, Ishimoto M, Sawa M (1988) Inheritance of resistance to infestation with azuki bean weevil in *Vigna sublobata* and successful incorporation to *V. radiata*. Jpn J Breed 38:459–464

Kooner BS, Cheema HK (2007) Screening of mungbean germplasm against whitefly, *Bemisia tabaci* and MYMV. Acta Hortic 752:307–310

Kooner BS, Chhabra KS, Arora BS (1997) Resistant sources in mungbean to manage whitefly, jassids and yellow mosaic virus. In: In: proceedings of third agricultural science congress. Ludhiana, PAU

Kpoviessi AD, Datinon B, Agbahoungba S, Agoyi EE, Chougourou DC, Sodedji FKA, Assogbadjo AE (2019) Source of resistance among cowpea accessions to bruchid, *Callosobruchus maculatus* F. Coleoptera: Chrysomelidae, in Benin. Afr Crop Sci J 28(1):49–65

Kpoviessi AD, Adoukonou-Sagbadja H, Agbahoungba S, Agoyi EE, Assogbadjo AE, Chougourou DC (2021) Inheritance and combining ability estimates for cowpea resistance to bruchid (*Callosobruchus maculatus* Fab.) in Benin cowpea. Ecol Genet Genom 18:100082

Kumar TNV, Srivastava CP, Sabuj G, Saswati P (2020) Evaluation of host–plant resistance of selected long duration pigeonpea genotypes to the gram pod borer, *Helicoverpa armigera* (Hübner). J Exp Zool India 23(2):1627–1631

Kumari J, Ahmad R, Chandra S, Jha GK (2009) Determination of morphological attributes imparting resistance against aphids (*Aphis craccivora* Koch) in lentil (*Lens culinaris* Medik). Arch Phytopathol Plant Protect 42(1):52–57

Kushwaha KS, Malik BPS (1988) Reaction of some pigeonpea genotypes to gram pod borer under field conditions. Indian J Ecol 15:194–195

Lal SS (1985) A review of insect pests of mungbean and their control in India. Int J Pest Manag 31(2):105–114

Lal SS (1987) Insect pests of mungbean, urd, cowpea, and pea and their management. In: Rao VM, Sithanantham S (eds) Plant protection in field crops. Plant Protection Association of India, Hyderabad, pp 185–202

Lambrides CJ, Imrie BC (2000) Susceptibility of mungbean varieties to the bruchid species *Callosobruchus maculatus* (F.), *C. phaseoli* (Gyll.), *C. chinensis* (L.), and *Acanthoscelides obtectus* (Say.) (Coleoptera: Chrysomelidae). Aust J Agric Res 51:85–90

Laosatit K, Somta P, Chen X, Srinives P (2020) Genomic approaches to biotic stresses. In: The mungbean genome. Springer, Cham, pp 133–167

Laserna-Ruiz I, De-Los-Mozos-Pascual M, Santana-Méridas O, Sánchez-Vioque R, Rodríguez-Conde MF (2012) Screening and selection of lentil (*Lens miller*) germplasm resistant to seed bruchids (*Bruchus* spp.). Euphytica 188(2):153–162

Lateef SS (1985) Gram pod borer (*Heliothis armigera*) (Hub.) resistance in chickpeas. Agric Ecosyst Environ 14(1–2):95–102

Lateef SS, Sachan JN (1990) Host-plant resistance to *Helicoverpa armigera* in different agroecological contexts. In: Chickpea in the nineties: proceedings of the second international workshop on chickpea improvement. ICRISAT Center, Hyderabad

Majhi PK, Mogali SC, Bhoi TK (2020) Towards development of mungbean [*Vigna radiata* (L.) Wilczek] genotypes with combined resistance to bruchid (*Callosobruchus maculatus*) and Mungbean yellow mosaic virus (MYMV). Ann Agric Crop Sci 5(2):1064

Malik SPS (1990) Comparative resistance of summer mungbean genotype to the thrips, *Megalurothrips distalis* (Karny.). Punjab Agricultural University, Ludhiana

Malo M, Hore J (2020) Pulse production in India: major constraints and way forward. Res Trends Multidiscip Res Dev 5:35–63

Mbeyagala KE, Amayo R, Obuo JP, Pandey AK, War AR, Nair RM (2017) A manual for mungbean (greengram) production in Uganda, vol 32. National Agricultural Research Organization (NARO)

Miyagi M, Humphry M, Ma ZY, Lambrides CJ, Bateson M, Liu CJ (2004) Construction of bacterial artificial chromosome libraries and their application in developing PCR-based markers closely linked to a major locus conditioning bruchid resistance in mungbean (*Vigna radiata* L. Wilczek). Theor Appl Genet 110:15–56

Nadeem S, Hamed M, Asghar MJ, Abbas G, Saeed NA (2014) Screening of mungbean (*Vigna radiata* (L.) Wilczek) genotypes against sucking insect pests under natural field conditions. Pak J Zool 46(3):863–866

Naresh JS, Sharma SS, Dahiya BS (1993) Assessment of losses caused by *Heliothis armigera* and *Melanagromyza* in eight varieties of pigeonpea in Hisar. Indian J Plant Prot 11:37–39

Neupane S, Subedi S, Darai R (2020) Field screening of lentil genotypes against aphid infestation in inner Tarai of Nepal. J Nepal Agric Res Counc 6:79–84

Ngugi-Dawit A, Hoang TML, Williams B, Higgins TJ, Mundree SG (2020) A wild *Cajanus scarabaeoides* (L.), Thouars, IBS 3471, for improved insect-resistance in cultivated pigeonpea. Agronomy 10(4):517

Njaci I, Ngugi-Dawit A, Oduor RO, Kago L, Williams B, Hoang LTM, Ghimire SR (2020) Comparative analysis delineates the transcriptional resistance mechanisms for pod borer resistance in the pigeonpea wild relative *Cajanus scarabaeoides* (L.) thouars. Int J Mol Sci 22(1):309

Ongom PO, Fatokun C, Togola A, Salvo S, Oyebode OG, Ahmad MS, Boukar O (2021) Molecular fingerprinting and hybridity authentication in cowpea using single nucleotide polymorphism based competitive allele-specific PCR assay. Front Plant Sci 12:734117

Ongom PO, Togola A, Fatokun C, Boukar O (2022) A genome-wide scan divulges key loci involved in resistance to aphids (*Aphis craccivora*) in Cowpea (*Vigna unguiculata*). Genes 13(11):2002

Oyewale RO, Bamaiyi LJ (2013) Management of cowpea insect pests. Sch Acad J Biosci (SAJB) 1(5):217–226

Patterson CA, Maskus H, Dupasquier C (2009) Pulse crops for health. Cereal Foods World (CFW) 54(3):108

Popelka JC, Gollasch S, Moore A, Molvig L, Higgins TJV (2006) Genetic transformation of cowpea (*Vigna unguiculata* L.) and stable transmission of the transgenes to progeny. Plant Cell Rep 25:304–312

Pratap A, Douglas C, Prajapati U, Kumari G, War AR, Tomar R, Dubey S (2020) Breeding progress and future challenges: biotic stresses. In: The mungbean genome. Springer, Cham, pp 55–80

Rajeswari E, Jayamani P (2020) Host plant resistance in pigeonpea (*Cajanus cajan* (L.) Millsp.) genotypes for root rot disease caused by *Rhizoctonia bataticola* (Taub.) Butler. Electron J Plant Breed 11(04):1148–1152

Rana DS, Dass A, Rajanna GA, Kaur R (2016) Biotic and abiotic stress management in pulses. Indian J Agron 61:S238–S248

Rathinam M, Roschitzki B, Grossmann J, Mishra P, Kunz L, Wolski W, Sreevathsa R (2020) Unraveling the proteomic changes involved in the resistance response of *Cajanus platycarpus* to herbivory by *Helicoverpa armigera*. Appl Microbiol Biotechnol 104:7603–7618

Reed W, Lateef SS (1990) Pigeonpea: pest management. In: Nene YL, Hall SD, Shiela VK (eds) The pigeonpea. CAB International, Wallingford, pp 349–374

Ruttoh EK, Mulwa RM, Ngode L, Gohole L, Towett B, Njogu NS, Kimurto PK (2013) Screening for host plant resistance to *Helicoverpa armigera* (Lepidoptera: Noctuidae) in selected chickpea (*Cicer arientinum* L.) genotypes in Kenya. Egerton J Sci Technol 13:39–55

Sahoo BK, Hota AK (1991) Field screening of greengram germplasm against insect pest and disease complex. Madras Agric J 78(1–4):84–86

Salimath PM, Patil SA, Gowda CLL, Sharma HC (2008) Biophysical and biochemical basis of host plant resistance to pod borer (*Helicoverpa armigera* Hubner) in chickpea (*Cicer arietinum* L.). Indian J Genet Plant Breed 68(03):320–323

Sarwar M, Ahmad N, Toufiq M (2009) Host plant resistance relationships in chickpea (*Cicer arietinum* L.) against gram pod borer (*Helicoverpa armigera* Hubner). Pak J Bot 41(6): 3047–3052

Sathish K, Jaba J, Katlam BP, Mishra SP, Rana DK (2020) Evaluation of chickpea, *Cicer arietinum*, genotypes for resistance to the pulse beetle, *Callosobruchus chinensis* (L.). J Entomol Zool Stud 8(3):1002–1006

Sedivy J (1972) The feeding activity of leaf weevil on varieties of lentil. Arch Pflanzenschutz 8: 209–217

Shanower TG, Romeis JMEM, Minja EM (1999) Insect pests of pigeonpea and their management. Annu Rev Entomol 44(1):77–96

Sharma HC (2016) Host plant resistance to insect pests in pigeonpea: potential and limitations. Legume Perspect 11:24–29

Sharma HC, Crouch JH (2004) Molecular marker-assisted selection: a novel approach for host plant resistance to insects in grain legumes

Sharma HC, Gowda CLL, Stevenson PC, Ridsdill-Smith TJ, Clement SL, Rao GR, El-Bouhssini M (2007) Host plant resistance and insect pest management in chickpea. In: Chickpea breeding and management. CABI, Wallingford, pp 520–537

Sharma HC, Gaur PM, Samineni S, Gowda CL (2014) Exploiting host plant resistance for pest management in chickpea. Legume Perspect 3:25–28

Singh SR, Allen DJ (1979) Cowpea pests and diseases, vol 2. International Institute of Tropical Agriculture, Ibadan

Singh SK, Singh PS (2014) Screening of mungbean (*Vigna radiata*) genotypes against major insects. Curr Adv Agric Sci 6(1):85–87

Singh SR, Jackai LEN, Dos Santos JHR, Adalla CB (1990) Insect pests of cowpea. In: Singh SR (ed) Insect pests of food legumes. Wiley, Chichester, pp 43–89

Sobda G, Atemkeng FM, Boukar O, Fatokun C, Tongoona PB, Ayertey J, Offei SK (2018) Generation mean analysis in cowpea [*Vigna unguiculata* (L.) Walp.] under flower thrips infestation. J Agric Sci 10(4):86

Somta P, Ammaranan C, Ooi PAC, Srinives P (2007) Inheritance of seed resistance to bruchids in cultivated mungbean (*Vigna radiata* L. Wilczek). Euphytica 155:47–55

Souleymane A, Aken'Ova ME, Fatokun CA, Alabi OY (2013) Screening for resistance to cowpea aphid (*Aphis craccivora* Koch) in wild and cultivated cowpea (*Vigna unguiculata* L. Walp.) accessions. Int J Sci Environ Technol 2:611–621

Sugawara F, Ishimoto M, Le-Van N, Koshino H, Uzawa J, Yoshida S, Kitamura K (1996) Insecticidal peptide from mungbean: a resistant factor against infestation with azuki bean weevil. J Agric Food Chem 44:3360–3364

Swaminathan R, Singh K, Nepalia V (2012) Insect pests of green gram *Vigna radiata* (L.) Wilczek and their management. In: Aflakpui G (ed) Agricultural science. IntechOpen, pp 197–222

Swarnalatha P (2007) Germplasm screening and insecticidal management of pest complex in greengram (*Vigna radiata* (L.) Wilczek). Ranga Agricultural University, Hyderabad

Talekar NS (1990) Agromyzid flies of food legumes in the tropics. New Delhi, Wiley Eastern Limited, p 299

Tamò M, Afouda L, Bandyopadhyay R, Bottenberg H, Cortada-Gonzales L, Murithi H, Ortega-Beltran A, Pittendrigh B, Sikirou R, Togola A, Wydra KD (2019) Identifying and managing plant health risks for key African crops: legumes. In: Critical issues in plant health: 50 years of research in African agriculture. Burleigh Dodds Science Publishing, pp 259–294. https://doi.org/10.19103/as.2018.0043.11

Togola A, Boukar O, Belko N, Chamarthi SK, Fatokun C, Tamo M, Oigiangbe N (2017) Host plant resistance to insect pests of cowpea (*Vigna unguiculata* L. Walp.): achievements and future prospects. Euphytica 213:1–16

Togola A, Boukar O, Chamarthi S, Belko N, Tamò M, Oigiangbe N, Fatokun C (2019) Evaluation of cowpea mini core accessions for resistance to flower bud thrips *Megalurothrips sjostedti* Trybom (Thysanoptera: Thripidae). J Appl Entomol 143(6):683–692

Togola A, Boukar O, Servent A, Chamarthi S, Tamo M, Fatokun C (2020) Identification of sources of resistance in cowpea mini core accessions to *Aphis craccivora* Koch (Homoptera: Aphididae) and their biochemical characterization. Euphytica 216(6):1–15

Vanambathina P, Rachaputi RC, Sultanbawa Y, Phan ADT, Henry RJ, Brier H (2021) Biochemical basis of resistance to pod borer (*Helicoverpa armigera*) in Australian wild relatives of pigeonpea. Legume Sci 3(4):e101

Yadav GS, Dahiya B (2000) Screening of some mungbean genotypes against major insect–pests and yellow mosaic virus. Ann Agri Bio Res 5:71–73

Yadav DK, Singh SK (2006) Forecast model of major insect pests of mungbean. Ann Plant Prot Sci 14(2):323–328

Plant Resistance to Insects in Oilseed Crops

8

Sheikh Salma Irshad, Neha Panwar, Lukesh Bansal, Sathya Thirumurugan, and Sarwan Kumar

Abstract

Oilseed crops are energy rich crops that are grown under energy deprived conditions. The demand for premium seed oils is increasing as the world's population grows. The attack of pests and diseases reduces the production potential of these crops. The cost of controlling insect pests by synthetic chemical insecticides in agriculture annually is billions of dollars worldwide. The overuse of insecticides has many associated ill effects. To maintain the ecological balance of the environment, it is imperative to develop alternative pest management systems that aim for minimal pesticide use and the conservation of natural enemies. Resistance in host plants is one such alternative strategy. Since one of the main bottlenecks in realization of full yield potential of oilseed crops is attack of insect pests, a traditional and effective alternative crop protection method serves as the fundamental tool of integrated pest management. Future development of insect-resistant oilseed crop cultivars will depend on the adoption of technologies and breeding techniques such as genomic selection, high-throughput phenomics, gene editing, and landscape genomics as well as the continuous use of sources of resistance from crop germplasm.

Keywords

Breeding · Defense · Insect–plant interactions · Phenotyping · Screening

S. S. Irshad · L. Bansal · S. Thirumurugan
Department of Entomology, Punjab Agricultural University, Ludhiana, India

N. Panwar
The Graduate School, Colorado State University, Fort Collins, CO, USA

S. Kumar (✉)
Department of Plant Breeding and Genetics, Punjab Agricultural University, Ludhiana, India
e-mail: sarwanent@pau.edu

S. Kumar, M. Furlong (eds.), *Plant Resistance to Insects in Major Field Crops*,
https://doi.org/10.1007/978-981-99-7520-4_8

8.1 Introduction

Oilseed crops are widely grown across the world due to their versatility and high economic value. Due to a rise in the demand for vegetable oil, animal feed, medicines, biofuels, and other oleochemical industrial uses, oilseeds have recently gained increased attention. Over the past 30 years, the increased interest has led to an 82% rise in the area under oilseed cultivation and a roughly 240% increase in global production (El-Hamidi and Zaher 2018). Some of the major oilseed crops grown in the world include seven edible oilseeds (soybean, rapeseed-mustard, groundnut/ peanut, sunflower, sesame, safflower, and niger) and two non-edible oilseeds (castor and linseed) due to their diverse agroecological conditions. More than 180 million hectares of crop land are used for oilseeds cultivation in the world (Oerke et al. 2006). These crops are important source of income for farmers and contribute significantly to the global food and energy supply. There are several limitations in successful cultivation of oilseed crops that can impact output, including climate, soil quality, water availability, pests, diseases, etc. Pests are a significant barrier to the global expansion of area under oilseed crops. Insect pests pose a serious threat to food security and sustainable development, demanding the creation of efficient plant protection technologies to prevent and reduce pest-related crop losses (Oerke et al. 2006). Chemical pesticides are farmers' first line of defense against insect pests, but their widespread use has led to several problems, such as the killing of beneficial insects, environmental pollution, problems with human and animal health, and the development of pest resistance to pesticides (Pedigo and Rice 2006; Stevens et al. 2012; Nderitu et al. 2020). One strategy for addressing this problem is Integrated Pest Management (IPM). Modern and more sustainable agriculture methods must be adopted to address these problems. Further, these negative off-target impacts have inspired scientists to develop brand-new, environmentally benign approaches to manage insect pests. Host plant resistance has the potential to serve as the foundation for pest management in many agricultural systems (Sharma 2007; Kumari et al. 2022). The term host plant resistance (HPR) has many different definitions. In general, "the proportional amount of inheritable plant traits to counteract the activities of insects to cause minimum percent drop in yield" can be used to define plant resistance. The development of insect-resistant cultivars based on a thorough understanding of the many host plant resistance mechanisms is one of the key milestones in the HPR initiative. Antixenosis, antibiosis, and tolerance are three types of resistance mechanisms in plants that confer resistance to insect pests. Plant traits that prevent or restrict insect pests from coming, landing, settling, feeding, or ovipositing are known as antixenosis, also known as non-preference. This type of mechanism is based on both physical (visual cues, morphological traits) and biochemical (the presence of chemical molecules that function as anti-deterrent, antifeedants, etc.) properties of plants. Antibiosis is brought on by traits that impair an insect's biology, growth, and reproduction. The basis for tolerance in plants is their ability to continue to develop normally and sustain minor yield losses in the face of insect assault or other damage (Palaniswamy 1996). The intensity of both current and novel pests and illnesses that affect oilseed crops will grow as oilseed production intensifies and scales up. Although great emphasis is paid to locating sources and analysing the intricate mechanisms of insect resistance, it is important to keep in mind that the end objective is reducing pest damage in future crop types. Resistance must be considered, carefully assessed, and chosen among other criteria of the variety-to-be during the developmental process (selection and breeding). Sadly, not enough is frequently done during breeding to develop traits in the new variety that make it resistant to insects. Our objective is to give readers a better grasp of host plant resistance for the creation of genuinely ecological IPM techniques in oilseed crops. An overview of research and development on several oilseed crops with regard to HPR is provided in this chapter.

8.1.1 Soybean

The leading agricultural crop in the world, soybean is used for a variety of things, including food for humans and other animals, edible oil, biofuel, industrial products, and cosmetics. Many pests, such as fungi, bacteria, and insects, affect the production of the crop in soybean growing areas (Hartman et al. 2011). The methods and input prices for soybean pest management have changed dramatically over time (Ragsdale et al. 2011; Hodgson et al. 2012). The peculiar grain legume known as soybean, *Glycine max*, is well known for serving as both a pulse and an oilseed, having 38–44% protein and 18–22% oil. Moreover, soybean is listed as an important ingredient in a wide range of industrial goods, including insulating foams, adhesives, inks, lubricants, and solvents. In a country like India, where a sizable portion of the population is vegetarian, soybean plays a significant role as a rich source of protein. India is one of the top exporters of soy meal to other Asian nations and the third-largest importer of soy oil in the world.

8.1.1.1 Damage by Insect Pests

About a dozen of the 275 species of insect pests that infest soybean in India has acquired the status of important pests, including the girdle beetle, tobacco caterpillar, green semilooper, Bihar hairy caterpillar, stem fly, aphids, jassids, and whitefly. A total of 380 different kinds of insects have been found on soybean crop across the globe (Patel and Rahul 2020). The soybean looper, velvet bean caterpillar, beet armyworm, bean leaf beetle, stem borer, Mexican bean beetle, and soybean leaf miner are examples of lepidopteran and coleopteran pests that feed on soybean in the world (Higley 1995). The threat posed by the soybean aphid and stink bugs has the potential to increase quickly as their geographic ranges expand. For example, the soybean aphid, which was first detected in Wisconsin, expanded to 30 US states and three Canadian provinces in less than 10 years (Ragsdale et al. 2011) while the brown-marmorated stink bug (BMSB) has already been found in 38 US states since it was first found in Pennsylvania in 1996 (Leskey et al. 2012).

8.1.1.2 Role of HPR

To lessen the harm caused by insect pests, an Integrated Pest Management (IPM) programme should take advantage of host plant resistance in soybean cultivars. We attempt to assess recent advancements in the analysis of soybean resistance to different pests in this chapter. Several crop plants have been developed that are resistant to hemipteran insects (Smith 2005; Smith and Boyko 2007). From a historical perspective, HPR has been quite effective in controlling hemipteran insect pests of soybean. Due to the presence of pubescence on plant leaves, soybeans are now resistant to the potato leaf hopper (*Empoasca fabae*), a pest that was once a severe problem for the crop (Hollowell and Johnson 1934). The potato leaf hopper was relegated to a non-pest category after the pubescence trait was subsequently included into commercial varieties of soybean (Boethel 1999). Many crops, including cereals, vegetables, fruits, and forages, have been found to have aphid-specific resistance genes (Smith 2005). The first aphid resistance genes were discovered at the beginning of the twenty-first century, and resistant cultivars have been sold for 10 years, but they have been terribly underutilized. Screening for resistance genes in soybean lines commenced very soon after the first report of soybean aphid in North America. Hill et al. (2004) investigated 1530 cultivars between 2001 and 2002 and identified *Rag1*, a single dominant gene that confers antibiosis, as the first of numerous *Rag* genes (resistance to *Aphis glycines*). The ability of insects to develop virulence that allows them to overcome the resistance is the biggest limiting factor to the long-term utility of resistant cultivars, as it is with any sort of host plant resistance (Kelley et al. 2021). Since then, screening by several research programs (mostly in the public sector) has uncovered at least 11 naturally occurring *Rag* genes as well as four quantitative trait loci, all in various stages of evaluation and development (Natukunda and MacIntosh 2020). Researchers discovered aphid biotypes with varying degrees of virulence against *Rag* genes in test plots before the restricted commercial distribution of *Rag*-containing cultivars. Four biotypes referred to as biotypes 1–4 have been identified to date in the US, and their survivability varies based on the biotype and resistance gene. We are aware of no records of virulent biotypes in Asia, the aphid's natural habitat. Despite being widespread in the soybean aphid's range in the US (e.g. in Ohio, Illinois, Wisconsin, and South Dakota), virulent biotypes have not yet become prevalent or numerous enough to pose a threat to the usefulness of organic *Rag* soybean varieties. This is possibly because these varieties make up a relatively small portion of the soybean acreage (Kelley et al. 2021). Using insecticide resistance management (IRM), it may be possible to control how aphid biotypes evolve, encouraging more active commercial development of these varieties. We need data, models, and tactics to accurately predict changes in the incidence of virulent aphids and respond to them for an IRM strategy to provide sustainable use of *Rag* soybeans. In this quest, the use of genetic markers to categorize aphid lineages and determine virulence would be extremely helpful, especially when confirmed using aphids collected in the field. To increase the durability and shorten the development time of hemiptera resistant soybean, a combined approach will probably be required. In a similar manner, *Anticarsia gemmatalis* Hübner causes serious economic harm to soybean. Use of

Bt transgenic soybeans and chemical control are the main management tactics for this bug. In conjunction with other tactics as part of an integrated pest management (IPM) approach, alternative strategies like host plant resistance are seen as an effective and less-aggressive method. Using oviposition, attractiveness, and food consumption tests, Ongaratto et al. (2021) assessed 30 soybean genotypes for antixenosis expression and chose 13 promising genotypes to confirm the potential for antibiosis. Their findings imply that the genotypes "TMG 133" RR, "TMG 1179" RR, "IAC 19", "IAC 17", "IAC 100", D75–10169, and IAC 78–2318 have antixenosis. The genotypes IAC 74–2832, "IAC 19", "IAC 17", "IAC 100", and "PI 274454" were found to have antixenosis and antibiosis, which were characterized by an effect on behaviour and a detrimental effect on larval survival. It was discovered that the "TMG 7062" IPRO provided antibiosis resistance by adversely affecting larval survival and development. Antixenosis was advised for "IAC 24" due to the larvae's decreased feeding consumption. While developing soybeans with *A. gemmatalis* resistance, breeding programmes should take these genotypes into account (Ongaratto et al. 2021).

While much of the current research has been on conventional and traditional host plant breeding and screening techniques, novel genomic and molecular technology-based research is beginning to emerge. The vast array of signalling molecules produced by the plant's immune system, as well as morphological barriers, all contribute to HPR. These plant-defensive responses are linked to either constitutive, therefore always present, or induced plant features that are shaped by an attack or elicitation in the morphology, physiology, and/or chemistry of the plant, which produces plant secondary metabolites (Romero et al. 2020; Bravo et al. 2019). Both strategies result in direct or indirect consequences through the attraction of natural enemies and are not mutually incompatible. Breeding initiatives must place a strong emphasis on tolerant cultivars. As was already established, tolerant cultivars can tolerate or recover from damage brought on by insect populations comparable to those that affect susceptible cultivars. The adoption of host plant resistance is encouraged more by the growth of insect pest populations that are resistant to pesticides in soybean. A more extensive commercialization of this effective pest control technique may be encouraged by recent research on the genetics and markers of insects, plant gene expression, and in-plant refuges. These findings point to important directions for insect resistance management (IRM).

8.1.2 Rapeseed-Mustard

The second-most significant and well-known group of winter oilseed crops in India is rapeseed-mustard after soybean. These crops are primarily grown in India's northern plains, while there are some cultivated areas in the country's eastern region as well. *Brassica juncea* (Indian mustard) is the predominant oilseed Brassica crop in India, while *B. rapa* (*Toria*, Yellow *Sarson*, Brown *Sarson*), *B. napus* (*Gobhi Sarson* popularly known as Canola), and *B. nigra* (Black Mustard) are some of the other crops included in the rapeseed-mustard group. The defatted cake is fed to animals.

Lack of high-quality seed, diseases, and insect pests have significant impact on mustard yield if not controlled during the crucial crop growth period. Dealing with pest infestations is one of the main challenges in oilseed Brassica cultivation. Insecticides are frequently used to control insect pests; however, the socioeconomic and, environmental issues and the growing incidence of insecticide-resistant populations argue against their exclusive use. At present, oilseed Brassica pests are primarily controlled by synthetic chemical pesticides. In addition to concerns about how insecticides impact people and other non-target species, the formation and spread of pesticide resistance in populations of the oilseed Brassica pests is another barrier to their usage. Plant resistance is a historic and reliable alternative crop protection technique that is the cornerstone of integrated pest management. The diamondback moth (*Plutella xylostella*) has developed resistance to all major classes of insecticides including Bt, while the insecticide resistance in cabbage stem flea beetle (*P. chrysocephala*) is quickly expanding in the population (Furlong et al. 2013). *Lipaphis erysimi,* mustard aphid, is one of the primary and most problematic pests of rapeseed-mustard in Indian subcontinent (Bakhetia et al. 2002; Blackman and Eastop 2000). Continuous feeding by large colonies of nymphs and adults on aboveground plant parts results in drastic yield reduction. The productivity losses of up to 83% in India's rapeseed-mustard crops are attributable to this pest alone (Mandal et al. 2002). Oilseed Brassicas have evolved a number of biophysical and biochemical defense systems to ward-off these pests (Mitchell et al. 2016). Plant tissues contain constitutive defense mechanisms that serve as preventative protection against future damaging circumstances (Mertens et al. 2021). In response to herbivore attack, defense mechanisms may also be dynamic and created, depending on the age, developmental stage, and genetic makeup of the plant (Howe and Jander 2008; Brunissen et al. 2009; Chandrasekhar et al. 2018; Batyrshina et al. 2020). Variation in these systems at the spatial level (individual sections inside a plant) may be one of the main variables determining pest proliferation and feeding behaviour (Awmack and Leather 2002; Karley et al. 2002; Jakobs and Müller 2018). How closely related the defense systems are, how they differ in, how well they prepare for attacks, and how well they defend against aphids have not been fully explored. Researchers have examined the associations between several physiological and biochemical traits and the resistance of the mustard variety to *Lipaphis erysimi*. Many physical and chemical barriers have a significant impact on mustard aphid resistance (Sandeep et al. 2017; Hervé 2018). In contrast to chemical defenses, which frequently include the production of low molecular weight natural chemicals known as secondary metabolites that may be toxic to the creatures attacking plants, biophysical defenses include structures like prickles, thorns, and cuticular waxes (Vanetten et al. 1994; Maffei et al. 2007a). The variation found in the aphid population counts on various varieties of different species is attributed to the specific host–insect interactions. Insects prefer specific genotypes/species for feeding, which is influenced by several factors including environmental and chemical interactions between host plant and insect–pests. Hopkins and Huner (2004) reported that plants challenged by insects respond through changes in the composition and physical properties of the cell wall as well as the biosynthesis of secondary metabolites. Defensive chemicals may be constitutively present in the plant before an insect attack (phytoanticipins), or they may be induced after attack (phytoalexins). Trichomes are present on a variety of oilseed plant species and stop aphid migration and stylet insertion. Trichomes may or may not have glands. Non-glandular trichomes, specialized epidermal hair-like structures, may affect aphid movement and reproduction rates (Riddick and Simmons 2014). Although the specific mechanism of how glandular trichomes function is unknown, plants use them to defend themselves against insect attacks. Genetic differences among varieties resulted in significant diversity in surface wax concentration, which can be phenotyped for resistance to insects in relation to wax content. Waxes differ widely among the plant species and even in different parts of the same plant, attesting to the genetic diversity and developmental influences (Rutledge and Eigenbrode 2003). There is a lot of evidence that trichomes decrease feeding and motility, but it is unclear how they affect other aspects. Additionally, the glandular trichomes exude some toxic exudates that capture and ultimately kill small insects including aphids. Research on plant mutants with varied surface properties provided the first evidence, in terms of structural qualities, of the significance of surface layers for host acceptance. Wax on the leaf epicuticular surface protects plants from pathogen infection, insect pest attack, and desiccation (Trdan et al. 2004; Frati et al. 2013; Ni et al. 2014). According to Paré and Tumlinson (1999), chemicals on leaf surfaces can cause diverse responses in insects. Moreover, the wax can physically obstruct an insect's movement across the surface of a leaf. According to Bjorkman et al. (2011), there is a negative relationship between the number of insects and the amount of epicuticular wax in the leaves (Bjorkman et al. 2011). Several hydrocarbons, long-chain fatty acids, primary and secondary alcohols, fatty aldehydes, ketone esters, and primary and secondary alcohols make up the chemical composition of epicuticular waxes (Jenks et al. 1994; Alcerito et al. 2002). There are fewer plant species that exhibit glossy phenotypes (types with less epicuticular wax crystals) (Kanno and Harris 2000; Steinbauer et al. 2004). Yet, for eating and oviposition, many insect species favour glossy leaves or leaves with the waxes removed over the leaves with normal or high epicuticular wax loads (Stork 1980; Bodnaryk 1992; Brennan et al. 2001; Cervantes et al. 2002).

8.1.2.1 Plant–Herbivore Interactions in Oilseed Brassica

It has been estimated that at least one species consumes at least one kind of plant due to the prevalence of herbivory (Schoonhoven et al. 2005). The plant develops several defensive strategies in response to herbivore attack. There are various pest control uses for this plant–herbivore interaction. Transcriptomics has shown the mechanisms by which insects respond to plant protection and the survival strategies they use (Vogel et al. 2014; Thorpe et al. 2016; Schweizer et al. 2017). In-depth understanding of these underlying mechanisms is required to develop innovative pest management techniques that either strengthen plant defenses or thwart pest attempts to get around those defenses. In response to infestation or infection, different phytohormone-dependent processes are activated. Salicylate (SA)-dependent responses are induced by several necrotrophic diseases (Thomma et al. 2001), and sucking and chewing insects (Maffei et al. 2007b), whereas ethylene

(ET) and jasmonate (JA) pathways are induced by biotrophic pathogens and chewing insects (Thomma et al. 2001). These interactions result in the production of several defense-related proteins and secondary metabolites with antixenotic or antibiotic properties. It indicated that aphid infestation resulted in an SA-dependent response. JA-dependent genes, however, were inhibited (Zhu-Salzman et al. 2004; Thompson and Goggin 2006; Gao et al. 2007; Walling 2008). All these responses result in the modification of plant metabolism to ensure beneficial interactions between insects and plants. Source germplasm for insect resistance genes (especially aphids) in Brassica spp. is still mostly unavailable despite multiple prior attempts (Sekhon and Åhman 1993; Bhadoria et al. 1995). During the laboratory-based screening, the wild crucifer *Brassica fruticulosa* reportedly displayed effective antibiosis against *L. erysimi* and gene(s) for aphid resistance have been successfully introgressed in *B. juncea* background (Atri et al. 2012). Several attempts to produce aphid-resistant transgenic mustard have led to varied success but the field testing of these transgenics is still awaited (Kanrar et al. 2002; Hossain et al. 2006). When looking for increased transcripts in response to taxonomically diverse plant–aphid interactions in the absence of specific R genes, many orthologous transcripts were discovered. These transcripts result in the production of proteins that are essential for cell maintenance, general plant defense and signalling, reactive oxygen species (ROS) production, hypersensitive response, and photosynthesis (Boyko et al. 2006; Koramutla et al. 2014). Recently, Singh and Dhillon (2022) reported some of the *B. juncea* lines to show resistance against *L. erysimi* but all the lines fell under susceptible to highly susceptible category given the aphid resistance index, thus making their claims self-contradictory.

Herbivorous insects use various visual, tactile, and chemical cues in host plant selection. It is important to choose the right host plant for oviposition because this will affect the progeny's survival and growth. A significant factor affecting how insects behave during oviposition is the presence of surface chemicals, plant volatiles, and trichomes (Hilker et al. 2002; Chamarthi et al. 2011). The most crucial defense mechanism for plants against insect herbivory is antixenosis for oviposition. The female moths use a variety of physical and chemical cues to choose an appropriate host plant for oviposition. Plants respond to insect oviposition by using direct and indirect defenses that try to kill or remove the insect eggs to prevent the larvae, that would hatch from them, from causing damage to the plant (Hilker and Meiners 2010). In reaction to oviposition, plants generate neoplasm at the egg deposition site, which lifts and lowers the eggs, as well as ovicidal substances, which kill the eggs (Doss et al. 2000; Seino et al. 1996; Yamasaki et al. 2003). Moreover, oviposition causes necrotic tissue to form at the oviposition site due to the plant tissues' hypersensitive reaction, which separates the eggs (Petzold-Maxwell et al. 2011). Salicylic acid (SA) and jasmonic acid (JA) are crucial phytohormones that help plants defend themselves against insect herbivory (Stotz et al. 2002; Traw and Bergelson 2003; Bruinsma et al. 2007; Zhao et al. 2009). They cause plants to produce harmful antinutritive chemicals and poisonous secondary metabolites, which stunt larval growth and development and prevent adult moths from ovipositing (Bruinsma et al. 2007). Many intermediary substances are released by the octadecanoid and phenylpropanoid routes, respectively, which are mediated by JA and SA. Some of these substances have an antibiotic effect on pest insects, while others have an antixenotic effect on oviposition (Dicke and van Poecke 2002; Bruinsma et al. 2007). A negative crosstalk is known to exist between the JA and SA pathways (Traw and Bergelson 2003). Significantly less number of *Pieris rapae* and *P. brassicae* eggs were laid on plants treated with JA compared to untreated plants (Bruinsma et al. 2007). In addition, eggs had more JA content than plant tissues or larval diets do (Hilker and Meiners 2010).

8.1.2.2 Perspectives for Integrated Insect Pest Protection in Oilseed Brassica Breeding

In the past, plant breeding efforts to breed for insect-resistant cultivar generally lacked the focus to develop resistant cultivar for incorporation in an IPM module. Plants breeders always looked for high level of resistance in plants and often ignored moderate resistance. Due to the lack of effective strategies for crop rotation, tillage, biocontrol, and cultivar tolerance, the management of insect pests in oilseed rape has relied on pesticides in recent years (Zheng et al. 2020). The frequent use of a limited number of insecticides, many with common modes of action, resulted in insect populations with resistance against various classes of insecticides (e.g. pyrethroids) within the last decade worldwide including populations of the diamondback moth (*Plutella xylostella*), pollen beetle (*Brassicogethess aeneus*), cabbage stem flea beetle (*Psylliodes chrysocephala*) and cabbage seed weevil (*Ceutorhynchus obstrictus,* syn. *assimilis*) (Hervé 2018). The 2013 European Union ban on neonicotinoid seed treatments led to increased insect damage and decreased oilseed rape yields. Moreover, fewer conventional crop protection products are getting new product approvals (McDougal 2018). Insect pests are projected to move into previously unaffected oilseed rape production areas (range expansion) because of changes in insect populations brought on by climate change (Bale et al. 2002). The cultivation of oilseed rape is now threatened by these trends. A major shift towards creative and integrated management approaches is urgently needed since insect pest management strategies that just rely on insecticides are not viable. A significant problem is maintaining environmentally sustainable rapeseed production amid growing dangers from insect predation. There are about 40 kinds of insects that prey on rapeseed globally, and only a few effective control measures have been developed. Rapeseed also exhibits relatively little resistance to insect predation. It is difficult to phenotype for host plant resistance because resistances are often quantitative in nature. By focusing on a variety of potential insect control techniques, new, comprehensive protection measures for rapeseed will emerge throughout the short, medium, and long term, maintaining the output of this important oil crop species (Obermeier et al. 2022) (Fig. 8.1).

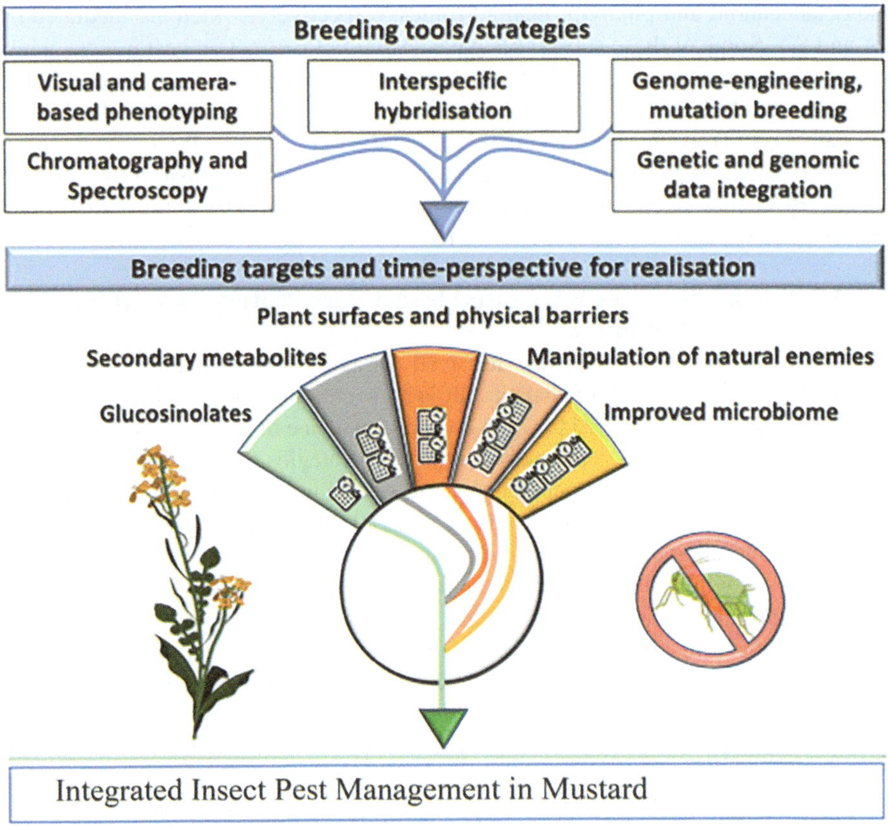

Fig. 8.1 Schematic representation of major breeding tools/strategies and breeding targets with time perspective for future realization and availability of insect-resistant breeding lines in integrated insect pest protection of mustard (Adapted from Obermeier et al. 2022)

8.1.3 Groundnut

A significant source of digestible proteins, cooking oil, and vitamins, groundnuts are a major oilseed crop in the world. Insect pests are the major biotic stressors in limiting peanut production. About 1.6 million tones of produce equal to US $ 30.9 million are lost each year in India due to groundnut production losses caused by insect infestations (Savage and Keenan 1994; Dhaliwal et al. 2010). Groundnut production during the *kharif* season is predicted to total 83.69 lakh tonnes in 2022–2023 as opposed to 83.75 lakh tonnes in 2021–2022 according to the first advance projections from the Government of India (Anonymous 2023). There are various biotic and abiotic stresses affecting the production of groundnut. Yield losses from pest attack are difficult to estimate because of the localized and sporadic nature of the outbreaks (Reddy and McDonald 1984). An estimated 15% of the groundnut crop's annual yield is lost to insect pest damage. According to Baskaran and Rajavel (2013), the tobacco caterpillar *Spodoptera litura* (Fabricius) is a significant groundnut pest that can lead to up to 47% production loss (Baskaran and Rajavel 2013). Larvae feed aggressively on vegetation at first, then completely defoliate the plants Furthermore, massive and careless use of pesticides has a negative impact on non-target organisms (predators and parasitoids), leads to pesticide residues in food, encourages the resurgence of pests, and leads to the development of insecticide resistance in insects in addition to adverse effects on human health and the environment (Sharma 2007). Given this, host plant resistance has the potential to offer sustainable solution to pest problems. Enhancing host plant defenses against insects will decrease losses from herbivores, reduce the need for insecticides, improve agricultural yields, and create a safer environment (Howe and Jander 2008). Host plant resistance is an important pest management tactic for *S. litura*, that significantly reduces groundnut harvests globally. Saleem et al. (2019) examined various biochemical (reducing, non-reducing and total sugars, proteins, and phenols content) and biophysical (trichome density, relative water content, specific leaf weight, epicuticular wax) parameters in the eight groundnut genotypes resistant to *S. litura* along with four controls and the wild species *Arachis monticola* L. There was a significant negative correlation of leaf damage with leaf weight, midrib, leaf lamina trichome density, wax content, proteins, and phenols content. ICG 928, one of the *S. litura* resistant genotypes, displayed higher leaf lamina (140 mm^{-2}) and midrib (180 mm^{-2}) trichome density, specific leaf weight (2.5 g dm^{-2}), wax content (0.36 g cm^{-2}), proteins (4.91 mg g^{-1}), and phenols content (3.77 mg g^{-1}) than the susceptible check JL 24, which displayed higher relative water content (73.3%), reducing sugars (11.32 mg g^{-1}). These biophysical and biochemical elements can be successfully used in groundnut breeding to develop cultivars resistant to *S. litura*.

During the time of seedling emergence and a few weeks thereafter, thrips feeding results in high damage. Early in the season, thrips feeding damage from high thrips pressure can cause yield loss and/or delayed maturity (Todd et al. 1995a; de Moraes et al. 2005; Funderburk et al. 2007). Thrips use a method of sex determination known as haplodiploidy, in which fertilized eggs develop into diploid females and unfertilized eggs into haploid males (Moritz 2002). They can quickly multiply due to their mode of reproduction and temperature-dependent life cycle. Breeding cultivars resistant to orthotospoviruses has received a lot of attention because it is difficult to control thrips, and the viral infection results in severe economic loss. Research with tomato spotted wilt virus (TSWV) in peanut undertaken in the United States and groundnut bud necrosis virus (GBNV) studies conducted in Asia provide a large portion of the knowledge on breeding for virus resistance. The Florunner and Southern runner cultivars were particularly vulnerable to the spotted wilt disease caused by TSWV in the Southern United States in the 1980s and early 1990s (Culbreath et al. 1992; Todd et al. 1995b). Most of the resistance was generated from a single genotype (PI 203363) that was brought from Brazil in 1953 because of screening and breeding activities (Culbreath et al. 2003). Since that the primary runner peanut cultivars contain a sizable percentage of PI203363 alleles, the introduction of this distinctive genotype had an immediate and considerable impact on American peanut breeding (Clevenger et al. 2017).

The current generation of 'third generation TSWV resistant' peanut cultivars has good level of field resistance to the disease, and losses caused by it have been reduced. Breeding initiatives have also identified some GBNV resistance in Asia (Amin et al. 1985; Dwivedi et al. 1995; Reddy et al. 2000; Kesmala et al. 2004; Mandal et al. 2012). Many of these lines have partial viral resistance and frequently involve the incorporation of other management techniques. Insecticides are still used, for instance, to lessen thrips feeding damage in early-season peanut (Mandal et al. 2012; Marasigan et al. 2016). Finding and implementing efficient thrips resistance in high yielding peanut cultivars can assist farmers financially and lessen the environmental impact of pesticide use.

Brar et al. (1994) attempted genetic transformation of peanut, but no resistance tests were carried out. Cry genes from *Bacillus thuringiensis* (Berliner), which confer resistance to a variety of insects, especially those of the order Lepidoptera, have been introduced into peanut (Krishna et al. 2015). However as per the available information, no transgenic peanut has been examined or found to be resistant to thrips. Transgenesis has been successful in transferring thrips resistance to other crops; for example, potato plants have been modified to resist the western flower thrips (Outchkourov et al. 2004). Stefin, potato cystatin, equistatin, and cystatin were combined, synthesized into a functional unit, and produced in potato plants as inhibitors of cysteine and aspartic proteases. Compared to non-transgenic control plants, the plants expressing these multidomain proteins had fewer larvae and adults (Outchkourov et al. 2004). Orthotospovirus resistance has been repeatedly attained in transgenic peanut (Li et al. 1997; Magbanua et al. 2000; Yang et al. 2004; Rao et al. 2013; Mallikarjuna et al. 2016). Valencia and runner genotypes were transformed using TSWV/GBNV N-gene or coat protein-based constructs in both the United States and India, and pathogen-derived resistance was attained in these transformants. Transformed peanut seedlings offered significant level of field or in vitro resistance against TSWV and GBNV (Li et al. 1997; Magbanua et al. 2000; Yang et al. 2004; Rao et al. 2013; Mallikarjuna et al. 2016). A transgenic peanut resistant to the peanut stem necrosis virus was also developed by Mehta et al. (2013). Hence, the possibility of viral resistance in peanut through transgenic technology cannot be ruled out.

RNAi is another technology that has the potential to offer resistance against viruses and insects (Whyard et al. 2009; Gan et al. 2010; Zhang et al. 2013). It involves the conversion of an invasive organism's dsRNA to short interfering (si) RNSs (20–23 nucleotides long) with the help of an enzyme complex. This siRNA, subsequently, prevents the translation of the mRNA (Fire et al. 1998; Pak and Fire 2007). In peanuts, the effectiveness of RNAi against viruses and/or thrips has not been proven. Yet, the effectiveness of RNAi has been established for *F. occidentalis,* where a considerable increase in reproductive fitness was seen after a double-stranded RNA (dsRNA) associated with an essential enzyme (vacuolar ATP-synthase or V-ATPase) was silenced (Badillo-Vargas et al. 2015). Developments in thrips genomics and transcriptomics have resulted in identification of several developmental genes linked to tobacco thrips and western flower thrips (Schneweis et al. 2017; Shrestha et al. 2017). It is important to first validate the use of these genes through in vitro assays before attempting to express and test the effectiveness of these genes against thrips in plants to assess the utility of RNAi as a management strategy. Although consumer preferences may prevent transgenic peanut from entering the market anytime soon, the transgenic technology that is currently accessible does present some extremely intriguing research opportunities. Plant-based transgene delivery could be avoided by making modifications to technologies like RNAi. In the first Genome Wide Association (GWA) investigation on seed composition attributes on the U.S. peanut small core collection, Zhang et al. (2021) found some important QTLs in peanut. GWAS and RNA-seq screening yielded a total of 282 differentially expressed genes (DEGs), and 16 genes substantially associated with seed composition features were also tested. These 16 genes' levels of expression were associated with five fatty acid and protein concentrations. Breeders looking to change the fatty acid ratios or boost the protein/oil content of peanut seeds may find the QTLs and genes discovered in this study valuable. These potential genes must be tested in the future against various biotic stressors, such as insect pests. In conclusion, groundnut plants have developed various biochemical and biophysical barriers to protect themselves against pests. Plants with high level of induced resistance to insect pests can offer resistance against insects which are not regular in occurrence and the attack is unpredictable (War et al. 2011). Groundnut develops numerous plant defense features after the exogenous application of jasmonic acid (JA) and salicylic acid (SA), which confer resistance to insect pests (War et al. 2011).

8.1.4 Sunflower

The consumer market recognizes sunflower as a significant oilseed crop which despite being used as oil crop also finds uses in roasted, confectionery, and bird feed products. Russian plant breeders increased the oil content of sunflower seeds, turning a common wayside plant into a well-known oilseed crop. The habitats in which sunflower species can be found include plains, deserts, and salt marshes. Because of their tolerance to a variety of ecological conditions, wild species can be seen as reservoirs of novel alleles to meet a variety of breeding objectives (Kantar et al. 2014). It has been widely documented that related species can be used in breeding programmes to produce different alleles (Seiler 1992, 2007), and this potential can be used to breed for pest and disease, drought, heat, and salinity resistance. Barriers made of phytomelanin and terpene repellents found in glandular trichomes are examples of known mechanisms for resistance in sunflower (Rogers et al. 1987). The presence of a phytomelanin layer may contribute to the pericarp's robustness (Seiler 1997; Rogers 2019). A single locus with a dominant allele has been postulated to be a genetic link to the phytomelanin layer, which has been investigated as a potential source of insect resistance (Johnson and Beard 1977). Development of resistant sunflower cultivars may be facilitated by identifying the genetics underlying the mechanisms. Phytomelanin barriers and terpene repellents located in glandular trichomes are two examples of sunflower host plant resistance features (Johnson and Beard 1977; Carlson 1971; Rogers and Kreitner 1983). Moreover, sunflower has morphological and physiological traits that may confer some tolerance. Based on different studies, it is hypothesized that the physical defense mechanisms of sunflowers against herbivory by sunflower moth late instar larval eating behaviour are significantly influenced by pericarp strength (Prasifka et al. 2014). More than 60 insects attack sunflowers in North America (Schulz 1978; Walker 1936; Phillips et al. 1973; Niide et al. 2006). The release of transgenics, particularly for weedy characteristics (herbicide resistance, illnesses, and insects), had an ecological impact due to the quick gene flow from cultivated to wild germplasm. The release of transgenic sunflowers in areas where wild populations coexist with cultivated fields carries a significant danger since natural gene flow between the two forms of sunflower happens quickly in these areas (Cantamutto and Poverene 2007; Gutierrez et al. 2010; Presotto et al. 2012).

8.1.5 Safflower

The safflower plant, *Carthamus tinctorius* L., is a bushy, herbaceous annual plant that is a member of the family Compositae. It is an important oilseed crop that is becoming more and more important in many countries of the world. The plant has historically been produced for its blossoms, which are used as a source of dye for colouring foods and textiles. Safflower oil has very high level of linoleic acid (78%). The safflower plant is known to be attacked by 101 different pests at various stages crop growth and development stages (Singh et al. 1999; Javed et al. 2013). Unfortunately, a variety of insect pests and diseases cause damage to safflower, resulting in significant productivity losses (Singh et al. 1999). The most devastating of these are insect pests, specifically the aphid, *Uroleucon compositae* (Theobald) and pod borer *Helicoverpa armigera* (Hubner) (Akashe et al. 2014). Nine different kinds of natural enemies and 20 insect pests have been identified on safflower in the Karnataka region (Mallapur et al. 2001). The aphid, *U. compositae* (Theobald) (Aphididae; Hemiptera), which attacks safflower, is one of the most destructive and persistent pests and infests crop, especially from its elongation stage until flowering phase (Akashe et al. 2014). Both nymphs and adults deplete the plants resources by sucking the cell sap from the underside of leaves and vulnerable shoots. Aphid performance on various genotypes can vary for a variety of reasons, including physical traits and the quality of the host plant. Since host plant suitability is influenced by the number of amino acids or nitrogen in the phloem sap and the secondary metabolites that affect aphid performance, host plant quality is a significant factor that contributes to the antibiotic resistance of plants (Gibson and Pickett 1983; Ave and Tingy 1986; Dixon 1998; Cisneros and Godfrey 2001; Karley et al. 2002). Different plant species exhibit variable suitability as hosts for diverse insects (Storer and van Emden 1995; Frei et al. 2003). Several genotypes of a plant species have various physical and chemical characteristics that influence its suitability as host (Ave and Tingy 1986). By analysing feeding deterrence and aphid settlement in a choice test, Saeidi (2020) studied safflower genotypes for antixenosis against *U. carthami* and found significant variations in the performance of the eight safflower genotypes examined. The genotypes with the fewest aphids present exhibit the highest antixenosis resistance. Thus, comparing the pest resistance of different genotypes considering the plant differences may provide crucial information about how suitable or unsuitable they are for insects.

Safflower fly, also known as shoot fly or capsule fly (*Acanthiophilus helianthi* Rossi), is another important insect pest of safflower in Asia and Europe (Talpur et al. 1995). This insect causes significant loss in most of the safflower producing areas in Iraq, Pakistan, and India in Asia (Al-Ali et al. 1977; Talpur et al. 1995; Verma et al. 1974). The insect heavily infests the plant throughout its reproductive phase with particular preference for egg laying during flowering and head development (Talpur et al. 1995). Hatched larvae break through the head bracts and consume the entire seed or the tissue that makes up the receptacle. According to Narayanan (1961), each larval or pupal stage lasts for 7 days, and eclosed flies can survive for up to 2 weeks before beginning a fresh cycle of oviposition. Significant reductions in seed weight, yield, and marketability are caused by larvae feeding on seeds (Ashri 1971). The most efficient and environment friendly approach to pest control should be to develop safflower cultivars that are resistant to different insects. Unfortunately, the lack of reliable sources of resistance has slowed down the development of breeding insect-resistant cultivars. It is important to note that out of 80 safflower germplasm accessions screened by Dambal and Patil (2016), 11 accessions reported foliage drying of 21–35% with aphid population on the central twig/plant. Ashri (1971) looked for fly resistance in more than 2000 cultivars of safflower, but no resistant genotype was found. Wild safflower species, on the other hand, demonstrated a moderate to high level of resistance to the safflower fly (Ashri 1971). Only two wild safflower cousin species, *Carthamus oxyacanthus* Bieb. and *Carthamus palaestinus* Eig., are easily crossable with the cultivated species and are resistant to safflower fly, making them suitable for safflower improvement (Ashri and Knowles 1960; Sabzalian et al. 2010). The development and deployment of resistant cultivars can be a successful technique to help decrease insect pest populations while minimizing the need for chemical crop treatments. Also, it can be combined with biological control and any other pest management technique used in IPM programmes.

8.1.6 Sesame

Sesame (*Sesamum indicum* Linn.), a member of the Pedaliaceae family, is a traditional and significant oilseed crop grown across the tropics and subtropics of India as well as other regions of the world. It was given a boost due to its high-quality edible oil. Sesame is also known as the "queen of oilseeds" and is a rich source of carbohydrates, protein, calcium, and phosphorus (Seegeler 1983). Moreover, it is used in the production of bread, cookies, cakes, margarine, and confections. Insecticides, cosmetics, fragrances, soaps, and pharmaceuticals are all made with the oil. The cake is also utilized as cattle feed (Mbah and Akueshi 2009). India has the largest area under sesame cultivation in the world, but yield is quite low (332 kg/ ha) compared to the global average (Singh et al. 2003). The major sesame-growing states in India are Rajasthan, Maharashtra, Gujarat, Madhya Pradesh, Andhra Pradesh, Karnataka, Uttar Pradesh, West Bengal, Orissa, Punjab, and Tamil Nadu. China, Myanmar, Sudan, Uganda, Nigeria, Pakistan, Ethiopia, and Bangladesh are the other sesame-producing countries (Ogbonna and Umar-Shaba 2012). Sesame yields in India are generally low for several reasons, but insect pests have a major role in reducing yield. The pest attack results in a significant decrease in seed production (25–90%) (Ahuja and Kalyan 2002). The crop is typically severely damaged by insect pests particularly at the flowering phase. Although the physical harm may be less than that caused by leaf pests, the impact on the production is enormous (Mahmoud 2013). The potential insect barrier to production from the seedling stage till maturity is the sesame leaf webber and capsule borer, *Antigastra catalaunalis* Duponchel (Lepidoptera: Pyraustidae) (Selvanarayanan and Baskaran 1996; Karuppaiah and Nadarajan 2011). Pesticide use has environmental concerns and unfavourable side effects on non-target organisms despite their effectiveness (Rai et al. 2002). For resource-constrained farmers, breeding for host plant resistance is thought to be the most effective, efficient, and practical control method (Bayoumi and El-Bramawy 2007). Karuppaiah et al. (2009) screened 21 sesame genotypes for resistance against *A. catalaunalis,* of which only four genotypes were less favoured for oviposition. The trichome density in the leaf had a positive correlation ($r = 0.749$) with the egg laying by *A. catalaunalis.* Comparison of growth index of this pest on different genotypes exhibited antibiosis mechanisms with significantly low growth index in resistant genotypes compared to susceptible ones. *A. catalaunalis* damage exhibited negative relationship with the total phenols content of leaves, flowers, and pods. Panday et al. (2021) identified several sources of resistance in sesame against leaf webber and capsule borer and found germplasm lines with inhibitory mechanisms of resistance to *Antigastra* that can be exploited to transfer the resistance into commercial varieties. Even cultivars that are only partially resistant could be able to offer sufficient control even when just sparingly using insecticides. By preventing the emergence of insecticide-resistant insect strains, it will aid in extending the usable commercial life of current pesticides. Choudhary et al. (2017) evaluated various sesame varieties and noted that the number of leaves, branches, capsules, and trichome density of these varieties had a negative relationship with the population of *A. catalaunalis.* The impacts of inducible defenses on phytophagous insects can include greater toxicity, a delay in larval development, or increased attack by insect parasitoids (Maleck and Dietrich 1999). Compared to constitutive defensive mechanisms, inducible defenses are thought to degrade plant fitness to a lesser extent and might be more robust (Agrawal 1998). The population dynamics of many herbivores depend heavily on the quality of the host plant, which is determined by the plant's nutritional condition (Sarwar et al. 2011). Several plant species have been demonstrated to benefit from salicylic acid in terms of the expression of dormant genes (Mahmoud 2013). The level of resistance in the cultivated sesame germplasm in India is quite low, and the available resistant germplasm has not been used to create resistant cultivars.

8.1.7 Castor

Castor is an important non-edible oilseed crop. Castor farming can be improved by growing it in areas with abundant rainfall (Cheema et al. 2013). Castor's genetic improvement has mostly been restricted to the exploitation of naturally occurring genetic variation present in the base population and limited to selection for high yield, desired branching type, non-shattering capsules, and seeds with better oil content. The types and hybrids have become more susceptible to a variety of biotic and abiotic challenges because of extensive production with high inputs, improper scientific management, and lack of crop rotation. About 100 species of insect pests have been reported on castor at different phonological stages, of which castor semilooper, capsule borer, *Spodoptera litura*, red hairy caterpillar, jassids, and whitefly are the most damaging (DOR 2005; Lakshminarayana and Sujatha 2005). According to yield loss estimates, loss in seed yield can range from 35 to 50% depending on the crop growth stage and the pest pressure. Nonetheless, complete losses have been reported in cases of significant outbreaks of red hairy caterpillar and castor semiloopers during the early stages of crop growth. Castor has the natural ability to withstand up to 25% of leaf damage without suffering a substantial loss in seed production; however, damage to the spikes and capsules causes a large yield loss (Sujatha et al. 2011). Breeders are constantly looking for new sources of pest resistance and stress tolerance. In contrast to plant disease resistance, resistance to insect pests is found to be transient and, for the most part, incomplete. There is evidence for both horizontal (polygenic) and vertical (single or few genes) disease resistance, although there is limited support for vertical resistance in terms of host–insect interactions. Insect resistance in castor has not yet been bred into any varieties or hybrids; breeding programmes aiming at introducing resistance to the main insect pests are still in their infancy. The production of cultivars for disease resistance is the focus of the parental-line and varietal front development programmes. Numerous reports on the resistance/tolerance of released varieties/hybrids—namely TMVCH-1, DCH-519, RHC1, DCH-32, GCH-4 to leaf hoppers; GCH-5 to jassids, whitefly and capsule borer; AKC1 to castor semilooper and capsule borer; TMV-6 to leaf hoppers, semilooper and capsule borer; and Kranti to semilooper—are based on field reaction to the pests. Except for the leaf miner, no systematic attempts are made to screen germplasm under consistent pest-load for the selection of promising genotypes. Castor germplasm has been tested for resistance to leaf miner, leading to the identification of the RG 1930 and RG 2008 lines, which have a purple morphotype and resistance to the miner with exceptionally low infestation compared to control plants (Anjani 2005). Resistance to leaf miner and its link to the purple colour phenotype showed uniparental inheritance (Anjani et al. 2007). Also, it was noted that cultivars SKI 73 and SKI 89 were very resistant to leaf miner infestation (Kapadia 1995). Low genetic variability for productivity attributes and sources of pest and disease resistance hinders the success of breeding for yield stability (Weiss 2000; Hegde et al. 2003). There are several opportunities for genetic enhancement of castor using traditional and biotechnological methods. Due to the lack of new sources of genes and germplasm, the main drawbacks of broad commercial production are unevenness and low yields. Castor genetic enhancement should focus on strengthening the composition against various biotic stresses as well as boosting castor productivity in a variety of conditions (Singh et al. 2015). For the establishment of newer lines and hybrids, molecular markers must be used in castor. However, the successful application of this information for development of DNA-based markers, understanding metabolic pathways of oil biosynthesis, toxins, biotic stress resistance, etc. still needs to be clarified.

8.2 Breeding for Host Plant Resistance

HPR offers a sustainable approach to manage insect pests (Stout 2014). This is especially true in underdeveloped nations (primarily tropical nations), where it is either impossible or inappropriate to use alternative control methods like pesticides. HPR is compatible with other management strategies, takes little from the farmer (other than the procurement of the proper variety of seed), has a cumulative effect frequently, and has no harmful effects on the environment. Additional forms of control must be integrated with HPR as a foundation. Nothing else will be required for economically viable crop production if the level is high enough. Development of pest-resistant cultivars may be the most effective and profitable to manage pests and diseases, but before making any research investments of this nature, several factors must be taken into account. First, think about how the pest or disease is spreading; if it is widespread, causes a considerable yield loss, and can only be successfully bred out from an agronomic standpoint, then that is a reasonable breeding goal. If an appropriate management strategy is in place, HPR is also considered a breeding goal, however it is costly and unreliable. For a pest or disease that is highly localized or temporally sporadic, has little impact on yield or quality, or is adequately controlled by conventional management measures, the time and expense of HPR breeding is not justified. The assessment of whether a resistant cultivar is the best method of controlling the pest or disease, however, must frequently be reviewed because the dynamics of the pest or disease and crop are constantly changing and because effective control chemicals can also be unexpectedly lost due to changes in government regulations. While breeding for HPR, it is important to consider the ability of the pest or pathogen to adapt to the developed lines, the improvement of the developed lines by additional recombination during breeding, and the changing environment from season to season. It frequently involves a high cost and a protracted timetable to justify the development of a resistant cultivar, especially from the secondary or tertiary gene pools. The normal process for developing resistant cultivars involves three steps: (1) identifying potential sources of resistance in the germplasm, (2) incorporating the source of resistance into an elite background, and (3) integrating the cultivar into current (or novel) crop management practices.

Every breeding programme has multiple competing objectives, so the selection criteria used to assess the resistance of germplasm must be stringent to ensure that only the most resistant material is advanced. How difficult it is to introduce new resistance into an elite background depends on the genetic distance between the source of the resistance and the cultivated species (i.e. whether it comes from primary, secondary, or tertiary germplasm), the genetic basis of the resistance, and any ploidy differences. Most of the polygenic resistance to a pest or pathogen occurs in oilseeds, although some, fortunately, are passed down by single genes. Finding germplasm with the proper level of resistance can be difficult because, like other polygenic traits, variation in resistance is typically continuous rather than discrete. It is also typical to find beneficial pest and disease resistance in the wild or unadapted genetics of plants. As a result, the process of eliminating poor alleles introduced by the resistant donor and enhancing the performance of the introgressed progeny is typically time-consuming when integrating resistance into top populations. Genetic markers connected to all the donor's key genomic regions that are sufficient to confer resistance are useful for assisting in the selection of lines with resistance and lowering the number of resources needed to create a resistant cultivar. The viability of molecular approaches has risen over the past 10 years with the adoption of low-cost, high-throughput sequencing technology known as 'next-generation sequencing'.

8.3 Genetic Engineering for Resistance to Insect Pests

In the current environment of a growing population and climate variations, increasing crop productivity is a crucial issue that needs to be addressed. Insect pests are one of the main biotic pressures that are holding back agriculture in the twenty-first century. The decrease in potential yields, which leads to stagnant productivity, is caused by insects. Despite the adoption of control measures, most of the commercially important crops experience a variety of yield losses. But, to combat insect pests, farmers have turned to chemical pesticides, their extensive usage has the potential to harm both people and the environment. Therefore, the main goal of modern agriculture has been to increase yields using the available resources and land to ensure food security for everyone throughout the world and agricultural sustainability. Genome editing is slowly and regularly becoming the tool of choice for laboratories all over the world for understanding gene function and its practical applications (Fig. 8.2). They are being used in a variety of crop development efforts to reduce biotic and abiotic stressors. Due to the editing strategies' great precision, effectiveness, cost, and time effectiveness, they have become useful tools. Significant phenotyping and enrichment of the resistance gene pool are crucial clues that demand novel approaches. To do this, it is necessary to evaluate the available germplasm, including wild relatives of crops, for pest response and to identify genes that respond to stress using multi-omics techniques. The details of genome editing for pest management involve multiplex editing of such identified resistance genes utilizing high-throughput transformation techniques. Targeted mutations

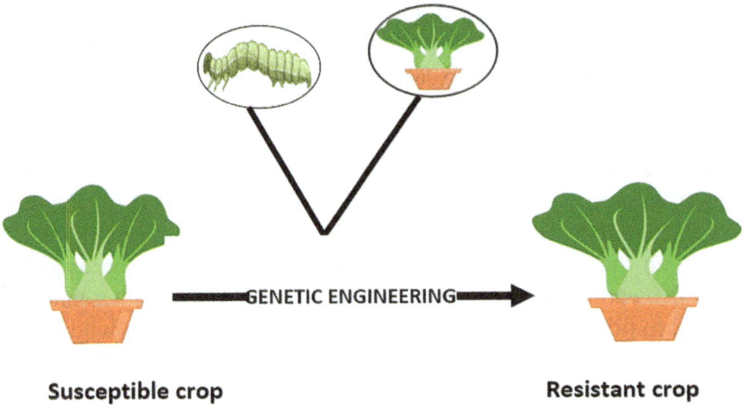

Fig. 8.2 Genetic engineering in plants

turning sensitive plants into ones that can control their pests are not far off in space or time with such studies presently in vogue. These factors, along with legislative restrictions on gene-edited crops, may help the technology succeed in terms of science.

Oilseed plants represent an important renewable source of fatty acids because they accumulate them in the form of triacylglycerol (TAG) as major storage components in seeds. In plants, two interconnected metabolic pathways—an acyl-CoA-dependent pathway and an acyl-CoA-independent pathway—lead to the export of processes for de novo fatty acid synthesis from plastids to the cytoplasm. The discovery and precise quantification of the expression of transcription factors and important genes associated with lipid metabolic pathways in soybean (Severin et al. 2010), *Jatropha curcas* (Costa et al. 2010), *Arabidopsis* (Beisson 2003), peanut (Gupta et al. 2016), and castor bean (Brown et al. 2012) have been made possible by high-quality RNA-seq data. Engineering plants to produce oilseed crops with higher yields and resilience to abiotic and biotic challenges make it possible to gather RNA-seq data and enhance plant transformation technology. To attain optimal metabolic flow, there are currently tactics that rely on more complicated methods that simultaneously overexpress or inhibit several genes. Knowing a metabolic network would enable predictable and beneficial natural product production and new chemical synthesis (Lu et al. 2006) that would help to improve the genetic makeup of oilseed crops. This is why, over the past 10 years, corporate and academic researchers have become interested in the metabolic engineering of oilseed plants.

8.4 Future Perspectives

Soon, oilseed crops and their products will account for a significant portion of global trade. These plants can be utilized for both food and non-food purposes. Gaining and maintaining future genetic gain require a thorough grasp of the development of oilseed crop breeding, including how breeding triumphs have been attained so far. The industry should feel confidence that our breeders can use a similar methodology for success in the future and the time limits involved if comparable obstacles are encountered. The effectiveness of using traditional breeding methods has been seen in oilseed breeding activities. The development of high-performing cultivars will depend on the native variation present in oilseed germplasm and potentially novel sources of additional genetically modified (GM) characteristics. Future traits must be able to take advantage of variation that is either already existing in germplasm or has been developed through recombination with closely related species or through transgenic methods. Recent developments in panomics suggest that these novel breeding tools should be kept in HPR oilseed breeding programmes together with new and existing crop management techniques and the ongoing search for new sources of resistance. Breeding for HPR will be crucial for future climates, even though crop management and precision agriculture are significant and should be considered if it is determined that a breeding solution is not the greatest path to success. For instance, using precision agricultural techniques enables farmers to adopt more effective management plans that are tailored to their individual circumstances by having a precise understanding of where a disease may emerge inside a field or throughout a farm. To lessen the impact of plant pests and diseases on the crop, techniques might range from using high level geospatial data to complex models of precise insecticide and fertilizer application (Shafi et al. 2019; Roberts et al. 2021).

8.5 Conclusion

One of the primary causes of falling agricultural output is insects. Farmers are more likely to use chemical insecticides because they offer an immediate solution to the insect pest problems. The potential deployment of alternative pest-control strategies has received increased attention because of the fast-growing public awareness of the risks indiscriminate pesticide usage poses to human and animal health as well as the environment. Host plant resistance is a pest management strategy that is safe for the environment and an important component of IPM (integrated pest management) programmes. Host plant resistance is an effective, economical, and environmentally friendly method of pest control. Because of its benefits to the economy, ecology, and environment, it has evolved into a crucial and important part of IPM methods against invertebrate pests in many agricultural systems (Teetes 1996). The most important feature of HPR is that farmers with limited resources practically never need technical know-how for application techniques and do not have to make any financial commitment. It has taken a lot of effort to find and develop agricultural cultivars that are resistant to the worst pests that affect different crops. It is essential to incorporate resistance genes into high yielding cultivars that can adapt to varied agroecosystems. Pest resistance should be one requirement for releasing varieties and hybrids for farmer use. The use of pesticide-resistant crop varieties can result in lower pesticide dosages and less frequent treatments, which lowers crop production costs and promotes biodiversity in the relevant agroecosystem. The prospects of using crop varieties resistant to invertebrate pests as a sole control method or as a supplement to other control strategies of IPM systems have also been demonstrated by empirical studies in oilseed crops, though more research is needed to elucidate the resistance mechanisms for the development and deployment of pest-resistant varieties in these crops. In fact, homogeneity is the worst thing from the standpoint of crop protection because it makes an area or time more vulnerable to stressors now or in the future (Grettenberger and Tooker 2015). It is now known that even the simplest form of diversity, such as the mixing of different genotypes in a field, can produce associational insect resistance in addition to several other advantages (Barot et al. 2017; Grettenberger and Tooker 2015; Tooker and Frank 2012). Consequently, it may be more effective (or long-lasting) to reconsider breeding techniques for complementary genotypes that produce associational resistance when grown in mixtures rather than aiming for a perfect genotype that would be entirely resistant in isolation. Future IPM strategies are also anticipated to heavily rely on genetically modified crops with improved insect resistance.

References

Agrawal AA (1998) Induced responses to herbivory and increased plant performance. Science 279: 1201–1202

Ahuja DB, Kalyan RK (2002) Losses in seed yield due to insect pests in different varieties of sesame, *Sesamum indicum* L. Ann Plant Soil Res 4(1):99–103

Akashe VB, Jadhav JD, Bavdekar VR, Shinde SK, Amrutsagar VM (2014) Predictive models for safflower aphid infestation for the scarcity zone of Maharashtra. J Agrometeorol 16(2):230–232

Al-Ali AS, Al-Neamy K, Abbas SA, Abdul-Masih AM (1977) On the life history of the safflower fly, *Acanthiophilus helianthi* Rossi (dip., Tephritidae) in Iraq. Z fur Angew Entomol 83:216–222

Alcerito T, Barbo FE, Negri G (2002) Foliar epicuticular wax of *Arrabidaea brachypoda*: flavonoids and antifungal activity. Biochem Syst Ecol 30:677–683

Amin PW, Singh KN, Dwivedi SL, Rao VR (1985) Sources of resistance to the jassid (*Empoasca kerri* Pruthi), thrips (*Frankliniella schultzei* (Trybom)) and termites (*Odontotermes* sp.) in groundnut (*Arachis hypogaea* L.). Peanut Sci 12:58–60. https://doi.org/10.3146/pnut.12.2.0002

Anjani K (2005) Purple-coloured castor (*Ricinus communis* L.)—a rare multiple resistant morphotype. Curr Sci 88:215–216

Anjani K, Pallavi M, Sudhakara Babu SN (2007) Uniparental inheritance of purple leaf and the associated resistance to leaf miner in castorbean. Plant Breed 126(5):515–520. https://doi.org/10.1111/j.1439-0523.2007.01395.x

Anonymous (2023) Groundnut outlook—2023. Agricultural Market Intelligence Centre, PJTSAU. https://agricoop.gov.in/

Ashri A (1971) Evaluation of the world collection of safflower, *Carthamus tinctorius* L. II. Resistance to the safflower fly, *Acanthiophilus helianthi* R. Euphytica 20:410–415

Ashri A, Knowles PF (1960) Cytogenetics of safflower (*Carthamus* L.) species and their hybrids. Agron J 52:11–17

Atri C, Kumar B, Kumar H, Kumar S, Sharma S, Banga SS (2012) Development and characterization of *Brassica juncea–fruticulosa* introgression lines exhibiting resistance to mustard aphid (*Lipaphis erysimi* Kalt). BMC Genet 13(1):1–9

Ave DA, Tingy WM (1986) Phenolic constituents of glandular trichomes on *Solanum berthaultii* and *Solanum polydenium*. Am J Potato Res 63:473–480

Awmack CS, Leather SR (2002) Host plant quality and fecundity in herbivorous insects. Annu Rev Entomol 47:817–844

Badillo-Vargas IE, Rotenberg D, Schneweis BA, Whitfield AE (2015) RNA interference tools for the western flower thrips, *Frankliniella occidentalis*. J Insect Physiol 76:36–46

Bakhetia DRC, Singh H, Chander H (2002) IPM for sustainable production of oilseeds: oilseeds and oils research and development needs. Indian Society of Oilseeds Research, Hyderabad, pp 184–218

Bale JS, Masters GJ, Hodkinson ID, Awmack C, Bezemer TM, Brown VK, Butterfeld J, Buse A, Coulson JC, Farrar J, Good JEG, Harrington R, Hartley S, Hefn Jones T, Lindroth RL, Press MC, Symrnioudis I, Watt AD, Whittake JB (2002) Herbivory in global climate change research: direct effects of rising temperature on insect herbivores. Glob Change Biol 1:1–16. https://doi.org/10.1046/j.1365-2486.2002.00451.x

Barot S, Allard V, Cantarel A, Enjalbert J, Gauffreteau A, Goldringer I, Porcher E (2017) Designing mixtures of varieties for multi-functional agriculture with the help of ecology—a review. Agron Sustain Dev 37(2):13

Baskaran RK, Rajavel DS (2013) Yield loss by major insect pests in groundnut. Ann Plant Sci 21(1):189–190

Batyrshina ZS, Yaakov B, Shavit R, Singh A, Tzin V (2020) Comparative transcriptomic and metabolic analysis of wild and domesticated wheat genotypes reveals differences in chemical and physical defense responses against aphids. BMC Plant Biol 20:19. https://doi.org/10.1186/s12870-019-2214-z

Bayoumi TY, El-Bramawy MAS (2007) Genetic analyses of some quantitative characters and fusarium wilt disease resistance in sesame. Afr Crop Sci Conf Proc 8:2198–2204

Beisson F (2003) Arabidopsis genes involved in acyl lipid metabolism. A 2003 census of the candidates, a study of the distribution of expressed sequence tags in organs, and a web-based database. Plant Physiol 132:681–697

Bhadoria NS, Jakhmola SS, Dhamdhere SV (1995) Relative susceptibility of mustard cultivars to *Lipaphis erysimi* in north West Madhya Pradesh (India). J Entomol Res 19:143–146

Bjorkman M, Klingen I, Birch AN et al (2011) Phytochemicals of Brassicaceae in plant protection and human health-Influences of climate, environment and agronomic practices. Phytochemistry 72:538–556

Blackman RL, Eastop VF (2000) Aphids on the World's crops: an identification and information guide, 2nd edn. Wiley, Chichester, p 466

Bodnaryk RP (1992) Leaf epicuticular wax, an antixenotic factor in Brassicaceae that affects the rate and pattern of feeding of flea beetles, *Phyllotreta cruciferae* (Goeze). Can J Plant Sci 72: 1295–1303

Boethel DJ (1999) Assessment of soybean germplasm for multiple insect resistance. In: Clement LL, Quiesenberry SS (eds) Global plant genetic resources for insect resistant crops. CRC Press, Boca Raton, FL, pp 101–129

Boyko EV, Smith CM, Thara VK, Bruno JM, Deng Y, Starkey SR, Klaahsen DL (2006) Molecular basis of plant gene expression during aphid invasion: wheat Pto- and Pti-like sequences are involved in interactions between wheat and Russian wheat aphid (Homoptera: Aphididae). J Econ Entomol 99(4):1430–1445

Brar GS, Cohen BA, Vick CL, Johnson GW (1994) Recovery of transgenic peanut (*Arachis hypogaea* L.) plants from elite cultivars utilizing ACCELL® technology. Plant J 5(5):745–753

Bravo RE, Chen G, Kim HK, Grosser K, van Dam NM, Leiss KA, Klinkhamer PGL (2019) Ultraviolet radiation exposure time and intensity modulate tomato resistance to herbivory through activation of jasmonic acid signaling. J Exp Bot 70(1):315–327. https://doi.org/10.1093/jxb/ery347

Brennan EB, Weinbaum SA, Rosenheim JA, Karban R (2001) Heteroblasty in *Eucalyptus globulus* (Myricales: Myricaceae) affects ovipositional and settling preferences of *Ctenarytaina eucalypti* and *C. spatulata* (Homoptera: Psyllidae). Environ Entomol 30:1144–1149

Brown AP, Kroon JTM, Swarbreck D et al (2012) Tissue-specific whole transcriptome sequencing in castor, directed at understanding triacylglycerol lipid biosynthetic pathways. PLoS One 7: e30100. [Epub ahead of print]. https://doi.org/10.1371/journal.pone.0030100

Bruinsma M, Van Dam NM, Van Loon JJA, Dicke M (2007) Jasmonic acid-induced changes in *Brassica oleracea* affects oviposition preference of two specialist herbivores. J Chem Ecol 33: 655–668

Brunissen L, Cherqui A, Pelletier Y, Vincent C, Giordanengo P (2009) Host–plant mediated interactions between two aphid species. Entomol Exp Appl 132:30–38. https://doi.org/10.1111/j.1570-7458.2009.00862.x

Cantamutto M, Poverene M (2007) Genetically modified sunflower release: opportunities and risks. Field Crop Res 101(2):133–144

Carlson EC (1971) New insecticides to control sunflower moth. J Econ Entomol 64(1):208–210

Cervantes DE, Eigenbrode S, Ding HJ, Bosque-Pérez NA (2002) Oviposition responses by Hessian fly, *Mayetiola destructor*, to wheats varying in surfaces waxes. J Chem Ecol 28:193–211

Chamarthi SK, Sharma HC, Vijay PM, Narasu LM (2011) Leaf surface chemistry of sorghum seedlings influencing expression of resistance to sorghum shoot fly, *Atherigona soccata*. J Plant Biochem Biotechnol 20:211–216

Chandrasekhar K, Shavit R, Distelfeld A, Christensen S, Tzin V (2018) Exploring the metabolic variation between domesticated and wild tetraploid wheat genotypes in response to corn leaf aphid infestation. Plant Signal Behav 13:e1486148. https://doi.org/10.1080/15592324.2018.1486148

Cheema NM, Farooq U, Shabbir G (2013) Prospects of castor bean cultivation in rain-fed tract of Pakistan. Pak J Bot 45(1):219–224

Choudhary MD, Kumawat KC, Yadav MK, Samota RG (2017) Host plant resistance in sesame genotypes to *Antigastra catalaunalis* Dup. J Entomol Zool Stud 6(1):1012–1015

Cisneros JJ, Godfrey LD (2001) Midseason pest status of the cotton aphid (Homoptera: Aphididae) in California cotton: is nitrogen a key factor? Environ Entomol 30:501–510

Clevenger J, Chu Y, Chavarro C (2017) Genome-wide SNP genotyping resolves signatures of selection and tetrasomic recombination in peanut. Mol Plant 10:309–322

Costa GGL, Cardoso KC, Del Bem LEV et al (2010) Transcriptome analysis of the oil-rich seed of the bioenergy crop *Jatropha curcas* L. BMC Genom 11:462

Culbreath AK, Todd JW, Demski JW (1992) Productivity of Florunner peanut infected with tomato spotted wilt virus. Peanut Sci 19:11–14

Culbreath AK, Todd JW, Brown SL (2003) Epidemiology and management of tomato spotted wilt in peanut. Annu Rev Phytopathol 41(1):53–75

Dambal GI, Patil RS (2016) Screening of safflower germplasm accessions for resistance against safflower aphid (*Uroleucon compositae* Theobald). Res J Agric Sci 7(1):128–130

Dhaliwal GS, Jindal V, Dhawan AK (2010) Insect pest problems and crop losses: changing trends. Indian J Ecol 37:1–7

Dicke M, Van Poecke R (2002) Signalling in plant–insect interactions: signal transduction in direct and indirect plant defense. In: Plant signal transduction. Oxford University Press, pp 289–316

Dixon AFG (1998) Aphid ecology. Chapman & Hall, London, p 300

DOR (2005) Integrated pest management in oilseed crops, 2nd edn. Directorate of Oilseeds Research, Hyderabad

Doss RP, Oliver JE, Proebsting WM, Potter SW, Kuy SR, Clement SL, Williamson RT, Carney JR, De Vilbiss ED (2000) Bruchins: insect derived plant regulators that stimulate neoplasm formation. Proc Natl Acad Sci U S A 97:6218–6223

Dwivedi SL, Nigam SN, Reddy DVR, Reddy AS, Ranga RGV (1995) Progress in breeding groundnut varieties resistant to peanut bud necrosis virus and its vector. In: Buiel AAM, Parlevliet JE, Lenne JM (eds) Recent studies on peanut bud necrosis disease: proceedings of a meeting. ICRISAT Asia Center, Nairobi, pp 35–40

El-Hamidi M, Zaher FA (2018) Production of vegetable oils in the world and in Egypt: an overview. Bull Natl Res Centre 42:19. https://doi.org/10.1186/s42269-018-0019-0

Fire A, Xu S, Montgomery MK, Kostas SA, Driver SE, Mello CC (1998) Potent and specific genetic interference by double-stranded RNA in *Caenorhabditis elegans*. Nature 391:806–811

Frati F, Salerno E, Conti E (2013) Cabbage waxes affect *Trissolcus brochymenae* response to short-range synomones. Insect Sci 20:763–762

Frei A, Gu H, Bueno JM, Cardona C, Dorn S (2003) Antixenosis and antibiosis of common beans to *Thrips palmi* Karny (Thysanoptera: Thripidae). J Econ Entomol 93:1577–1584

Funderburk J, Diffie S, Sharma J, Hodges A, Osborne L (2007) Thrips of ornamentals in the southeastern US. ENY-845 (IN754), Entomology and Nematology Department, Florida Cooperative Extension Service, Institute of Food and Agricultural Sciences, vol 2008. University of Florida

Furlong MJ, Wright DJ, Dosdall LM (2013) Diamondback moth ecology and management: problems, progress, and prospects. Annu Rev Entomol 58(1):517–541. https://doi.org/10.1146/annurev-ento-120811-153605

Gan D, Zhang J, Jiang H, Jiang T, Zhu S, Cheng B (2010) Bacterially expressed dsRNA protects maize against SCMV infection. Plant Cell Rep 29:1261–1268

Gao LL, Anderson JP, Klingler JP, Nair RM, Edwards OR, Singh KB (2007) Involvement of the octadecanoid pathway in bluegreen aphid resistance in *Medicago truncatula*. Mol Plant Microbe Interact 20(1):82–93

Gibson RW, Pickett JA (1983) Wild potato repels aphids by release of aphid alarm pheromone. Nature 302:608–609

Grettenberger IM, Tooker JF (2015) Moving beyond resistance management toward an expanded role for seed mixtures in agriculture. Agric Ecosyst Environ 208:29–36

Gupta K, Kayam G, Faigenboim-Doron A et al (2016) Gene expression profiling during seed-filling process in peanut with emphasis on oil biosynthesis networks. Plant Sci 248:116–127

Gutierrez A, Carrera A, Basualdo J et al (2010) Gene flow between cultivated sunflower and *Helianthus petiolaris* (Asteraceae). Euphytica 172(1):67–76

Hartman GL, West ED, Herman TK (2011) Crops that feed the world 2. Soybean - worldwide production, use, and constraints caused by pathogens and pests. Food Sec 3:5–17. https://doi.org/10.1007/s12571-010-0108-x

Hegde DM, Sujatha M, Singh NB (2003) Castor in India. Directorate of Oilseeds Research, Hyderabad

Hervé MR (2018) Breeding for insect resistance in oilseed rape: challenges, current knowledge, and perspectives. Plant Breed 137:27–34. https://doi.org/10.1111/pbr.12552

Higley LG (1995) Handbook of soybean insect pests. Entomological Society of America, p 136

Hilker M, Meiners T (2010) How do plants 'notice' attack by herbivorous arthropods? Biol Rev 85:267–280

Hilker M, Kobs C, Varma M, Schrank KK (2002) Insect egg deposition induces *Pinus sylvestris* to attract egg parasitoids. J Exp Biol 205:455–461

Hill CB, Li Y, Hartman GL (2004) Resistance of glycine species and various cultivated legumes to the soybean aphid (Homoptera: Aphididae). J Econ Entomol 97:1071–1077

Hodgson EW, McCornack BP, Tilmon K, Knodel JJ (2012) Management recommendations for soybean aphid (Hemiptera: Aphididae) in the United States. J Integr Pest Manage 3(1):1–10. https://doi.org/10.1603/IPM11019

Hollowell EA, Johnson HW (1934) Correlation between rough hairy pubescence in soybean and freedom from injury by *Empoasca fabae*. Phytopahology 24:12

Hopkins WG, Huner NPA (2004) Introduction to plant physiology. Wiley, New York, pp 479–481

Hossain MA, Maiti MK, Basu A, Sen S, Ghosh AK, Sen SK (2006) Transgenic expression of onion leaf lectin in Indian mustard offers protection against aphid colonization. Crop Sci 46:2022–2032

Howe GA, Jander G (2008) Plant immunity to insect herbivores. Annu Rev Plant Biol 59:41–66. https://doi.org/10.1146/annurev.arplant.59.032607.092825

Jakobs R, Müller C (2018) Effects of intraspecific and intra-individual differences in plant quality on preference and performance of monophagous aphid species. Oecologia 186:173–184. https://doi.org/10.1007/s00442-017-3998-x

Javed H, Iqbal J, Khan TM (2013) Studies on population dynamics of insect pests of safflower *Carthamus tinctorius* L. Pak J Zool 45(1):213–217

Jenks MA, Joly RJ, Peters PJ, Rich PJ, Axtell JD, Ashworth EN (1994) Chemically induced cuticle mutation affecting epidermal conductance to water vapor and disease susceptibility in *Sorghum bicolor* (L.) Moench. Plant Physiol 105:1239–1245

Johnson AL, Beard BH (1977) Sunflower moth damage and inheritance of the Phytomelanin layer in sunflower achenes. Crop Sci 17(3):369–372

Kanno H, Harris MD (2000) Physical features of grass leaves influence the placement of eggs within the plant by the Hessian fly. Entomol Exp Appl 96:69–80

Kanrar S, Venkateswari J, Kirti PB, Chopra VL (2002) Transgenic Indian mustard (*Brassica juncea*) with resistance to the mustard aphid (*Lipaphis erysimi* Kaltenbach). Plant Cell Rep 20:976–981

Kantar MB, Betts K, Michno JM et al (2014) Evaluating an interspecific *Helianthus annuus* × *Helianthus tuberosus* population for use in a perennial sunflower breeding program. Field Crop Res 155:254–264

Kapadia MN (1995) Varietal preference of the castor leaf miner, *Liriomyza trifolii* (Burgess) and its parasitoids. Int J Trop Agric 13:269–271

Karley AJ, Douglas AE, Parker WE (2002) Amino acid composition and nutritional quality of potato leaf phloem sap for aphids. J Exp Biol 205:3009–3018

Karuppaiah V, Nadarajan L (2011) Evaluation of sesame genotypes for resistance to sesame leaf roller and capsule borer, *Antigastra catalaunalis* Duponchel (Pyraustidae: Lepidoptera). Arch Phytopathol Plant Protect 44(9):882–887

Karuppaiah V, Nadarajan L, Kumar K (2009) Mechanism of resistance in sesame genotypes to *Antigastra catalaunalis* Dup. Ann Plant Prot Sci 17(2):337–340

Kelley JT, Andy M, O'Neal ME (2021) Aphid resistance is the future for soybean production, and has been since 2004: efforts towards a wider use of host plant resistance in soybean. Curr Opin Insect Sci 45:53–58

Kesmala T, Jogloy S, Wongkaew S, Akkasaeng C, Vorasoot N, Patanothai A (2004) Heritability and phenotypic correlation of resistance to Peanut bud necrosis virus (PBNV) and agronomic traits in peanut. Songklanakarin J Sci Technol 26:129–138

Koramutla MK, Kaur A, Negi M, Venkatachalam P, Bhattacharya R (2014) Elicitation of jasmonate-mediated host defense in *Brassica juncea* (L.) attenuates population growth of mustard aphid *Lipaphis erysimi* (Kalt.). Planta 240:177–194

Krishna G, Singh BK, Kim EK, Morya VK, Ramteke PW (2015) Progress in genetic engineering of peanut (*Arachis hypogaea* L.)—a review. Plant Biotechnol J 13(2):147–162

Kumari P, Jasrotia P, Kumar D, Kashyap PL, Kumar S, Mishra CN, Kumar S, Singh GP (2022) Biotechnological approaches for host plant resistance to insect pests. Front Genet 13:914029

Lakshminarayana M, Sujatha M (2005) Toxicity of *Bacillus thuringiensis* var. *kurstaki* strains and purified crystal proteins against *Spodoptera litura* (Fabr.) on castor, *Ricinus communis* (L.). J Oilseeds Res 22:433–434

Leskey TC, Hamilton GC, Nielsen AL et al (2012) Pest status of the brown marmorated stink bug *Halyomorpha halys* in the USA. Outlooks Pest Manag:218–226. https://doi.org/10.1564/23oct07

Li Z, Jarret RL, Demski JW (1997) Engineered resistance to tomato spotted wilt virus in transgenic peanut expressing the viral nucleocapsid gene. Transgenic Res 6:297–305

Lu C, Fulda M, Wallis JG (2006) A high-throughput screen for genes from castor that boost hydroxy fatty acid accumulation in seed oils of transgenic Arabidopsis. Plant J 45:847–856

Maffei ME, Mithöfer A, Boland W (2007a) Insects feeding on plants: rapid signals and responses preceding the induction of phytochemical release. Phytochemistry 68:2946–2959

Maffei ME, Mithöfer A, Boland W (2007b) Before gene expression: early events in plant–insect interaction. Trends Plant Sci 12(7):310–316

Magbanua ZV, Wilde HD, Roberts JK, Chowdhury K, Abad J, James WM, Hazel YW, Wayne A (2000) Field resistance to tomato spotted wilt virus in transgenic peanut (*Arachis hypogaea* L.) expressing an antisense nucleocapsid gene sequence. Mol Breed 6:227–236

Mahmoud MF (2013) Induced plant resistance as a pest management tactic on piercing sucking insects of sesame crop. Arthropods 2(3):137–149

Maleck K, Dietrich RA (1999) Defense on multiple fronts: how do plants cope with diverse enemies? Trends Plant Sci 4:215–219

Mallapur CP, Kubasad VS, Hulihalli UK, Gulaganji GG (2001) Revalidation of recommended insecticides against safflower aphid. Karnataka J Agric Sci 14(3):671–672

Mallikarjuna G, Rao TS, Kirti PB (2016) Genetic engineering for peanut improvement: current status and prospects. Plant Cell Tissue Organ Cult 125:399–416

Mandal S, Yadav S, Singh R, Begum G, Suneja P, Singh M (2002) Correlation studies on oil content and fatty acid profile of some cruciferous species. Genet Res Crop Evol 49:551–556

Mandal B, Jain RK, Krishnareddy M, Krishna Kumar NK, Ravi KS, Pappu HR (2012) Emerging problems of Tospoviruses (Bunyaviridae) and their management in the Indian sub-continent. Plant Dis 96:468–479. https://doi.org/10.1094/PDIS-06-11-0520

Marasigan K, Toews M, Kemerait JR, Abney MR, Culbreath A, Srinivasan R (2016) Evaluation of alternatives to carbamate and organophosphate insecticides against thrips and tomato spotted wilt virus in peanut production. J Econ Entomol 109:544–557. https://doi.org/10.1093/jee/tov336

Mbah MC, Akueshi CO (2009) Aflatoxin in mould infested sesame seeds. Afr J Biotechnol 8(3): 391–394

McDougal P (2018) Evolution of the crop protection industry since 1960. CropLife International. https://croplife.org/wpcontent/uploads/2018/11/Phillips-McDougall-Evolution-ofthe-Crop-Protection-Industry-since-1960-FINAL.pdf. Assessed 18 June 2021

Mehta R, Radhakrishnan T, Kumar A, Yadav R, Dobaria JR, Thirumalaisamy PP, Jain RK, Chigurupati P (2013) Coat protein-mediated transgenic resistance of peanut (*Arachis hypogaea* L.) to peanut stem necrosis disease through *Agrobacterium*-mediated genetic transformation. Indian J Virol 24:205–213

Mertens D, Boege K, Kessler A, Koricheva J, Thaler JS, Whiteman NK et al (2021) Predictability of biotic stress structures plant defense evolution. Trends Ecol Evol 36:444–456. https://doi.org/10.1016/j.tree.2020.12.009

Mitchell C, Brennan RM, Graham J, Karley AJ (2016) Plant defense against herbivorous pests: exploiting resistance and tolerance traits for sustainable crop protection. Front Plant Sci 7:1132. https://doi.org/10.3389/fpls.2016.01132

Moraes ARAD, Lourenção AL, Godoy IJD, Teixeira GDC (2005) Infestation by *Enneothrips flavens* Moulton and yield of peanut cultivars. Sci Agric 62:469–472

Moritz G (2002) The biology of thrips is not the biology of their adults: a developmental view. In: Thrips and Tospoviruses: Proceedings of the 7th international symposium on Thysanoptera. CSIRO, Canberra, pp 259–267

Narayanan ES (1961) Insect pests of safflower and methods of their control. In: Chavan VM (ed) Niger and safflower. Indian Central Oilseeds Committee, Hyderabad, pp 123–127

Natukunda MI, MacIntosh GC (2020) The resistant soybean–*Aphis glycines* interaction: current knowledge and prospects. Front Plant Sci 11:1223

Nderitu PW, Jonsson M, Arunga E, Otieno M, Muturi JJ, Wafula GO (2020) Combining host plant resistance, selective insecticides, and biological control agents for integrated management of *Tuta absoluta*. Adv Agric 2020:1–8. https://doi.org/10.1155/2020/6239491

Ni Y, Guo NJ, Wang J, Xia RE, Wang XQ, Ash G, Li JN (2014) Responses of physiological indexes and leaf epicuticular waxes of *Brassica napus* to *Sclerotinia sclerotiorum* infection. Plant Pathol 63:174–184

Niide T, Bowling RD, Pendleton BB (2006) Morphometric and mating compatibility of *Dectes texanus texanus* (Coleoptera: Cerambycidae) from soybean and sunflower. J Econ Entomol 99 (1):48–53. https://doi.org/10.1093/jee/99.1.48

Obermeier C, Mason AS, Meiners T, Petschenka G, Rostás M, Will T, Wittkop B, Austel N (2022) Perspectives for integrated insect pest protection in oilseed rape breeding. Theor Appl Genet 135(11):3917–3946

Oerke EC, Dehne HW, Schönbeck F, Weber A (2006) Crop production and crop protection: estimated losses in major food and cash crops. Elsevier Science BV, Amsterdam. https://doi.org/10.1017/S0021859605005708

Ogbonna PE, Umar-Shaba YG (2012) Time of sowing affects the growth and yield of sesame in a derived savanna agroecology of southeastern Nigeria. Philipp Agric Sci 95(2):153–159

Ongaratto S, Silveira CM, Santos MC, Gorri JER, Sartori MMP, Hunt TE, Lourenção AL, Baldin ELL (2021) Resistance of soybean genotypes to *Anticarsia gemmatalis* (Lepidoptera: Erebidae): antixenosis and antibiosis characterization. J Econ Entomol 114(6):2571–2580

Outchkourov NS, De Kogel WJ, Wiegers GL, Abrahamson M, Jongsma MA (2004) Engineered multidomain cysteine protease inhibitors yield resistance against Western flower thrips (*Frankliniella occidentalis*) in greenhouse trials. Plant Biotechnol J 2(5):449–458

Pak J, Fire A (2007) Distinct populations of primary and secondary effectors during RNAi in *C. elegans*. Science 315:241–244

Palaniswamy P (1996) Host plant resistance to insect pests of cruciferous crops with special reference to flea beetles feeding on canola - a review. Acta Hortic 407:469–481

Panday AK, Sundaria MM, Chandrasekaran M, Bisen R (2021) Identification of sources of resistance in sesame against leaf webber and capsule borer (*Antigastra catalaunalis* Dup.). Bangladesh J Bot 50(3):565–575

Paré PW, Tumlinson JH (1999) Plant volatiles as a defense against insect herbivores. Plant Physiol 121:325–331

Patel S, Rahul SN (2020) Insect pests of soybean and their management. Popul Kheti 8(4):58–61

Pedigo LP, Rice ME (2006) Economic decision levels for pest populations. In: Entomology and pest management. Pearson Prentice Hall, Columbus, pp 253–284

Petzold-Maxwell J, Wong S, Arellano C, Gould F (2011) Host plant direct defense against eggs of its specialist herbivore, *Heliothis subflexa*. Ecol Entomol 36:700–708

Phillips RL, Randolph NM, Teetes GL (1973) Seasonal abundance and nature of damage of insects attacking cultivated sunflowers. Texas A&M University, College Station

Prasifka JR, Hulke BS, Seiler GJ (2014) Pericarp strength of sunflower and its value for plant defense against the sunflower moth, *Homoeosoma electellum*. Arthropod Plant Interact 8:101–107

Presotto A, Ureta MS, Cantamutto M, Poverene M (2012) Effects of gene flow from IMI resistant sunflower crop to wild *Helianthus annuus* populations. Agric Ecosyst Environ 146(1):153–161

Ragsdale DW, Landis DA, Brodeur J, Heimpel GE, Desneux N (2011) Ecology and management of soybean aphid in North America. Annu Rev Entomol 56:375–399. https://doi.org/10.1146/annurev-ento-120709-144755

Rai HS, Gupta MP, Verma ML (2002) Efficacy of insecticides against the incidence of leaf webber and capsule borer, *Antigastra catalaunalis* Dup. in sesame. Ann Plant Prot Sci 10(1):55–58

Rao SC, Bhatnagar-Mathur P, Kumar PL, Reddy AS, Sharma KK (2013) Pathogen-derived resistance using a viral nucleocapsid gene confers only partial non-durable protection in peanut against peanut bud necrosis virus. Arch Virol 158:133–143

Reddy LJ, McDonald D (1984) Research strategies for increased groundnut production in India. In: Paper presented at the national seminar on strategies of increased production of oilseed and edible oils. National Productivity Council, New Delhi., during (mimeographed), p 23

Reddy AS, Reddy LJ, Mallikarjuna N, Abdurahaman MD, Reddy YV, Bramel PJ et al (2000) Identification of resistance sources to peanut bud necrosis virus (PBNV) in wild Arachis germplasm. Ann Appl Biol 137:135–139. https://doi.org/10.1111/j.1744-7348.2000.tb00045.x

Riddick EW, Simmons AM (2014) Do plant trichomes cause more harm than good to predatory insects? Pest Manag Sci 70:1655–1665. https://doi.org/10.1002/ps.3772

Roberts DP, Short NM, Sill J, Lakshman DK, Hu X, Buser M (2021) Precision agriculture and geospatial techniques for sustainable disease control. Indian Phytopathol 74:287–305. https://doi.org/10.1007/s42360-021-00334-2

Rogers CE (2019) Entomology of indigenous *helianthus* species and cultivated sunflowers. In: The entomology of indigenous and naturalized systems in agriculture. CRC Press, Boca Raton, pp 1–38

Rogers CE, Kreitner GL (1983) Phytomelanin of sunflower achenes: a mechanism for pericarp resistance to abrasion by larvae of the sunflower moth (Lepidoptera: Pyralidae). Environ Entomol 12:277–285. https://doi.org/10.1093/ee/12.2.277

Rogers CE, Gershenzon J, Ohno N, Mabry TJ, Stipanovic RD, Kreitner GL (1987) Terpenes of wild sunflowers (*Helianthus*): an effective mechanism against seed predation by larvae of the sunflower moth, *Homoeosoma electellum* (Lepidoptera: Pyralidae). Environ Entomol 16 (3):586–592

Romero B, Dillon FM, Zavala JA (2020) Different soybean cultivars respond differentially to damage in a herbivore-specific manner and decrease herbivore performance. Arthropod Plant Interact 14:89–99

Rutledge CE, Eigenbrode SD (2003) Epicuticular wax on pea plants decreases instantaneous search rate of *Hippodamia convergens* larvae and reduces their attachment to leaf surfaces. Can Entomol 135:93–101

Sabzalian MR, Saeidi G, Mirlohi A, Hatami B (2010) Wild safflower species (*Carthamus oxyacanthus*): a possible source of resistance to the safflower fly (*Acanthiophilus helianthi*). Crop Prot 29(6):550–555

Saeidi K (2020) Plant resistance to the safflower aphid, *Uroleucon carthami* (Theobald) (Homoptera: Aphididae) in safflower genotypes. Mun Ent Zool 15(1):189–196

Saleem MA, Gopalakrishna NK, Tippannavar PS, Nadaf HL (2019) Biophysical and biochemical mechanisms of resistance to *Spodoptera litura* in groundnut (*Arachis hypogaea* L.). J Entomol Zool Stud 7(4):86–96

Sandeep RR, Thakur SV, Hussen VS, Joshi MN, Tyagi SN, Bagatharia SB (2017) Development of PR genes panel for screening aphid tolerant cultivars in Brassica juncea. Biotech 7:129. https://doi.org/10.1007/s13205-017-0785-7

Sarwar M, Ahmad N, Tofique (2011) Impact of soil potassium on population buildup of aphid (Homoptera: Aphididae) and crop yield in canola (*Brassica napus* L.) field. Pak J Zool 43(1): 15–19

Savage GP, Keenan JI (1994) The composition and nutritive value of groundnut kernels. In: Smartt J (ed) The groundnut crop: a scientific basis for improvement. Chapman and Hall, London, pp 173–213

Schneweis DJ, Whitfield AE, Rotenberg D (2017) Thrips developmental stage-specific transcriptome response to tomato spotted wilt virus during the virus infection cycle in *Frankliniella occidentalis*, the primary vector. Virology 500:226–237

Schoonhoven LM, Van Loon JA, Dicke M (2005) Insect–plant biology. Oxford University Press, New York

Schulz JT (1978) Insect pests. Sunflower Sci Technol 19:169–223

Schweizer F, Heidel-Fischer H, Vogel H, Reymond P (2017) *Arabidopsis* glucosinolates trigger a contrasting transcriptomic response in a generalist and a specialist herbivore. Insect Biochem Mol Biol 85:21–31. https://doi.org/10.1016/j.ibmb.2017.04.004

Seegeler CJP (1983) Oil plants in Ethiopia: their taxonomy and agricultural significance. Wageningen University and Research, Wageningen

Seiler GJ (1992) Utilization of wild sunflower species for the improvement of cultivated sunflower. Field Crop Res 30(3):195–230

Seiler GJ (1997) Anatomy and morphology of sunflower. In: Schneiter AA (ed) Agronomy monographs. https://doi.org/10.2134/agronmonogr35.c3

Seiler GJ (2007) The potential of wild sunflower species for industrial uses. Helia 30(46):175–198

Seino Y, Suzuki Y, Sogawa K (1996) An ovicidal substance produced by rice plants in response to oviposition by the whitebacked planthopper, *Sogatella furcifera* (Horvath) (Homoptera: Delphacidae). Appl Entomol Zool 31:467–473

Sekhon BS, Åhman I (1993) Insect resistance with special reference to mustard aphid. In: Breeding oilseed brassicas. Springer, Berlin, pp 206–221

Selvanarayanan V, Baskaran P (1996) Varietal response of sesame to the shoot webber and capsule borer, *Antigastra catalaunalis* Duponchel (Lepidoptera: Pyraustidae). Int J Pest Manag 42 (4):335–336

Severin AJ, Woody JL, Bolon YT et al (2010) RNA-seq atlas of *Glycine max*: a guide to the soybean transcriptome. BMC Plant Biol 10:160

Shafi U, Mumtaz R, García-Nieto J, Hassan SA, Zaidi SAR, Iqbal N (2019) Precision agriculture techniques and practices: from considerations to applications. Sensors 19:3796. https://doi.org/10.3390/s19173796

Sharma HC (2007) Host plant resistance to insects: modern approaches and limitations. Ind J Plant Prot 35(2):179–184

Shrestha A, Champagne DE, Culbreath AK, Rotenberg D, Whitfield AE, Srinivasan R (2017) Transcriptome changes associated with tomato spotted wilt virus infection in various life stages of its thrips vector, *Frankliniella fusca* (Hinds). J Gen Virol 98(8):2156–2170

Singh N, Dhillon MK (2022) Nucleo-cytoplasmic interactions affecting biological performance of *Lipaphis erysimi* in *Brassica juncea*. Front Plant Sci 13:971606. https://doi.org/10.3389/fpls.2022.971606

Singh V, Prasad YG, Lakshminarayana M (1999) Insect pests of safflower and their management. In: Rajeev K, Upadhyay K, Mukerji G, Rajak RL (eds) IPM systems in agriculture. Adiya Books Pvt Ltd, India

Singh BP, Prakash O, Singh PK (2003) Dry-matter accumulation in weeds and qualitative characters of sesame (*Sesamum indicum*) as influenced by nitrogen levels and weed-control measures. Indian J Agron 48(2):120–123

Singh AS, Kumari S, Modi AR, Gajera BB, Narayanan S, Kumar N (2015) Role of conventional and biotechnological approaches in genetic improvement of castor (*Ricinus communis* L.). Ind Crops Prod 74:55–62

Smith CM (2005) Plant resistance to arthropods: molecular and conventional approaches. Springer, p 423

Smith CM, Boyko EV (2007) The molecular bases of plant resistance and defense responses to aphid feeding: current status. Entomol Exp Appl 122:1–16. https://doi.org/10.1111/j.1570-7458.2006.00503.x

Steinbauer MJ, Schiestl FP, Davies NW (2004) Monoterpenes and epicuticular waxes help female autumn gum moth differentiate between waxy and glossy eucalyptus and leaves of different ages. J Chem Ecol 30:1117–1142

Stevens J, Dunse K, Fox J, Evans S, Anderso M (2012) Biotechnological approaches for the control of insect pests in crop plants. In: Soundararajan RP (ed) Pesticides: advances in chemical and botanical pesticides. IntechOpen, pp 269–308. https://doi.org/10.5772/46233

Storer JR, van Emden HF (1995) Antibiosis and antixenosis of chrysanthemum cultivars to the aphid *Aphis gossypii*. Entomol Exp Appl 77:307–314

Stork NE (1980) Role of waxblooms in preventing attachment to Brassicas by the mustard beetle, *Phaedon cochleariae*. Entomol Exp Appl 28(1):100–107. https://doi.org/10.1111/j.1570-7458.1980.tb02992.x

Stotz HU, Koch T, Biedermann A, Weniger K, Boland W, Mitchell-Olds T (2002) Evidence for regulation of resistance in *Arabidopsis* to Egyptian cotton worm by salicylic and jasmonic acid signaling pathways. Planta 214:648–652

Stout MJ (2014) Host plant resistance in pest management. In: Abrol DP (ed) Integrated pest management: current concepts and ecological perspective. Academic Press, pp 1–21. https://doi.org/10.1016/B978-0-12-398529-3.00002-6

Sujatha M, Devi PV, Reddy TP (2011) Insect pests of castor (*Ricinus communis* L) and their management strategies. CRC Press, Boca Raton
Talpur MA, Hussan T, Rustamani MA, Gaad MA (1995) Relative resistance of safflower varieties to safflower shootfly, *Acanthiophilus helianthi* Rossi (Diptera: Tephritidae). Proc Pak Congr Zool 15:177–181
Teetes GL (1996) Plant resistance to insects: a fundamental component of IPM. In: Radcliffe EB, Hutchison WD, Cancelado RE (eds) Radcliffe's IPM world textbook. University of Minnesota, St Paul. http://ipmworld.umn.edu/chapters/teetes.htm
Thomma BP, Tierens KF, Penninckx IA, Mauch-Mani B, Broekaert WF, Cammue BP (2001) Different micro-organisms differentially induce Arabidopsis disease response pathways. Plant Physiol Biochem 39(7–8):673–680
Thompson GA, Goggin FL (2006) Transcriptomics and functional genomics of plant defense induction by phloem-feeding insects. J Exp Bot 57(4):755–766
Thorpe P, Cock PJA, Bos J (2016) Comparative transcriptomics and proteomics of three different aphid species identifies core and diverse effector sets. BMC Genom 17(1):172. https://doi.org/10.1186/s12864-016-2496-6
Todd JW, Culbreath AK, Brown SL (1995a) Dynamics of vector populations and progress of spotted wilt disease relative to insecticide use in peanuts. Tospoviruses Thrips Floral Veg Crops 431:483–490
Todd JW, Culbreath AK, Chamberlin JR, Beshear RJ, Mullinix BG (1995b) Colonization and population dynamics of thrips in peanuts in the southern United States. In: Thrips biology and management. Springer, Boston, pp 453–460
Tooker JF, Frank SD (2012) Genotypically diverse cultivar mixtures for insect pest management and increased crop yields. J Appl Ecol 49(5):974–985
Traw BM, Bergelson J (2003) Interactive effects of jasmonic acid, salicylic acid, and gibberellins on induction of trichomes in *Arabidopsis*. Plant Physiol 133:1367–1375
Trdan S, Žnidarčič D, Zlatič E, Jerman J (2004) Correlation between epicuticular wax content in the leaves of early white cabbage (*Brassica oleracea* L. var. *capitata*) and damage caused by *Thrips tabaci* Lindeman (Thysanoptera: Thripidae). Acta Phytopathol Entomol Hungarica 39:173–185
VanEtten HD, Mansfield JW, Bailey JA, Farmer EE (1994) Two classes of plant antibiotics: phytoalexins versus phytoanticipins. Plant Cell 6:1191–1192
Verma AN, Singh R, Mehratra N (1974) *Acanthiophilus halianthi* Rossi, a serious pest of safflower in Haryana. Indian J Entomol 34(4):364–365
Vogel H, Musser RO, de la Paz C-MM (2014) Transcriptome responses in herbivorous insects towards host plant and toxin feeding. In: Voelckel C, Jander G (eds) Annual plant reviews: insect–plant interactions. Wiley, Chichester, pp 197–233
Walker FH (1936) Observations on sunflower insects in Kansas. J Kansas Entomol Soc 9(1):16–25
Walling LL (2008) Avoiding effective defenses: strategies employed by phloem-feeding insects. Plant Physiol 146(3):859–866
War AR, Paulraj MG, War MY, Ignacimuthu S (2011) Jasmonic acid mediated induced resistance in groundnut (*Arachis hypogaea* L.) against *Helicoverpa armigera* (Hubner) (Lepidoptera: Noctuidae). J Plant Growth Regul 30:512–523
Weiss EA (2000) Castor. Oilseed crops. Blackwell Science, Oxford, pp 13–52
Whyard S, Singh AD, Wong S (2009) Ingested double-stranded RNAs can act as species-specific insecticides. Insect Biochem Mol Biol 39(11):824–832
Yamasaki M, Yoshimura A, Yasui H (2003) Genetic basis of ovicidal response to whitebacked planthopper (*Sogatella furcifera* Horvath) in rice (*Oryza sativa* L.). Mol Breed 12:133–143
Yang H, Ozias-Akins P, Culbreath AK, Gorbet DW, Weeks JR, Mandal B, Pappu HR (2004) Field evaluation of tomato spotted wilt virus resistance in transgenic peanut (*Arachis hypogaea*). Plant Dis 88(3):259–264
Zhang J, Li D, Ge P, Yang M, Guo Y, Zhu KY, Zhang J (2013) RNA interference revealed the roles of two carboxylesterase genes in insecticide detoxification in *Locusta migratoria*. Chemosphere 93(6):1207–1215

Zhang H, Wang ML, Dang P, Jiang T, Zhao J, Lamb M, Chen C (2021) Identification of potential QTLs and genes associated with seed composition traits in peanut (*Arachis hypogaea* L.) using GWAS and RNA-Seq analysis. Gene 769:145215

Zhao LY, Chen JL, Cheng DF, Sun JR, Liu Y, Tian Z (2009) Biochemical and molecular characterizations of *Sitobion avenae*-induced wheat defense responses. Crop Prot 28:435–442

Zheng X, Koopmann B, Ulber B, von Tiedemann A (2020) A global survey on diseases and pests in oilseed rape-current challenges and innovative strategies of control. Front Agron 2:590908

Zhu-Salzman K, Salzman RA, Ahn JE, Koiwa H (2004) Transcriptional regulation of sorghum defense determinants against a phloem-feeding aphid. Plant Physiol 134(1):420–431

Host Plant Resistance to Insects in Vegetable Crops

9

Sunil Kumaraswamy, Saifullah Omar Nasif, Abu Bakar Siddique, Ashok Karuppannasamy, Rajesh Chowdary, Muthugounder Mohan, and Alamuru Krishna Chaitanya

Abstract

Host plant resistance is an effective approach to manage insect pests in vegetable crops. This approach involves the use of plant traits that deter or reduce insect feeding and reproduction involving the mechanisms of antibiosis, antixenosis, and tolerance. These traits affect plant–insect interactions through the production of metabolites and volatiles, which can modify insect behavior and physiology. Tri-trophic interactions between plants, insects, and natural enemies also protect plants from herbivory and increase insect parasitization. The use of transgenics and genome editing provides new opportunities to enhance host plant resistance

S. Kumaraswamy (✉)
University of Agricultural Sciences, Bangalore, India

ICAR-Indian Institute of Horticultural Research, Bengaluru, India

S. O. Nasif
Global Centre for Environmental Remediation (GCER), ATC Building, The University of Newcastle, Callaghan, NSW, Australia

A. B. Siddique
Tasmanian Institute of Agriculture, University of Tasmania, Prospect, Tasmania, Australia

A. Karuppannasamy
ICAR-Indian Institute of Horticultural Research, Bengaluru, India

Tamil Nadu Agricultural University, Coimbatore, India

R. Chowdary
Acharya N.G. Ranga Agricultural University, Lam, Guntur, Andhra Pradesh, India

M. Mohan
ICAR-National Bureau of Agricultural Insect Resources, Bengaluru, India

A. K. Chaitanya
Centre for Crop Health, University of Southern Queensland, QLD, Australia

S. Kumar, M. Furlong (eds.), *Plant Resistance to Insects in Major Field Crops*,
https://doi.org/10.1007/978-981-99-7520-4_9

by introducing or modifying genes involved in the production of insect-resistant traits. Understanding the mechanisms of host plant resistance and their interactions with insect pests and natural enemies is crucial for the development of effective pest management strategies that are environmentally friendly and economically sustainable. Overall, host plant resistance offers a promising approach to reducing reliance on synthetic pesticides in vegetable crop production, but continued studies are required to fully realize its potential.

Keywords

Antibiosis · Antixenosis · Tolerance · Plant-insect interactions · Constitutive and induced metabolites · Priming · Volatiles · Tri-trophic interactions · RNAi

9.1 Introduction

In nature, plants are exposed to a number of biotic and abiotic stresses that limit their productivity. In response to these stresses, they produce a variety of resistance mechanisms that allow them to avoid, withstand, or recover from the consequences of pest infestations (Hanley et al. 2007). Plant traits that influence herbivore biology, such as physical defenses on the plant's surface (hairs, trichomes, thorns, spines, and thicker leaves), are what mediate direct defenses and affect the herbivore growth (Samadia and Haldhar 2019). In order to reduce the likelihood of insects using a host plant for oviposition, food, or refuge, they exhibit antixenosis, which refers to potential plant qualities or attributes that affect or alter insect behavior (War et al. 2012).

Host plant resistance is considered the core component of any IPM module (Horgan et al. 2020; Mansour and Eryan 2022). Antibiosis, antixenosis (non-preference), and tolerance are three modalities of resistance that affect how an insect and a plant interact (Koch et al. 2016; Iqbal et al. 2018). Antibiosis resistance affects the pest biology, which ultimately leads to reduction in plant damage. According to Smith (2005), antixenosis resistance affects a pest's behavioral features by preventing it from favoring a resistant plant (Painter 1951). According to Smith (2005), tolerance is a form of resistance where a plant can withstand or recover from harm brought on by a pest population.

India has a relatively low share in vegetable exports due to a number of factors including high domestic consumption, more than 100,000 diseases caused by viruses, bacteria, fungi, and other microorganisms (Hall 1995; Dhaliwal et al. 2007). According to estimates by Pimentel and Levitan (1986), the total global food and preharvest losses resulting from insect pests and weeds were around 45% (of the total food production) and 30%, respectively. In order to manage insect pests in horticultural crops, such plant resistance mechanisms have been widely and successfully exploited (War et al. 2012; Haldhar et al. 2015). Long-term use of both indigenous and imported genetic materials from arid horticulture crops for breeding variations in the nation led to the creation of numerous new genotypes/lines through selection and, to a lesser extent, hybridization (Haldhar et al. 2017). By expediting the development of novel pest-resistant varieties in vegetable crops and even underutilized crops, modern biotechnology offered the finest possibilities for diversifying agricultural production (Abah et al. 2010). The systematic use of wild gene pool in strategic breeding for the development of genotypes with biotic and abiotic stress resistance or tolerance, as well as their conservation as related species, is therefore urgently needed (Samadia and Haldhar 2017).

Host plant resistance has occasionally provided a straightforward method for managing insect pests and insect vector transmissible diseases on a number of agricultural and horticultural crops. In this chapter, we have reviewed the plant–insect interactions, tri-trophic interactions, herbivore-induced metabolites/volatiles, transgenics, and genome editing approaches in vegetable crops against insect pests.

9.2 Importance of Host Plant Resistance

Synthetic pesticides have been widely used in agriculture for decades to control pests and increase crop yields. However, the indiscriminate use of these chemicals has led to several negative impacts on the environment, human health, and biodiversity (Aktar et al. 2009). To address these issues, host plant resistance has emerged as a viable alternative to synthetic pesticides, which is a natural defense mechanism that plants have developed over thousands of years to protect themselves from insect pests (Sharma and Ortiz 2002). By incorporating host plant resistance into pest management strategies, farmers can reduce their dependence on synthetic pesticides, thereby reducing the negative impact of these chemicals on the environment and human health. Moreover, host plant resistance has been shown to be effective against a wide range of insect pests, including those that have developed resistance to synthetic pesticides. This is because host plant resistance is a dynamic defense mechanism that can evolve and adapt to changing pest pressures over time (Huot et al. 2013). As such, it offers a sustainable and long-term solution to pest management that can help ensure the continued productivity of agricultural systems.

9.2.1 Mechanisms of Host Plant Resistance

Plants have evolved several mechanisms including antibiosis, antixenosis, and tolerance to defend themselves against insect pests. Each of these mechanisms involves different strategies that plants use to reduce the damage caused by insect pests. This section describes the mechanisms of host plant resistance to insect pests with reported examples in vegetable crops.

9.2.1.1 Antibiosis

Antibiosis is a term used to describe an organism's ability to restrict the growth or survival of another organism. This can be accomplished through the production of toxic or inhibitory substances, such as antibiotics or toxins. In relation to insect pests, antibiosis refers to the ability of specific plants to produce chemical compounds that can exterminate or suppress insect growth. Plants employ antibiosis as a defense mechanism, producing compounds that are harmful to insect pests, thus altering their growth, development, and survival (Bischoff et al. 2023). One example of antibiosis in plants is the production of toxic proteins including lectins produced by various plants such as beans, which can be toxic to insect pests such as aphids and caterpillars (Li et al. 2023). They bind to the digestive tract of the insects, causing damage and inhibiting their growth and survival. Another example of antibiosis is the production of protease inhibitors by plants. Protease inhibitors are compounds that inhibit the digestion of protein in insects, causing them to starve and ultimately die (Casaretto and Corcuera 1995). For instance, soybean plants produce a protease inhibitor known as Bowman–Birk inhibitor, which has been found to be effective against several insect pests such as corn earworm, tobacco budworm, and fall armyworm (Azzouz et al. 2005). The Bowman–Birk inhibitor targets digestive enzymes in the gut of these insects, which leads to reduced food digestion and eventually their death.

9.2.1.2 Antixenosis

Antixenosis refers to the ability of one organism to reduce the attraction or suitability of a habitat for another organism. In the context of insect pests, antixenosis is the defensive mechanism employed by plants to insect pests involving the production of chemical or physical barriers that discourage insects from feeding or ovipositing on the plant (Stenberg and Muola 2017). Essentially, it is the plant's ability to make its habitat less attractive or less suitable for the insects, which can be achieved through both physical and chemical barriers. Trichomes, spinescence, waxy cuticle, and sclerophyll in leaf and stem in plants act as physical barriers for insects (Mitchell et al. 2016; Xing et al. 2017). Leaves, stems, and even fruits of some plants showed an increase in trichome density and prevented the insects to attack the plants. Besides, the glandular trichomes secrete viscous fluids, which also prevent the insect's attack (Wheeler and Krimmel 2015). Due to the presence of trichomes, some plants also manipulate the host selectivity of herbivores. The presence of trichomes in kidney beans resulted in lowering the attack and life span of insects (Xing et al. 2017). For instance, some plants have developed physical barriers such as thorns, spines, or tough leaves to prevent insects from feeding or laying eggs on them. Others have developed chemical barriers such as the production of toxic or unpalatable substances that deter or repel the insects. For example, the presence of glandular trichomes on cucumber leaves can deter feeding by spider mites and whiteflies (Xue et al. 2019). Similarly, the production of volatile compounds by tomatoes and carrots can repel insect pests from laying eggs on the plant. For instance, tomato plants produce the volatile compound methyl salicylate, which has been shown to repel the whitefly, *Bemisia tabaci*, and the tomato fruit worm, *Helicoverpa zea* (Shi et al. 2016).

9.2.1.3 Plant Tolerance to Insects

Plant tolerance is one of the most important aspects of resistance against pests and pathogens where plants show their ability to withstand even though there are enough insects to cause severe damage to other susceptible plants. Tolerance enables the plant to continue its growth, and plants can complete their life cycle under insect attack. Plant alters its morphological features as a part of its defense against insect pests. Tolerance mechanisms employed by vegetable crops in defending insects are presented in Table 9.1. Besides the alteration in morphological changes, plants change their growth characters to reduce the damage caused by insect infestation. Delaying flowering, reduced growth, and delayed fruit setting are some of the tolerance characteristics in some plants to pass the insect attack (Stenberg and Muola 2017). However, the opposite characters are also observed in some plants such as higher number of branches, tillers, and higher number of roots and shoots to continue its growth under the insect attack (Koch et al. 2016). Breaking the dormancy under insect attack results in increase in the number of branches (Koch et al. 2016).

The continuous and increased growth of plants under insect attack has been correlated with the physiological basis of plant tolerance to insects. The increase in photosynthetic rate is common in several plants to adjust the growth under insect infestation, which also sustain the growth rate (Weintraub et al. 2018). Tolerant plants can increase the photosynthetic rate by increasing the photosynthetic pigments including the chlorophyll a, b and carotenoids and increase the enzymatic activities related to carbon capture and light harvesting systems such as RuBP and ATPase (Koch et al. 2016; Mitchell et al. 2016). An increase in photosynthesis under the insect infestation results in ensuring plant growth with higher vigor. Besides, the increase of carbon translocation to shoots from roots increases the sink strength to continue its growth and in some cases regrowth under insect infestation.

Over-accumulation of reactive oxygen species (ROS) under insect attack has been reported, which acts as signaling molecules for several defense systems. Though ROS lead to cell death, they activate several defense systems including hormonal regulation, metabolism, growth development, and gene expression having a negative impact on insects (Rashid and Chung 2017; Wani et al. 2022). Plants increase their antioxidative compounds including SOD, POD, APX, and proline to suppress the detrimental effect of ROS (Mansoor et al. 2022). However, the hormonal regulation, especially JA and SA translocate from the infested leaf to a distant leaf and activate the defense mechanisms, resulting in the activation of systemic acquired tolerance to insects (Divekar et al. 2022; Kumari et al. 2015; Zhang et al. 2019a, 2022a; Sahu et al. 2022). Besides, these events include the gene expression in systemic leaves, which are linked to plant defense against the insect.

Increase in metabolic biosynthesis is one of the tolerant mechanisms to insect attack. The most common metabolites are isoflavonoids, terpenoids, phenolics, alkaloids, and phytoalexins, which showed negative effect on insect performance and decrease in insect attack (Jeckel et al. 2022; Jha and Mohamed 2022; Divekar et al. 2022; War et al. 2012). Some of them have properties like feeding deterrence, releasing volatile compounds, direct toxins, and repellents having an overall

Table 9.1 Tolerance mechanisms to different insects in vegetable crops

Crop	Insect	Tolerance mechanism	Reference
Potato	*Leptinotarsa decemlineata*	Tolerance to beetle was attributed with higher doses of nitrogen which changed the leaf morphological features including leaf expansion, increased photosynthetic rate which led to tolerate defoliation.	Weintraub et al. (2018)
	L. decemlineata	Transgenic plant showed reduction in EcR transcripts in larvae.	Hussain et al. (2019)
	Myzus persicae	Tolerant potato cultivars showed lower infestation rate with higher leaf chlorophyll, photosynthesis, leaf hair density, and leaf hair number.	Quandahor et al. (2019)
	L. decemlineata	Plants overexpressed with Cry3A gene showed higher tolerance with higher mortality rate of insect, lower foliage consumption, and lower survival rates of insect.	Mi et al. (2015)
	Halyomorpha halys	Plants showed tolerance to herbivore insect through the increase biosynthesis of ROS.	Sperdouli et al. (2022)
Tomato	*Manduca sexta*	Changes in primary and secondary metabolites with a higher concentration in sink tissues showing the higher utilization to alleviate the attack.	Gomez et al. (2012)
	Helicoverpa armigera	Tolerance was attributed through the JA signaling pathway with an increase of photsynthetic rate, sucrose phosphate synthase, and flower and fruit biomass.	Guo et al. (2012)
	H. armigera	Heat shock protein (GroEL) enhanced the tolerance through the minimization of photo-oxidation of chlorophyll by increasing the activity of antioxidants and lower cell death.	Kumari et al. (2015)
	H. armigera	Overexpressed tomato lines with proteinase inhibitor gene showed antibiosis effect on insect with delayed growth rate.	Tanpure et al. (2017)
	Tuta absoluta	Cry1AC gene in plant showed higher mortality rate of insect.	Selale et al. (2017)
	Heliothis peltigera; *Spodoptera littoralis*	Overexpression of BnFAD3 enhanced the tolerance to both chewing insects through the JA signaling pathway.	Zhang et al. (2019a)
	H. zea	Tolerant plants showed higher trichomes and volatile organic compounds.	Paudel et al. (2019)
	T. absoluta	Tolerant tomato genotypes showed higher level of secondary metabolism, volatile organic compounds and higher level of SNP variation.	D'Esposito et al. (2023)
	Bemisia tabaci	siRNA transgenic lines showed higher mortality rate of whitefly.	Pizetta et al. (2022)

(continued)

Table 9.1 (continued)

Crop	Insect	Tolerance mechanism	Reference
	M. persicae and *B. tabaci*	Tolerant plants showed higher density of trichome with thick stem and increased concentration of phenols and tannins.	Anu et al. (2021)
Pepper	*B. tabaci*	Tolerant genotype was found to have higher level of resistance compounds with lower levels of nutrients.	Jiao et al. (2018)
	Spodoptera litura	Plant overexpressed with Cry2Aa2 gene showed tolerance to insect by increasing the photosynthetic pigments and rate, proline, soluble sugars, ROS with better growth and yield.	Zhu et al. (2015)
Broccoli	*Ascia monuste*	Tolerant character was associated with methyl jasmonate, and PR1 gene expression.	Venegas-Molina et al. (2020)
Cucumber	*Acalymma vitattum*	Tolerant plants showed higher germination rates, higher number of flowers and biomass, and higher JA and SA.	Rivera-Vega et al. (2022)
Common bean	*M. persicae*	PeBC1 elicitor protein triggered the biosynthesis of JA and SA by increasing the associated genes which resulted in lethality to aphids.	Basit et al. (2019)
Eggplant	*M. persicae*	Least susceptible plants showed lower number of adult aphids, lowest reduction in height and dry weight with greater plant resistance indices.	Raeyat et al. (2021)

negative effect on insects by disruption of their growth and development, ultimately leading to reduction in their number (Zhu et al. 2015).

Volatile organic compounds (VOCs) are plant-derived chemicals having deterrent effect on plant herbivores (Zhou and Jander 2022). They are the intermediate products of different metabolic pathways, which have diverse roles in plant defense against both biotic and abiotic stresses. VOCs play roles in ROS sequestration by membrane stabilization, inhibiting pathogen growth, attraction of parasitoids, and induction of systemic resistance (Brilli et al. 2019; Russo et al. 2022; MacDougall et al. 2022). A large number of vegetable crops including onion, potato, tomato, carrot, and other vegetables have been reported to contain several VOCs (Wang et al. 2019; Tiwari et al. 2020). A recent study showed that tomato and pepper increased the volatile synthesis after whitefly attack leading to disturbance of insect population (Ghosh et al. 2022).

A diverse range of proteins, transcripts, and gene expressions contribute to plant tolerance (War et al. 2012). Plant resistance proteins have been reported in vegetable crops including tomato, potato, pepper, and cucumber (Abuyusuf et al. 2022; Li et al. 2019; Leybourne and Aradottir 2022; Wani et al. 2022), which showed tolerance to insects such as *Manduca sexta*, *Spodoptera littoralis*, and *Macrosiphum euphorbiae*. Some proteins have functions like protease inhibitors, catalase, chitinase, and peroxidase, which suppress the growth and establishment of insects (Wang et al. 2018). Many of them are also related to the signaling cascade to activate systemic tolerance by activating the JA and SA signaling. Gene expression due to the insect attack is one of the most specific molecular defense mechanisms in plants including vegetables. Tomato overexpressed with Cry1A genes enabled it to reduce the damage against a diverse range of insects (Murata et al. 2021). Genes that encode the proteinase inhibitor, phytolectin, amylase inhibitor, and chitinase have been reported in plants, which contributed to the natural defense against insects and microbes (Wang et al. 2018, 2021a).

Crop improvement focusing on the tolerance mechanism has been used as a part of integrated pest control. Identification of metabolites, hormonal regulation, and plant volatile compounds for general and specific targets have great challenges to include them in pest control strategy. Due to their no effect on selection and reduction in the number of insect pests, tolerance can be a sustainable approach that can help to promote the effects of beneficial arthropods in agricultural practices.

9.2.2 Pseudo Resistance

Plants possess several traits that make them less vulnerable to insect attack. These mechanisms can occur at both spatial and temporal levels. In terms of spatial escape, plants may not be detectable by insects due to the plant alterations/changes in morphological features, whereas the temporal escape occurs due to lack of the synchronization of most susceptible stage(s) of plants with the peak pest activity period (Stenberg and Muola 2017). As a part of their indirect defense against insect pests, plants use spatial and temporal escape to sustain their growth and development. While host escape occurs at the individual level, host evasion happens at the population level. Plants pass through the most susceptible stage quickly or at a time when insects are less numerous. Plants may also alter their growth patterns to be asynchronous with the number of insect pests, further reducing their susceptibility to insect attacks.

Under insect pressure, plants modify their morphology, physiology, and biochemical synthesis, resulting in the activation of their defense systems. Escape can be attributed to morphological, physiological, and biochemical alterations in response to plant–insect interactions (Wani et al. 2022). For instance, plants may alter their leaf, flower, and vegetative growth to keep insects away. Changes in plant height and stature can also be considered escape traits. Mimicry can also be effective, whereby a plant may have leaves that are similar in shape and morphology to those of another species but differ in their chemical composition, helping the plant escape insect attacks (Stenberg and Muola 2017). Plants may also exhibit escape traits that increase their tolerance to insects. For example, some plants camouflage themselves to avoid herbivores by altering the appearance of their leaves. They may also change leaf traits such as trichome density, trichome number, and leaf thickness to prevent insect attack (Dominguez et al. 2017).

Associational resistance, also known as escape, can enable plants to remain undetected by insects and direct them toward neighboring plants. Additionally, some plants produce volatile organic compounds (VOCs), which serve as long-distance signaling molecules and act as insect repellents while also triggering plant immunity against insects (Conboy et al. 2020; Piesik et al. 2022; Nalam et al. 2019). Tomato plants, for instance, produce methyl salicylate (MeSA), which serves as both a repellent and an activator of plant defense mechanism when the plant is attacked by insects. VOCs have also been reported in other plant species such as potato, broccoli, eggplant, cucumber, and onion, among others, and have shown their efficiency in helping plants escape insect attacks (Darshanee et al. 2017; Steglińska et al. 2022; Murakami et al. 2019; Ahmed et al. 2021; Jo et al. 2022; Zhang et al. 2022a; Wang et al. 2019).

9.3 Plant–Insect Interactions

Plants interact with their environment, and the interactions with insects are regulated by a complex interplay between the systems that enable plants to detect insect attacks and coordinate the downstream signaling processes that activate plant defenses. In response to herbivory, plant defenses typically decrease the fitness of insects by reducing their survival and reproduction by triggering a cascade of resistance mechanisms. Plant–insect interaction is a two-way process, i.e., plants initiate a specific defense response after insect injury to protect themselves with a variety of induced chemical defenses that can act either directly or indirectly on herbivores.

9.3.1 Constitutive and Induced Defenses

Plants defend insects through constitutive and/or induced defenses. The constitutive defenses are the inherent defenses that are present regardless of herbivory/attack, and these defenses include the structures and compounds of primary and secondary metabolism (Bar and Shtein 2019). The constitutive defenses can work independently of one another, or they can act synergistically. For instance, a structural and chemical defense system is formed by glandular trichomes and their secretory canals (Glas et al. 2012; Wang et al. 2021b). Constitutive chemical defenses may be poisonous, repulsive, or antifeedant, and they may affect the growth of herbivores by altering their resistance, fecundity, and digestive capabilities.

By contrast, induced defenses are activated in response to herbivory or other biotic and abiotic stresses and consist of the structures, products, and compounds of the plant's secondary metabolism that are subsequently primed (Kersch-Becker et al. 2019). A sophisticated phytohormone system controls the layered set of responses known as induced defenses (Schuman and Baldwin 2016; Howe et al. 2018). The induced cellular mechanisms caused by interacting herbivores change the plant's metabolome, and insect herbivory activates defense metabolites to prevent pest invasion and the emergence of plant resistance (Maag et al. 2015; Jiang et al.

2019). Plants that have been damaged perceive a range of endogenous chemicals as danger signals, which are referred to as damage-associated molecular patterns (DAMPs), which also include secondary metabolites. Green-leaf volatiles are also DAMPs, and they respond to herbivory by activating both short-term and long-term defense mechanisms (Quintana-Rodriguez et al. 2018; Santamaria et al. 2018). Both constitutive and induced defenses can be direct and indirect, the direct ones include the buildup of toxins or the thickening of cell walls, which have an impact on the growth and performance of herbivore (Lin et al. 2020; Yang et al. 2020). Indirect defenses influence the behavior of herbivores, predators, or parasitoids by attracting them to the infested plant and/or to neighboring plants by sending out alarm signals (Coppola et al. 2017). Induced metabolites/volatiles of vegetable crops and their defense function against herbivores are presented in Table 9.2.

9.3.2 Herbivore-Induced Metabolites

The induced defenses are often controlled by metabolic pathways such as jasmonic acid (JA), salicylic acid (SA), and ethylene (ET)-mediated signaling, which may have a direct or indirect impact on the host choice by insect (Bonaventure 2012). For instance, spider mite (*Tetranychus urticae*) infestation induces jasmonic acid (JA) and salicylic acid (SA) signaling as well as the synthesis of terpenoids and flavonoids in pepper (Zhang et al. 2020). In cassava, the phenylpropanoid super pathway encodes the production of monolignol, flavonoid, and lignan, which reduce the performance of whiteflies (Perez-Fons et al. 2019). Multiple metabolic pathways and biomarkers like galactaric acid, hydroxycinnamic acids, and quinic acid promote resistance to whitefly-mediated tomato curly stunt virus (ToCSV) (Rossouw et al. 2019).

Plant responses result in morphological or metabolic changes that influence indirect interactions with other community members (Han et al. 2020). In cucumber, spider mite infestation leads to changes in terpene, green-leaf volatiles, and cucurbitacin biosynthesis that are potentially involved in the regulation of induced direct and indirect defenses (He et al. 2020). Secondary metabolites have been studied extensively in the context of plant–insect interactions and may have an impact on insect establishment. An initial herbivore attack can strengthen the plant's defense profile (Poelman and Dicke 2018). For example, *Phthorimaea absoluta* (*Tuta absoluta*) infestation induces various secondary metabolites and volatile organic compounds (VOCs), which enhance the resistance of tomato plants (Chen et al. 2021b; Roumani et al. 2022). Signaling genes from PR1b1, NPR1, NPR3, MAPKs, and ANP1 families associated with salicylic acid in eggplant mediate plant immunity against *T. absoluta* (Chen et al. 2021a).

The polyphenol oxidase (PPO) and protease inhibitors (PIs) in plant tissues may also be impacted by *T. absoluta* attack, which would hinder larval development (D'Esposito et al. 2021). Similarly, *Spodoptera litura* herbivory in pepper-induced toxic phenolic compounds like vanillic acid and syringic acid. The metabolite profiles of pepper containing capsianosides are associated with western flower thrips

Table 9.2 Herbivore-induced plant metabolites/volatiles and their defense function in vegetable crops

Crop	Insect interaction	Induced metabolite/ volatile	Defense function	Reference
Tomato	*Phthorimaea absoluta* (*Tuta absoluta*)	Accumulation of phenolamides, spermine, dihydrocinnamic acid derivatives and caffeoylputrescine	Antibiosis	Roumani et al. (2022)
	T. absoluta	Aldehydes, alcohols, sugars, aromatics, amines, terpenoids, ketones, phenolics and olefins	Antibiosis	Chen et al. (2021b)
	T. absoluta	Monoterpenes	Attract predator *Nesidiocoris tenuis*	Ayelo et al. (2021)
	T. absoluta and *Bemisia tabaci*	Chorismate-derived compounds, C_{18}-fatty acids and cyclic sesquiterpenes	Attract natural enemies	Silva et al. (2017)
	T. absoluta and *Tetranychus urticae*	(Z)-3-hexenyl propanoate [(*Z*)-3-HP]	Host plant resistance	Pérez-Hedo et al. (2021)
	B. tabaci	Increased levels of galactaric acid, hydroxycinnamic acids and quinic acid	Antibiosis	Rossouw et al. (2019)
	B. tabaci	Flavonoids	Reduce the ability of *Bemisia tabaci* to land, settle, probe, and phloem-feeding	Yao et al. (2019)
	B. tabaci	7-Epizingiberene	Repellent	Rosen et al. (2015)
	Trialeurodes vaporariorum	Methyl salicylate (MeSA)	Intracellular defense signaling	Conboy et al. (2020)
	Macrosiphum euphorbiae	Isoprenoids	Attract the parasitoid *Aphidius ervi*	Coppola et al. (2019)
	Frankliniella occidentalis	α-Tomatine and phenolic compound	Defensive compounds	Bac-Molenaar et al. (2019)
	Thrips, whiteflies, aphids and	Acylsugars and myricetin	Inhibit oviposition and antibiosis	Leckie et al. (2016)

(continued)

Table 9.2 (continued)

Crop	Insect interaction	Induced metabolite/ volatile	Defense function	Reference
	Spodoptera exigua			Vosman et al. (2018)
	S. litura	Chlorogenic acid and catechin	Redistribution of primary and secondary metabolites in local and systemic leaves	Kundu et al. (2018)
	S. exigua	Physostigmine, 4-oxododecanedioic acid, and azelaic acid	Antibiosis	Rivero et al. (2021)
	S. litura	(*Z*)-3-hexenol and glycoside	Feeding inhibition	Sugimoto et al. (2014)
Potato	*P. operculella*	Jasmonic acid (JA) and abscisic acid (ABA)	Activation of early defense response	Mao et al. (2022)
	P. operculella	α-Chaconine and glycoalkaloids	Larval feeding inhibition	Wang et al. (2020)
	Tecia solanivora *S. exigua* *S. frugiperda*	Chlorogenic acid, α-solanine and α-chaconine	Antibiosis	Kumar et al. (2016)
	Macrosiphum euphorbiae	Cis-jasmone (CJ)	Affect settling and performance	Sobhy et al. (2017) Sobhy et al. (2020)
	Myzus persicae	β-Bisabolene, (*E*)-β-farnesene, trans-α-bergamotene,	Attract the parasitoid, *Diaeretiella rapae*	Ali et al. (2022a)
Eggplant	*T. absoluta*	Signaling genes associated with salicylic acid	Plant immunity	Chen et al. (2021a)
	F. occidentalis	Quinic acid	Host plant resistance	Liu et al. (2022)
Pepper	*T. urticae*	Terpenoids and flavonoids	Jasmonic acid (JA) and salicylic acid (SA) signaling	Zhang et al. (2020)
	S. litura	Vanillic acid and syringic acid	Induced resistance	Movva and Pathipati (2017)
	F. occidentalis	Capsianosides	Host plant resistance	Macel et al. (2019)

(continued)

Table 9.2 (continued)

Crop	Insect interaction	Induced metabolite/ volatile	Defense function	Reference
Pepper and cabbage	*M. persicae*	α-Pinene, decanal phthalic acid, and isophorone	Attracts the endoparasitoid *Aphelinus varipes*	Ali et al. (2022b)
Cucumber	*T. urticae*	Terpene, green-leaf volatiles and cucurbitacin	Induced direct and indirect defenses	He et al. (2020)
Common bean	*Liriomyza huidobrensis*	Methyl salicylate and (*E*,*E*)-4,8,12-trimethyl-1,3,7,11-tridecatetraene (TMTT)	Attracts its parasite *Opius* sp.	Yang et al. (2021)

(*Frankliniella occidentalis*) resistance (Movva and Pathipati 2017; Macel et al. 2019). Bac-Molenaar et al. (2019) identified α-tomatine and a phenolic compound as potential defensive compounds against western flower thrips in tomato. The higher amounts of quinic acid in thrips-resistant eggplant improved the resistance of susceptible eggplant to western flower thrips on external application (Liu et al. 2022).

Following herbivory, induced defenses cause localized and systemic elevations of toxic secondary compounds above constitutive levels, which affect herbivore preference, performance, and feeding activity (Perkins et al. 2013; Zalucki et al. 2017) (Fig. 9.1). *S. litura* herbivory in tomato redistributes primary and secondary metabolites in local and systemic leaves, which substantially slow down larval growth (Kundu et al. 2018). Tomato plants exhibit higher levels of resistance following herbivore attack, as evidenced by the decrease in larval development of insects that feed on damaged leaves (Escobar-Bravo et al. 2017; Hamza et al. 2018). Metabolites can also affect the insect behavior and interrupt their host choice; for instance, tomato flavonoids reduce the ability of *Bemisia tabaci* to land, settle, probe, and phloem-feeding efficiency and prevent the spread of the tomato yellow leaf curl virus (TYLCV) (Yao et al. 2019). Plants mediate interactions between aboveground and belowground herbivores via plant secondary metabolites. Aboveground herbivory by *Phthorimaea operculella* in potato increased α-chaconine and glycoalkaloids in tubers and negatively affected larvae feeding on tubers (Wang et al. 2020). On the other hand, localized tuber damage by Guatemalan tuber moth (*Tecia solanivora*) induces aboveground resistance against *Spodoptera exigua* and *S. frugiperda* by increased foliar abundance of chlorogenic acid, α-solanine, and α-chaconine (Kumar et al. 2016).

The acyl sugars, sesquiterpenes, and methyl ketones are the three main allelochemicals present in tomato. Acylglycosis and acylsucrose are important acylsugars in the leaf trichomes of wild tomato species (Da Silva et al. 2016). Acylsugars and myricetin from *Solanum pennellii* and *S. galapagense* demonstrate resistance against thrips, whiteflies, aphids, and *S. exigua* (Leckie et al. 2016; Vosman et al. 2018). Similarly, *S. pennellii* and *S. habrochaites* lines predominated by acyl sugars are linked to whitefly resistance (Marchant et al. 2020; Kortbeek et al.

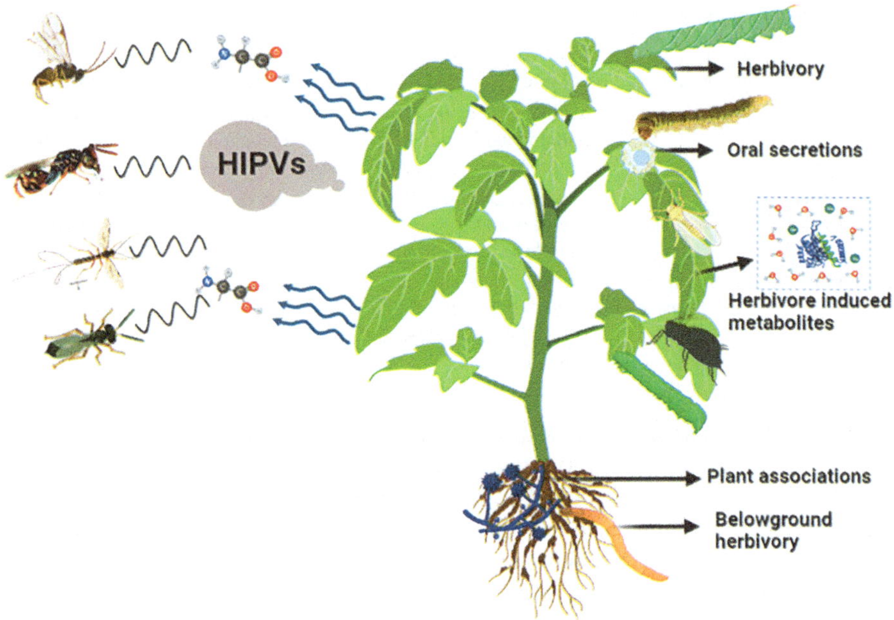

Fig. 9.1 Plant–insect interactions in vegetable plants involving defense elicitors such as herbivory, oral secretions, plant associations, and belowground herbivory, as well as induced defense responses of plants against herbivores that include herbivore-induced metabolites and herbivore-induced plant volatiles (HIPVs) attracting natural enemies

2021). Tolerant tomato genotypes display high levels of acyl sugars and other metabolites after *T. absoluta* infestation (De Falco et al. 2019).

Plants must be able to recognize the difference between physical injury and insect feeding in order to launch an insect-specific defense. The saliva of chewing and phloem-feeding insects contains chemical compounds and particular proteins that have been linked to herbivore manipulation of plant defenses (Villarroel et al. 2016). Various elicitors in insect saliv/oviposition fluid modulate plant's defense. These elicitors are recognized by plants and play a crucial role in downstream signaling cascades linked to defense (Bonaventure 2018) (Fig. 9.1). Based on their molecular makeup and structure, elicitors are categorized into six groups: enzymes, fatty acid amino acid conjugates (FACs), fatty acids, peptides, esters, and benzyl cyanide (Zunjarrao et al. 2020). In oral secretions, enzymes, fatty acids, and peptides are produced; however, during egg laying, esters and benzyl cyanide are released in oviposition fluids (Aljbory and Chen 2018; Bonaventure 2018).The microbes in the oral secretions of *P. operculella* regulate jasmonic acid (JA) and abscisic acid (ABA) during herbivory, which activate early defense response in potato (Mao et al. 2022).

Plant associations with other organisms help them to withstand various stresses and have an impact on the expression of defenses. For instance, tomato plants primed with *Trichoderma harzianum* and arbuscular mycorrhizal fungi (AMF)

enhance the defense against aphids and *S. exigua* by increased accumulation of isoprenoids, physostigmine, 4-oxododecanedioic acid, and azelaic acid (Coppola et al. 2019; Rivero et al. 2021). In some cases, insect herbivory can increase plant susceptibility and make the host more alluring to the pest. For example, Cucurbit chlorotic yellows virus (CCYV) infection in cucumber attracts whiteflies by reducing protective flavonoids and terpenoids (Zhang et al. 2022b). Similarly, *B. tabaci* infestation decreases the production of repellent flavonoids and monoterpenes (-α-phellandrene and α-terpinene), which enhanced the oviposition of whiteflies on tomato (Su et al. 2018).

9.3.3 Herbivore-Induced Plant Volatiles (HIPVs)

Upon herbivore attack, plants emit blends of volatile organic compounds (VOCs) that differ from those that are constitutively emitted. Herbivore-induced plant volatiles (HIPVs) are a class of small-molecule volatile substances that are produced by plants (Pérez-Hedo et al. 2021). These include terpenoids, volatile fatty acid derivatives, phenylpropanoids, volatile amino acid derivatives, glucosinolates, benzoxazinoids, and aromatic compounds (Bouwmeester et al. 2019). HIPVs form an indirect defense strategy as they are used as cues by natural enemies of plant-feeding insects to locate their prey or hosts, and also they are used to activate pre-infestation defense mechanisms by adjacent healthy plants (Pérez-Hedo et al. 2017; Silva et al. 2017) (Fig. 9.1). HIPVs may differ quantitatively and qualitatively upon damage by different species, for example, tomato volatiles were dominated by chorismate-derived compounds and C_{18}-fatty acids when infested with *T. absoluta*, whereas *B. tabaci* herbivory induced cyclic sesquiterpenes; these volatiles attract species-specific natural enemies (Silva et al. 2017). Plant volatiles are useful indicators of ongoing herbivore attack because they are released during insect feeding. Such signals are detected by plants, which are then used to intensify their innate or induced defense mechanisms. For instance, exogenously applied green leaf volatile ((Z)-3-hexenol) can prime the production of defense signaling molecule jasmonic acid in tomato (Su et al. 2020). Utilization of such HIPVs might be a more environmentally sound choice. In order to effectively manage begomovirus diseases, volatile terpenes, such as 7-epizingiberene from *S. habrochaites*, can be used as repellents against *B. tabaci* (Rosen et al. 2015).

Some green leaf volatiles can be transformed into toxic compounds or have defensive qualities of their own when perceived by plants. For instance, in tomato uptake of (Z)-3-hexenol, conversion of it into a glycoside inhibits the ability of *S. litura* to feed on recipient plants (Sugimoto et al. 2014). Similarly, tomato plants artificially induced with (Z)-3-hexenyl propanoate [(Z)-3-HP] were less susceptible to the attack of *T. absoluta* and *T. urticae* (Pérez-Hedo et al. 2021). Application of elicitor, methyl salicylate (MeSA) triggers intracellular defense signaling on un-infested tomato plants, which reduced the greenhouse whitefly (*Trialeurodes vaporariorum*) population (Conboy et al. 2020).

Herbivore-damaged plants emit volatile organic compounds (VOCs) that can alert neighbors and boost their resistance. VOCs released by herbivore-induced potato plants boosted resistance in receiver plants against *S. exigua* (Vázquez-González et al. 2023). Sweet potato weevil (*Cylas formicarius*) feeding induced volatiles, (Z)-3-hexenyl acetate (z3HAC) and allo-ocimene, which prime neighboring plants to defend themselves against sweet potato weevil (Xiao et al. 2023). Similarly, potato plants primed with defense activator cis-jasmone (CJ) significantly result in elevated emission of aphid defense-related volatiles, which negatively affect aphid (*Macrosiphum euphorbiae*) settling and performance (Sobhy et al. 2017, 2020). HIPVs not only draw in natural enemies but also influence the movement of conspecific herbivores. For example, tomato and eggplant volatiles ((*Z*)-3-hexen-1-ol and terpenoids) mediate the preference of greenhouse whitefly to its hosts (Darshanee et al. 2017). Similarly, Zhang et al. (2019b) showed that HIPVs from tomato plants infested with whiteflies can make nearby plants more susceptible by inhibiting JA-dependent defenses.

9.3.4 Tri-trophic Interactions

The increased release of volatiles from a plant following an insect attack is thought to be an induced indirect defense. Tri-trophic interactions are reliant on communicating the location of herbivores on infested plants to parasitoids and predators of the herbivores (Turlings and Erb 2018) (Fig. 9.1). Understanding the evolution of plant-natural enemy relationships requires knowledge of these mechanisms (Stahl et al. 2018). HIPVs can induce plants to repel herbivores or draw pest's natural enemies (Ayelo et al. 2021; Naselli et al. 2016; Pérez-Hedo et al. 2015a). For instance, aphid (*Myzus persicae*) induces α-pinene, decanal phthalic acid, and isophorone in pepper and cabbage, which attract its endoparasitoid *Aphelinus varipes* (Ali et al. 2022b). Similarly, aphid (*M. euphorbiae*) infestation on tomato plants treated with *Trichoderma harzianum* induced aphid mortality and attracted the aphid parasitoid *Aphidius ervi* (Coppola et al. 2019). *T. absoluta*-induced monoterpenes attract its predator *Nesidiocoris tenuis* in tomato (Ayelo et al. 2021). Pea leafminer (*Liriomyza huidobrensis*) induced methyl salicylate, and (*E,E*)-4,8,12-trimethyl-1,3,7,11-tridecatetraene (TMTT) in bean leaf attracts its parasite *Opius* sp. (Yang et al. 2021). In response to herbivory, plants synthesize and release variable mixtures of herbivore-induced plant volatiles (HIPVs) as indirect defense traits that attract enemies of the herbivores. For example, HIPVs emitted by insect-infested cucumber and potato were used as cues by the predatory ant *Formica pratensis* to locate its prey (Schettino et al. 2017). Volatiles from wild potato (*Solanum stoloniferum*) accessions significantly attracted the parasitoid, *Diaeretiella rapae* against *M. persicae* (Ali et al. 2022a).

Apart from herbivores, the predator activity on host plants can also prime defense against insects. For instance, green lacewing larvae (*Chrysoperla carnea*) activity in tomato induces myrcene and α-terpinene, which aid in plant defense (Errard et al. 2016). Zoophytophagous insect, *N. tenuis* feeding punctures on tomato stimulate the jasmonic and abscisic acid (ABA) pathways, making the plants less attractive to *T. absoluta* and *B. tabaci* but more attractive to the whitefly parasitoid *Encarsia formosa* (Pérez-Hedo et al. 2015b). In tomato plants, defenses are primed both locally and systemically in response to attacks from predators (*Macrolophus pygmaeus* and *tenuis*), (Z)-3-hexenyl propanoate induced by *N. tenuis* damage can indirectly prime the defenses of healthy plants (Pappas et al. 2015; Pérez-Hedo et al. 2018, 2021).

9.4 Transgenics and Genome Editing

Insect pests are the primary biotic stressors that cause significant crop losses worldwide by directly feeding on and transferring plant diseases (Douglas 2018). Biotechnological interventions in insect pest management to protect crop output have received increased focus, ranging from insect resistance breeding through transgenic introgression of novel genes to the genome editing (Doudna and Charpentier 2014). Some biotechnological approaches include gene transformation, genome editing, RNA interference, marker-assisted selection, anther culture, embryo culture, protoplast fusion, somaclonal variations etc. (Talakayala et al. 2020). Introducing a specific DNA segment or gene into crop plants to provide resistance against insect pests is known as gene transformation or genetic engineering of crops for insect resistance (Gatehouse 2013). The DNA segment that is inserted normally encodes an insecticidal protein. Plants obtain resistance to specific insect pests by expressing an insecticidal protein found in the inserted DNA segment (Birkett and Pickett 2014).

9.4.1 Cry Genes

Insecticidal activity in insect-resistant Bacillus thuringiensis (Bt) crops is expressed by the genes coding for parasporal crystal protoxins (Panwar et al. 2018). Insecticidal crystal protoxins produced by transgenic plants have significantly impacted the successful evolution of insect resistance (Paul and Das 2021). The crystal involves a protoxin protein, which gets solubilized in the larval midgut due to alkaline pH to produce more toxic protein that leads to gut paralysis, cessation of feeding and mortality within 2/3 days (Rajadurai et al. 2018). Soliman et al. (2021) reported that tomato leaf miner, *T. absoluta*, was susceptible to a synthetic Cry1Ab gene inserted into tomato, with 100% insect mortality in T_0 generation within 4–5 days. Similarly, Cry1Aa gene expression in sweet potatoes conferred resistance to *Spodoptera litura* (Zhong et al. 2019).

9.4.2 Lectins

Plant lectins are proteins that bind to carbohydrates and have an increased affinity for certain sugars found in glycoproteins and glycolipids in the cell membrane. Lepidoptera, Coleoptera, and Hemiptera insects of various orders have been demonstrated to perform worse when exposed to lectins in fake diets and their production in transgenic plants (Camaroti et al. 2018). Fitches et al. (2004) found increased toxicity of GNA-spider-venom toxin I (SFI1) fusion protein to larvae of the tomato moth (*L. oleracea*) and the peach potato aphid (*M. persicae*).

9.4.3 Fusion Proteins

Two recent techniques to address potential limits in conventional transgenic insect pest control are the stacking or pyramiding of numerous transgenes in the same transgenic plant and the use of hybrid toxins against insect pests (Fig. 9.2) (Boddupally et al. 2018; Talakayala et al. 2020). The feeding and growth of the tomato moth were found to be inhibited by a GNA-neuropeptide-allatostatin fusion protein (*L. oleracea*) (Fitches et al. 2004). A fusion protein from the South Indian red scorpion (*Mesobuthus tamulus*) that contains a GNA-lepidopteran-specific toxin (ButalT) was found to be more harmful to tomato moth larvae (Trung et al. 2006).

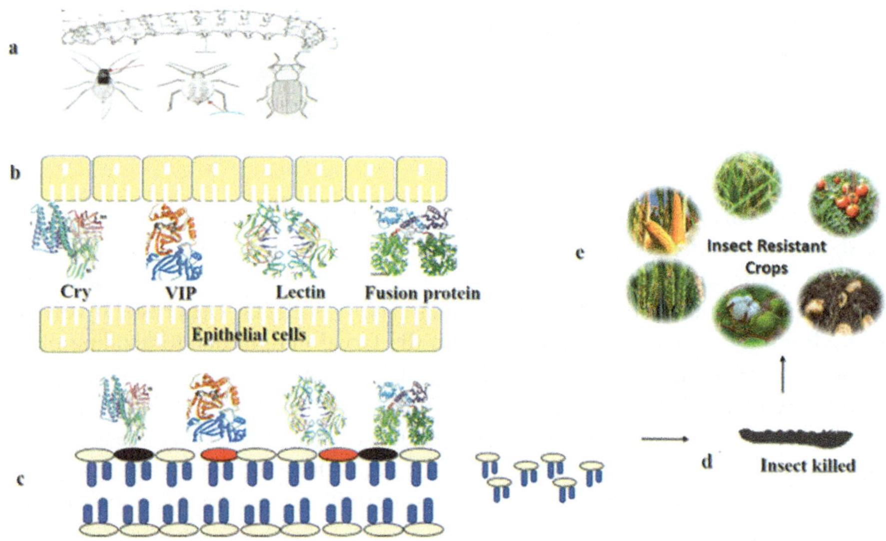

Fig. 9.2 Schematic representation of the mode of action of insecticidal proteins like Cry proteins, VIP proteins, lectins, and other fusion proteins in the gut of lepidopterans, aphids, and other insects. (**a**) Different insects that feeds on major crops (**b**) mode of action of different proteins in the gut epithelial cells (**c**) cell lysis by ICP on epithelial cells (**d**) death of the insect (**e**) insect-resistant crops (Source: Talakayala et al. 2020)

9.4.4 RNAi for Plant Resistance to Insect

RNAi, referred to as co-concealment, post-transcriptional gene silencing (PTGS), and suppressing, is a technique for reducing gene expression by reducing particular regions (Kamthan et al. 2015). Planning the (distribution) strategy and deciding whether a transformative or non-transformative RNAi-plant protection technique might be preferable requires knowledge about the target insect's eating patterns. For chewing insects, dsRNA can be supplied through a foliar spray, such as a typical plant protection chemical, and picked up straight from leaves. Products based on RNAi that can be sprayed are currently being developed and will soon hit the market. However, the spraying strategy would not affect the piercing-sucking bugs that consumed phloem sap, insects that fed on the root system, or stem borer pests that consumed plant stems (Fig. 9.3). The delivery of dsRNA through the phloem sap is necessary for sap-sucking insects, and this can be done by irrigation water, trunk injection for perennial trees, in planta dsRNA creation (transgenic or transplastomic plants), or recombinant plant viruses (Joga et al. 2016). In order to develop insect resistance in tobacco and tomatoes, the chitinase gene (HaCHI) in *H. armigera* was silenced using HI-RNAi. This resulted in the downregulation of the target gene

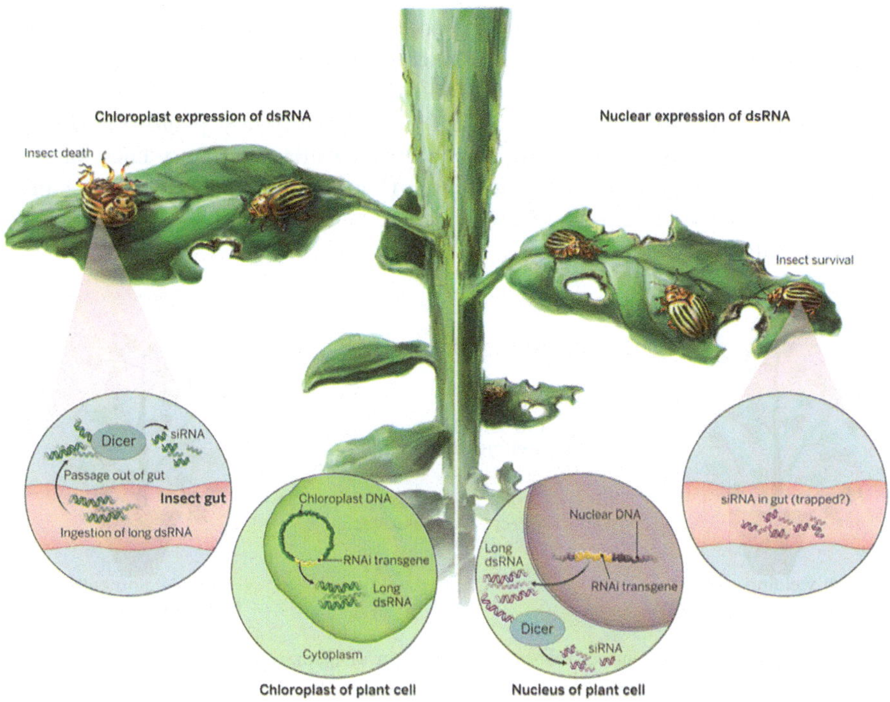

Fig. 9.3 The transgenic potato plants producing dsRNA in the plant cell cytoplasm do not kill potato beetles, whereas the plants bred to express insect-specific long dsRNA in chloroplasts do. (Source: Zhang et al. 2015)

transcripts, which led to mortality and developmental deformities at the larval, pupal, and adult stages (Mamta and Rajam 2017). The tobacco whitefly, *Bemisia tabaci*, has recently been identified as having no impact on tomato plants using a dsRNA that targeted a gene expressing a phenolic glucoside malonyltransferase, detoxifying phenolic glycosides (Xia et al. 2021). These results indicate that RNAi is one of the most efficient ways to produce insect-resistant plants. Although the technology is still in its infancy, its current drawbacks make it less practical to manage insect pests (Scott et al. 2013).

9.4.5 Genome Editing

The process of "genome editing," also known as "gene editing," includes adding, removing, or replacing DNA bases in a target DNA sequence of the genome in order to successfully change a gene's function through the cell's inherent mechanisms (Bortesi and Fischer 2015; Asokan et al. 2022; Ashok et al. 2023). The most recent and technically straightforward tool for creating insect pest resistant cultivars is CRISPR/Cas9 (Fig. 9.4). It has been successfully employed by eliminating the gene or causing missense mutations in the target gene to stop the accumulation of specific gene products in various crops (Gao 2021). In the wild tomato *S. pimpinellifolium*, CRISPR/Cas9 was utilized to target six loci related to tomato yield and efficiency (Zsögön et al. 2018). *Plutella xylostella* had increased levels of resistance to cry1Ac protoxin following the CRISPR/Cas9 tool's knockout of two ABC transporters, PxABCC2 and PxABCC3 (Guo et al. 2019). To identify host plants, the majority of polyphagous insects rely on the plant's own volatiles, gustatory cues, outward appearance, oviposition sites, and interactions (Larsson et al. 2004). Genome editing can change plant volatile mixtures, which could be an alternative pest management strategy in future.

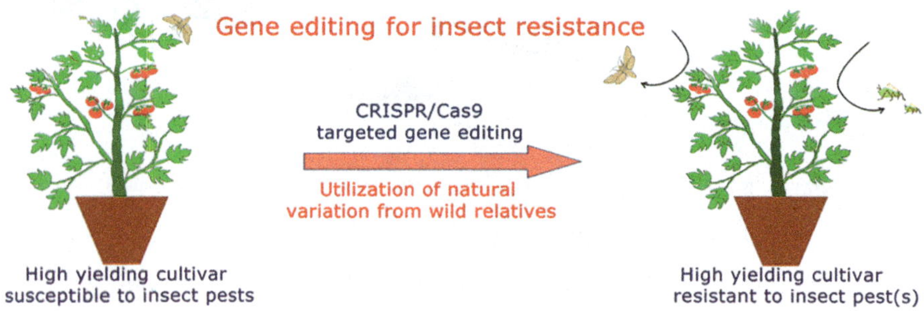

Fig. 9.4 Gene editing for insect resistance: utilization of natural variation for improvement of susceptible cultivars for resistance against insect pest(s) through CRISPR. (Source: Tyagi et al. 2020)

9.5 Critical Gaps, Challenges, and Priorities

The dynamic interplay of molecular and physiological responses that probably develop between plants and insects throughout time is not well captured by many current techniques. There are now more options for pest control, thanks to the modern biotechnological techniques, which offer excellent chances to build a long-lasting, multi-mechanistic resistance to insect pests. Although current technology is used in crops to help them become resistant to a range of insect pests, some agricultural pests regularly acquire a resistance to insecticidal toxins, which has a disastrous effect on crop yield. The research teams should focus on the challenges to understanding the plant–insect interactions.

Plant–insect interaction research has historically taken a stance, approaching issues from either an insect- or a plant-centric standpoint. Particularly with regard to molecular reactions, both partners are rarely taken into account at once (e.g., tomato against hornworm, Havko et al. 2020). Most studies frame concerns about defensive systems from the viewpoint of plants, with insects being portrayed as having countermeasures or adaptations to their hosts (Heidel-Fischer and Vogel 2015; Wouters et al. 2016; War et al. 2018). As a result, our understanding of host plant defenses and the molecular mechanisms behind insect virulence is far from perfection (Bansal and Michel 2015; Yates-Stewart et al. 2020).

Although numerous genes or pathways are likely involved in HPR in reality, quantitative approaches are required, much as those that are gaining popularity for creating pathogen resistance (Pilet-Nayel et al. 2017). The emergence of various biotechnology-based technologies, including marker-assisted selection, somaclonal variation, anther culture, embryo culture, genetic transformation, and protoplast fusion, may hasten the development of insect-resistant crops in the present and in the future. Future HPR breeding efforts will depend mainly on the ability to reliably identify fresh sources of resistance and conduct large-scale screening for that resistance. Resistance cannot be introduced into the pipeline for cultivar development if it cannot be discovered or screened for.

9.5.1 Phenomics

The measurement of groups of phenotypes, frequently physical or biochemical properties, is known as phenomics. Phenomics is rapidly being included into commercial HPR breeding projects because of its enormous potential to reduce the high cost of field phenotyping of plant resistance responses. One of the main benefits of phenomics is that the classification of resistance reactions is often noninvasive and standardized across fields, reducing human error (Mahlein et al. 2019). However, applicability across crops and diseases/pests remains a challenge (Varshney et al. 2021).

9.5.2 Genome Editing for Insect Resistance

This fascinating tool has emerged as an imperative contribution that allows modifications in the genome by adding/editing/deleting particular stretch of DNA sequences, thereby providing opportunities for utility in plants, animals, and humans. In the present scenario of constricted agricultural fields and increased load of insect pests on crop plants, genome editing will serve as a potential tool to combat insect pests. The latest revolutionary technology for genome editing is based on the RNA-guided engineered nucleases called CRISPR (clustered regularly interspaced short palindromic repeats)/Cas9, which hold great promise because of their specificity, simplicity, efficiency, and versatility by addressing key challenges posed by other genome editing tools (Tyagi et al. 2020). Genome editing in insect pest management for diamondback moth in cabbage for body segmentation trait with target gene (*abdominal-A* gene knockout) was done (Sun et al. 2017). Despite success in economically important crops to combat pathogens, their use in insect management has not been exploited to the fullest.

Though editing insects is an intriguing option, it requires caution in the selection of traits that need to be environmentally friendly so that the food chain is not affected. The major setback has been the lack of availability of target genes in contrast to other stresses. Therefore, emphasis needs to be given by scientists globally to the identification of resistance sources that can form a platform for insect management.

9.5.3 Volatilomics (Volatile Organic Compounds-VOCs)

Comparing the volatilomics of plants helps us to find out how they respond to different kinds of herbivory. The plants are also known to communicate with each other and share information among them. Volatilomics deals with the studies on how biological systems emit volatile organic compounds. To date, VOCs are applied in agriculture solely for the "push-pull strategy," where the crop of interest is both intercropped with plant species that emit VOCs able to repel ("push") herbivores, and surrounded with plants emitting VOCs that simultaneously attract ("pull") herbivores away from the field (Stenberg et al. 2015; Picket and Khan 2016). VOCs are not more intensively used in the host plant resistance as the experiments on the efficacy in the laboratory cannot be simulated and correlated with the open field for conferring the resistance. There is a need to generate baseline information on the inheritance of resistance to insect pests and the host plant–insect–environment interactions to understand the genetic control of different mechanisms of resistance for the development of suitable strategies to increase the levels and diversify the bases of resistance. There is a necessity to break the linkage between the parameters conferring resistance to the target insect pests and the low-yield trait that results in susceptibility and at the same time does not have a negative effect on the quality of the product (Sharma et al. 2017).

9.6 Conclusion

Vegetable crops are influenced by interactions with their environment and insects in terms of quality, nutrients, and yield. In natural ecosystems, it is very likely that numerous herbivores would attack plants sequentially or simultaneously. The primary chemical components of cells that directly or indirectly affect the interactions between plants and insects in vegetables are metabolites and volatiles. These chemical cues mediate tri-trophic interactions influencing plant resistance specificity and local adaptation to natural enemies. Therefore, it is necessary to investigate the metabolites and volatiles that are involved in the resistance mechanism in order to enhance the innate potential of vegetable crops to withstand insect damage. The sustainable management of vegetable plant health can be achieved by understanding constitutive and induced candidate metabolites and/or volatiles that function as phytohormones, signaling molecules, communication agents, elicitors, inhibitors, repellents, and antimicrobials involving numerous biosynthetic pathways.

It has been discovered that plant metabolites and HIPVs are an efficient and environmentally friendly way to manage insects with the potential to lead to the development of new pest management technologies. Vegetable plant associations with beneficial microbes influence metabolic changes that govern plant resistance and attract natural enemies of herbivores. Developing insect-resistant cultivars reduces the number of pests in a stable and cumulative manner while having no negative environmental effects. There has been a substantial advancement in the identification of insect resistant sources in different crops. Unfortunately, because of the entanglement of quantitative features at numerous loci, the production of insect-resistant crop varieties using standard approaches is slow and challenging to achieve. Developing a global database and incorporating it into molecular breeding programs for improved resistance and crop health could be aided by identifying the metabolites involved in plant–insect interactions in various vegetable systems and validating them with insects with different modes of feeding. Development of a long-lasting, multi-mechanistic resistance to insect pests has been possible because of new prospects presented by more advanced biotechnological techniques. This is a new and promising approach to future sustainable, environmentally friendly vegetable production.

References

Abah J, Ishaq MN, Wada AC (2010) The role of biotechnology in ensuring food security and sustainable agriculture. Afr J Biotechnol 9(52):8896–8900

Abuyusuf M, Rubel MH et al (2022) Glucosinolates and biotic stress tolerance in Brassicaceae with emphasis on cabbage: a review. Biochem Genet. https://doi.org/10.1007/s10528-022-10269-6

Ahmed Q, Agarwal M et al (2021) Evaluation of volatile organic compounds from Broccoli plants infested with *Myzus persicae* and parasitoids *Aphidius colemani* attraction. IOS Conf Ser Earth Environ Sci 761:012029

Aktar W, Sengupta D, Chowdhury A (2009) Impact of pesticides use in agriculture: their benefits and hazards. Interdiscip Toxicol 2(1):1

Ali J, Sobhy IS, Bruce TJA (2022a) Wild potato ancestors as potential sources of resistance to the aphid *Myzus persicae*. Pest Manag Sci 78(9):3931–3938. https://doi.org/10.1002/PS.6957

Ali MY, Naseem T, Zhang J, Pan M, Zhang F, Liu TX (2022b) Plant volatiles and herbivore induced plant volatiles from chili pepper act as attractant of the aphid parasitoid *Aphelinus varipes* (Hymenoptera: Aphelinidae). Plants 11(10):1350. https://doi.org/10.3390/PLANTS11101350/S1

Aljbory Z, Chen MS (2018) Indirect plant defense against insect herbivores: a review. Insect Sci 25(1):2–23. https://doi.org/10.1111/1744-7917.12436

Anu BC, Saha T, Akhtar S, Kumari K (2021) Morphological and biochemical constituents influencing aphids and whiteflies tolerance in tomato genotypes. Bangladesh J Bot 50(3): 483–489. https://doi.org/10.3329/bjb.v50i3.55826

Ashok K, Bhargava CN, Babu KP, Rohan W, Manamohan M, Rai A et al (2023) First report on CRISPR/Cas9 mediated editing of the eye colour gene, tryptophan 2,3-dioxygenase in egg plant shoot and fruit borer *Leucinodes orbonalis* Guenée (Lepidoptera: Crambidae). J Asia Pacific Entomol 26(1):102031

Asokan R, Rai A, Dash S, Manamohan M, Ashok K, Bhargava CN et al (2022) Application of genome editing in entomology. Indian J Entomol 84:96–103

Ayelo PM, Yusuf AA, Pirk CWW, Chailleux A, Mohamed SA, Deletre E (2021) Terpenes from herbivore-induced tomato plant volatiles attract *Nesidiocoris tenuis* (Hemiptera: Miridae), a predator of major tomato pests. Pest Manag Sci 77(11):5255–5267. https://doi.org/10.1002/PS.6568

Azzouz H, Cherqui A, Campan EDM, Rahbé Y, Duport G, Jouanin L et al (2005) Effects of plant protease inhibitors, oryzacystatin I and soybean Bowman–Birk inhibitor, on the aphid *Macrosiphum euphorbiae* (Homoptera, Aphididae) and its parasitoid *Aphelinus abdominalis* (Hymenoptera, Aphelinidae). J Insect Physiol 51(1):75–86

Bac-Molenaar JA, Mol S, Verlaan MG, van Elven J, Kim HK, Klinkhamer PGL, Leiss KA, Vrieling K (2019) Trichome independent resistance against western flower thrips in tomato. Plant Cell Physiol 60(5):1011–1024. https://doi.org/10.1093/pcp/pcz01

Bansal R, Michel A (2015) Molecular adaptations of aphid biotypes in overcoming host-plant resistance. In: Raman C, Goldsmith MR, Agunbiade TA (eds) Short views on insect genomics and proteomics. Springer International Publishing, Cham, pp 75–93

Bar M, Shtein I (2019) Plant trichomes and the biomechanics of defense in various systems, with Solanaceae as a model. Botany 97(12):651–660. https://doi.org/10.1139/CJB-2019-0144

Basit A, Hanan A, Nazir T, Majeed M, Qiu D (2019) Molecular and functional characterization of elicitor pebc1 extracted from *Botrytis cinerea* involved in the induction of resistance against green peach aphid (*Myzus persicae*) in common beans (*Phaseolus vulgaris* L.). Insects 10(2):35. https://doi.org/10.3390/insects10020035

Birkett MA, Pickett JA (2014) Prospects of genetic engineering for robust insect resistance. Curr Opin Plant Biol 19:59–67

Bischoff AM, Araujo ES, Benatto A, Zimmermann RC, de Oliveira MC, da Rosa JM et al (2023) Evidence of antibiosis resistance of four strawberry cultivars against *Duponchelia fovealis* (Lepidoptera: Crambidae). Crop Prot 168:106213

Boddupally D, Tamirisa S, Gundra SR, Vudem DR, Khareedu VR (2018) Expression of hybrid fusion protein (Cry1Ac: ASAL) in transgenic rice plants imparts resistance against multiple insect pests. Sci Rep 8(1):8458

Bonaventure G (2012) Perception of insect feeding by plants. Plant Biol 14(6):872–880. https://doi.org/10.1111/J.1438-8677.2012.00650.X

Bonaventure G (2018) Plants recognize herbivorous insects by complex signalling networks. In: Annual plant reviews online. Wiley, pp 1–35

Bortesi L, Fischer R (2015) The CRISPR/Cas9 system for plant genome editing and beyond. Biotechnol Adv 33(1):41–52

Bouwmeester H, Schuurink RC, Bleeker PM, Schiestl F (2019) The role of volatiles in plant communication. Plant J 100(5):892–907

Brilli F, Loreto F, Baccelli I (2019) Exploiting plant volatile organic compounds (VOCs) in agriculture to improve sustainable defence strategies and productivity of crops. Front Plant Sci 10:264. https://doi.org/10.3389/fpls.2019.00264

Camaroti JRSL, de Almeida WA, do Rego Belmonte B, de Oliveira APS, de Albuquerque Lima T, Ferreira MRA et al (2018) *Sitophilus zeamais* adults have survival and nutrition affected by *Schinus terebinthifolius* leaf extract and its lectin (SteLL). Ind Crops Prod 116:81–89

Casaretto JA, Corcuera LJ (1995) Plant proteinase inhibitors: a defensive response against insects (review). Biol Res 28(4):239–249

Chen LM, Li X, He W, Jun T, Li P, LiuY ZS, Xing W, Cong Q, Chen T, Ting L, Bin Y, Hou YM (2021a) Comparative biochemical and transcriptome analyses in tomato and eggplant reveal their differential responses to *Tuta absoluta* infestation. Genomics 113(4):2108–2121

Chen L, Li X, Zhang J, He T, Huang J, Zhang Z, Wang Y, Hafeez M, Zhou S, Ren X, Hou Y, Lu Y (2021b) Comprehensive metabolome and volatilome analyses in eggplant and tomato reveal their differential responses to *Tuta absoluta* infestation. Front Plant Sci 12:2231

Conboy NJA, McDaniel T, George D, Ormerod A, Edwards M, Donohoe P, Gatehouse AMR, Tosh CR (2020) Volatile organic compounds as insect repellents and plant elicitors: an integrated pest management (IPM) strategy for glasshouse whitefly (*Trialeurodes vaporariorum*). J Chem Ecol 46(11–12):1090–1104. https://doi.org/10.1007/S10886-020-01229-8

Coppola M, Cascone P, Madonna V, Di Lelio I, Esposito F, Avitabile C, Romanelli A, Guerrieri E, Vitiello A, Pennacchio F, Rao R, Corrado G (2017) Plant-to-plant communication triggered by systemin primes anti-herbivore resistance in tomato. Sci Rep 7(1):1–13. https://doi.org/10.1038/s41598-017-15481-8

Coppola M, Diretto G, Digilio MC, Woo SL, Giuliano G, Molisso D, Pennacchio F, Lorito M, Rao R (2019) Transcriptome and metabolome reprogramming in tomato plants by *Trichoderma harzianum* strain t22 primes and enhances defense responses against aphids. Front Physiol 10: 745

D'Esposito D, Manzo D, Ricciardi A, Garonna AP, De Natale A, Frusciante L, Pennacchio F, Ercolano MR (2021) Tomato transcriptomic response to *Tuta absoluta* infestation. BMC Plant Biol 21(1):1–14. https://doi.org/10.1186/s12870-021-03129-9

D'Esposito D, Guadango A et al (2023) Genomic and metabolic profiling of two tomato contrasting cultivars for tolerance to *Tuta absoluta*. Planta 257:47. https://doi.org/10.1007/s00425-023-04073-8

Da Silva AA, Andrade MC, Carvalho RC, Neiva IP, Santos DC, Maluf WR (2016) Resistência à *Helicoverpa armigera* em genótipos de tomateiro obtidos do cruzamento de *Solanum lycopersicum* com *Solanum galapagense*. Pesquisa Agropecuária Brasileira 51(7):801–808

Darshanee HLC, Ren H, Ahmed N, Zhang ZF, Liu YH, Liu TX (2017) Volatile-mediated attraction of greenhouse whitefly *Trialeurodes vaporariorum* to tomato and eggplant. Front Plant Sci 8: 1285. https://doi.org/10.3389/fpls.2017.01285

De Falco B, Manzo D, Incerti G, Garonna AP, Ercolano M, Lanzotti V (2019) Metabolomics approach based on NMR spectroscopy and multivariate data analysis to explore the interaction between the leafminer *Tuta absoluta* and tomato (*Solanum lycopersicum*). Phytochem Anal 30(5):556–563

Dhaliwal GS, Dhawan AK, Singh R (2007) Biodiversity and ecological agriculture: Issues and perspectives. Indian J Ecol 34(2):100–109

Divekar PA, Narayana S et al (2022) Plant secondary metabolites as defense tools against herbivores for sustainable crop protection. Int J Mol Sci 23:2690. https://doi.org/10.3390/ijms23052690

Dominguez E, Heredia-Guerrero JA et al (2017) The plant cuticle: old challenges, new perspective. J Exp Bot 68(19):5251–5255. https://doi.org/10.1093/jxb/erx389

Doudna JA, Charpentier E (2014) The new frontier of genome engineering with CRISPR-Cas9. Science 346(6213):1258096

Douglas AE (2018) Strategies for enhanced crop resistance to insect pests. Ann Rev Plant Biol 69: 637–660

Errard A, Ulrichs C, Kühne S, Mewis I, Mishig N, Maul R, Drungowski M, Parolin P, Schreiner M, Baldermann S (2016) Metabolite profiling reveals a specific response in tomato to predaceous *Chrysoperla carnea* larvae and herbivore(S)-predator interactions with the generalist pests *Tetranychus urticae* and *Myzus persicae*. Front Plant Sci 7:1256. https://doi.org/10.3389/FPLS.2016.01256

Escobar-Bravo R, Klinkhamer PGL, Leiss KA (2017) Induction of jasmonic acid-associated defenses by thrips alters host suitability for conspecifics and correlates with increased trichome densities in tomato. Plant Cell Physiol 58(3):622. https://doi.org/10.1093/PCP/PCX014

Fitches E, Wilkinson H, Bell H, Bown DP, Gatehouse JA, Edwards JP (2004) Cloning, expression and functional characterisation of chitinase from larvae of tomato moth (*Lacanobia oleracea*): a demonstration of the insecticidal activity of insect chitinase. Insect Biochem Mol Biol 34(10): 1037–1050

Gao C (2021) Genome engineering for crop improvement and future agriculture. Cell 184(6): 1621–1635

Gatehouse JA (2013) In: Christou P, Savin R, Costa-Pierce BA, Misztal I, Whitelaw CBA (eds) Genetic engineering of crops for insect resistance in sustainable food production. Springer, New York

Ghosh S, Didi-Cohen S, Cna'ani A, Kontsedalov S, Lebedev G, Tzin V, Ghanim M (2022) Comparative analysis of volatiles emitted from tomato and pepper plants in response to infection by two whitefly-transmitted persistent viruses. Insects 13:840. https://doi.org/10.3390/insects13090840

Glas JJ, Schimmel BCJ, Alba JM, Escobar-Bravo R, Schuurink RC, Kant MR (2012) Plant glandular trichomes as targets for breeding or engineering of resistance to herbivores. Int J Mol Sci 13(12):17077–17103. https://doi.org/10.3390/ijms131217077

Gomez S, Steinbrenner AD et al (2012) From shoots to roots: transport and metabolic changes in tomato after simulated feeding by a specialist lepidopteran. Entomol Exp Appl 144:101–111. https://doi.org/10.1111/j.1570-7458.2012.01268.x

Guo H, Sun Y, Ren Q, Zhu-Salzman K, Kang L et al (2012) Elevated CO_2 reduces the resistance and tolerance of tomato plants to *Helicoverpa armigera* by suppressing the JA signalling pathway. PLoS One 7(7):e41426. https://doi.org/10.1371/journal.pone.0041426

Guo Z, Li Y, Ding SW (2019) Small RNA-based antimicrobial immunity. Nat Rev Immunol 19(1): 31–44

Haldhar SM, Choudhary BR, Bhargava R, Meena SR (2015) Antixenotic and allelochemical resistance traits of watermelon against *Bactrocera cucurbitae* in a hot arid region of India. Florida Entomol 93:827–834

Haldhar SM, Samadia DK, Bhargava R, Singh D (2017) Host plant genotypes determine bottom-up effect of *Cucumis melo* var. callosus against melon fruit fly. Crop Prot 98:157–165

Hall R (1995) Challenges and prospects of integrated pest management. In: Novel approaches to integrated pest management. Lewis Publishers, pp 1–99

Hamza R, Pérez-Hedo M, Urbaneja A, Rambla JL, Granell A, Gaddour K, Beltrán JP, Cañas LA (2018) Expression of two barley proteinase inhibitors in tomato promote endogenous defensive response and enhances resistance to *Tuta absoluta*. BMC Plant Biol 18(1):1–14. https://doi.org/10.1186/S12870-018-1240-6

Han P, Becker C, Le Bot J, Larbat R, Lavoir AV, Desneux N (2020) Plant nutrient supply alters the magnitude of indirect interactions between insect herbivores: from foliar chemistry to community dynamics. J Ecol 108(4):1497–1510. https://doi.org/10.1111/1365-2745.13342

Hanley ME, Lamont BB, Fairbanks MM, Rafferty CM (2007) Plant structural traits and their role in anti-herbivore defence. Pers Plant Ecol Evol Syst 8(4):157–178

Havko NE, Das MR, McClain AM, Kapali G, Sharkey TD, Howe GA (2020) Insect herbivory antagonizes leaf cooling responses to elevated temperature in tomato. Proc Natl Acad Sci U S A 117:2211–2217

He J, Bouwmeester HJ, Dicke M, Kappers IF (2020) Transcriptional and metabolite analysis reveal a shift in direct and indirect defences in response to spider-mite infestation in cucumber

(*Cucumis sativus*). Plant Mol Biol 103(4–5):489–505. https://doi.org/10.1007/S11103-020-01005-Y

Heidel-Fischer HM, Vogel H (2015) Molecular mechanisms of insect adaptation to plant secondary compounds. Curr Opin Insect Sci 8:8–14

Horgan FG, Garcia CPF, Haverkort F, de Jong PW, Ferrater JB (2020) Changes in insecticide resistance and host range performance of planthoppers artificially selected to feed on resistant rice. Crop Prot 127:104963

Howe GA, Major IT, Koo AJ (2018) Modularity in jasmonate signaling for multistress resilience. Annu Rev Plant Biol 69:387–415. https://doi.org/10.1146/ANNUREV-ARPLANT-042817-040047

Huot OB, Nachappa P, Tamborindeguy C (2013) The evolutionary strategies of plant defenses have a dynamic impact on the adaptations and interactions of vectors and pathogens. Insect Sci 20(3): 297–306

Hussain T, Aksoy E et al (2019) Transgenic potato lines expressing hairpin RNAi construct of molting-associated EcR gene exhibit enhanced resistance against Colorado potato beetle (*Leptinotarsa decemlineata*, Say). Transgenic Res 28:151–164. https://doi.org/10.1007/s11248-018-0109-7

Iqbal AM, Waqar Hassan M, Rehman MH, Aslam M, Jamil M (2018) Evaluation of different wheat varieties for resistance against aphid, *Schizaphis graminum* R. (Homoptera: Aphididae) under laboratory conditions. Asian J Agric Biol 6(4):549–555

Jeckel AM, Beran F et al (2022) Metabolization and sequestration of plant specialized metabolites in insect herbivores: current and emerging approaches. Front Physiol 3:1001032. https://doi.org/10.3389/fphys.2022.1001032

Jha Y, Mohamed HI (2022) Plant secondary metabolites as a tool to investigate biotic stress tolerance in plants: a review. Gesunde Pflanzen 74:771–790. https://doi.org/10.1007/s10343-022-00669-4

Jiang Y, Zhang CX, Chen R, He SY (2019) Challenging battles of plants with phloem-feeding insects and prokaryotic pathogens. Proc Natl Acad Sci U S A 116(47):23390–23397. https://doi.org/10.1073/PNAS.1915396116

Jiao X, Xie W et al (2018) Lack of correlation between host choice and feeding efficiency for the B and Q putative species of *Bemisia tabaci* on four pepper genotypes. J Pest Sci 91:133–143. https://doi.org/10.1007/s10340-017-0906-4

Jo HE, Song K, Kim JG, Lee CH (2022) Non-targeted metabolomic analysis for the comparative evaluation of volatile organic compounds in 20 globally representative cucumber lines. Front Plant Sci 13:1028735. https://doi.org/10.3389/fpls.2022.1028735

Joga MR, Zotti MJ, Smagghe G, Christiaens O (2016) RNAi efficiency, systemic properties, and novel delivery methods for pest insect control: what we know so far. Front Physiol 7:553

Kamthan A, Chaudhuri A, Kamthan M, Datta A (2015) Small RNAs in plants: recent development and application for crop improvement. Front Plant Sci 6:208

Kersch-Becker MF, Thaler JS, Mônica Kersch-Becker CF (2019) Constitutive and herbivore-induced plant defences regulate herbivore population processes. J Anim Ecol 88(7): 1079–1088. https://doi.org/10.1111/1365-2656.12993

Koch KG, Chapman K, Louis J, Heng-Moss T, Sarath G (2016) Plant tolerance: a unique approach to control hemipteran pests. Front Plant Sci 7:1363. https://doi.org/10.3389/fpls.2016.01363

Kortbeek RWJ, Galland MD, Muras A, van der Kloet FM, André B, Heilijgers M, van Hijum SAFT, Haring MA, Schuurink RC, Bleeker PM (2021) Natural variation in wild tomato trichomes; selecting metabolites that contribute to insect resistance using a random forest approach. BMC Plant Biol 21(1). https://doi.org/10.1186/s12870-021-03070-x

Kumar P, Ortiz EV, Garrido E, Poveda K, Jander G (2016) Potato tuber herbivory increases resistance to aboveground lepidopteran herbivores. Oecologia 182(1):177–187. https://doi.org/10.1007/S00442-016-3633-2

Kumari P, Mahapatro GK et al (2015) Ectopic expression of GroEL from *Xenorhabdus nematophila* in tomato enhances resistance against *Helicoverpa armigera* and salt and thermal stress. Transgenic Res 24:859–873. https://doi.org/10.1007/s11248-015-9881-9

Kundu A, Mishra S, Vadassery J (2018) Spodoptera litura-mediated chemical defense is differentially modulated in older and younger systemic leaves of *Solanum lycopersicum*. Planta 248(4): 981–997. https://doi.org/10.1007/s00425-018-2953-3

Larsson MC, Domingos AI, Jones WD, Chiappe ME, Amrein H, Vosshall LB (2004) Or83b encodes a broadly expressed odorant receptor essential for *Drosophila* olfaction. Neuron 43(5):703–714

Leckie BM, D'Ambrosio DA, Chappell TM, Halitschke R, de Jong DM, Kessler A, Kennedy GG, Mutschler MA (2016) Differential and synergistic functionality of acylsugars in suppressing oviposition by insect herbivores. PLoS One 11(4). https://doi.org/10.1371/journal.pone.0153345

Leybourne DJ, Aradottir GI (2022) Common resistance mechanisms are deployed by plants against sap-feeding herbivorous insects: insights from a meta-analysis and systematic review. Sci Rep 12:17836. https://doi.org/10.1038/s41598-022-20741-3

Li Z, Kund G et al (2019) Effects of high-level acyl sugar-producing tomato lines on the development of tomato Psyllids (Hemiptera: Triozidae). J Econ Entomol 112(4):1926–1931. https://doi.org/10.1093/jee/toz070

Li Y, Fan H, Si Y, Guo Y, Wu Z, Du C (2023) The phloem lectin PP2-A1 enhances aphid resistance by affecting aphid behavior and maintaining ROS homeostasis in cucumber plants. Int J Biol Macromol 229:432–442

Lin PA, Peiffer M, Felton GW (2020) Induction of defensive proteins in Solanaceae by salivary glucose oxidase of *Helicoverpa zea* caterpillars and consequences for larval performance. Arthropod Plant Interact 14(3):317–325

Liu Y, Wang X, Luo S, Ma L, Zhang W, Xuan S, Wang Y, Zhao J, Shen S, Ma W, Gu A, Chen X (2022) Metabolomic and transcriptomic analyses identify quinic acid protecting eggplant from damage caused by western flower thrips. Pest Manag Sci 78(12):5113–5123. https://doi.org/10.1002/PS.7129

Maag D, Erb M, Glauser G (2015) Metabolomics in plant-herbivore interactions: challenges and applications. Entomol Exp Appl 157(1):18–29. https://doi.org/10.1111/EEA.12336

MacDougall S, Bayansal F, Ahmadi A (2022) Emerging methods of monitoring volatile organic compounds for detection of plant pests and diseases. Biosensors 12:239. https://doi.org/10.3390/bios12040239

Macel M, Visschers IGS, Peters JL, Kappers IF, de Vos RCH, van Dam NM (2019) Metabolomics of thrips resistance in pepper (*Capsicum* spp.) reveals monomer and dimer acyclic diterpene glycosides as potential chemical defenses. J Chem Ecol 45(5–6):490–501. https://doi.org/10.1007/s10886-019-01074-4

Mahlein AK, Alisaac E, Al Masri A, Behmann J, Dehne HW, Oerke EC (2019) Comparison and combination of thermal, fluorescence, and hyperspectral imaging for monitoring Fusarium head blight of wheat on spikelet scale. Sensors 19(10):2281. https://doi.org/10.3390/s19102281

Mamta B, Rajam MV (2017) RNAi technology: a new platform for crop pest control. Physiol Mol Biol Plants 23:487–501

Mansoor S, Ali Wani O et al (2022) Reactive oxygen species in plants: from source to sink. Antioxidants 11:225. https://doi.org/10.3390/antiox11020225

Mansour MR, Eryan NL (2022) Effects of certain weather, biotic factors and chemical components on the population of aphids in Egyptian wheat fields. Egyptian Acad J Biol Sci A Entomol 15(1):1–13

Mao Z, Ge Y, Zhang Y, Zhong J, Munawar A, Zhu Z, Zhou W (2022) Disentangling the potato tuber moth-induced early-defense response by simulated herbivory in potato plants. Front Plant Sci 13:902342

Marchant WG, Legarrea S, Smeda JR, Mutschler MA, Srinivasan R (2020) Evaluating acylsugars-mediated resistance in tomato against *Bemisia tabaci* and transmission of tomato yellow leaf curl virus. Insects 11(12):1–18. https://doi.org/10.3390/insects11120842

Mi X, Ji X, Yang J et al (2015) Transgenic potato plants expressing cry3A gene confer resistance to Colorado potato beetle. C R Biol 338:443–450. https://doi.org/10.1016/j.crvi.2015.04.005

Mitchell C, Brennan RM, Graham J, Karley AJ (2016) Plant defence against herbivorous pests: exploiting resistance and tolerance traits for sustainable crop protection. Front Plant Sci 7:1132. https://doi.org/10.3389/fpls.2016.01132

Movva V, Pathipati UR (2017) Feeding-induced phenol production in *Capsicum annuum* L. influences *Spodoptera litura* F. larval growth and physiology. Arch Insect Biochem Physiol 95(1). https://doi.org/10.1002/ARCH.21387

Murakami Y, Iwabuchi H, Ohba Y, Fukami H (2019) Analysis of volatile compounds from chili peppers and characterization of habanero volatiles. J Oleo Sci 68(12):1251–1269. https://doi.org/10.5650/jos.ess.19155

Murata M, Konno K, Wasano N, Mochizuki A, Mitsuhara I (2021) Expression of a gene for an MLX56 defense protein derived from mulberry latex confers strong resistance against a broad range of insect pests on transgenic tomato lines. PLoS One 16(1):e0239958. https://doi.org/10.1371/journal.pone.0239958

Nalam V, Louis J, Shah J (2019) Plant defense against aphids, the pest extraordinaire. Plant Sci 279: 96–107. https://doi.org/10.1016/j.plantsci.2018.04.027

Naselli M, Urbaneja A, Siscaro G, Jaques JA, Zappalà L, Flors V, Pérez-Hedo M (2016) Stage-related defense response induction in tomato plants by *Nesidiocoris tenuis*. Int J Mol Sci 17(8): 1210

Painter RH (1951) Insect resistance in crop plants. University of Kansas Press, Lawrence, KS

Panwar BS, Ram C, Narula RK, Kaur S (2018) Pool deconvolution approach for high-throughput gene mining from *Bacillus thuringiensis*. Appl Microbiol Biotechnol 102:1467–1482

Pappas ML, Steppuhn A, Geuss D, Topalidou N, Zografou A, Sabelis MW, Broufas GD (2015) Beyond predation: the zoophytophagous predator *Macrolophus pygmaeus* induces tomato resistance against spider mites. PLoS One 10(5):e0127251. https://doi.org/10.1371/JOURNAL.PONE.0127251

Paudel S, Lin P et al (2019) Induced plant defenses against herbivory in cultivated and wild tomato. J Chem Ecol 45:693–707. https://doi.org/10.1007/s10886-019-01090-4

Paul S, Das S (2021) Natural insecticidal proteins, the promising bio-control compounds for future crop protection. Nucleus 64:7–20

Perez-Fons L, Bohorquez-Chaux A, Irigoyen ML, Garceau DC, Morreel K, Boerjan W, Walling LL, Becerra Lopez-Lavalle LA, Fraser PD (2019) A metabolomics characterisation of natural variation in the resistance of cassava to whitefly. BMC Plant Biol 19(1):1–14. https://doi.org/10.1186/S12870-019-2107-1

Pérez-Hedo M, Bouagga S, Jaques JA, Flors V, Urbaneja A (2015a) Tomato plant responses to feeding behavior of three zoophytophagous predators (Hemiptera: Miridae). Biol Control 86: 46–51

Pérez-Hedo M, Urbaneja-Bernat P, Jaques JA, Flors V, Urbaneja A (2015b) Defensive plant responses induced by *Nesidiocoris tenuis* (Hemiptera: Miridae) on tomato plants. J Pest Sci 88(3):543–554

Pérez-Hedo M, Suay R, Alonso M, Ruocco M, Giorgini M, Poncet C, Urbaneja A (2017) Resilience and robustness of IPM in protected horticulture in the face of potential invasive pests. Crop Prot 97:119–127

Pérez-Hedo M, Arias-Sanguino ÁM, Urbaneja A (2018) Induced tomato plant resistance against *tetranychus urticae* triggered by the phytophagy of *Nesidiocoris tenuis*. Front Plant Sci 9. https://doi.org/10.3389/FPLS.2018.01419

Pérez-Hedo M, Alonso-Valiente M, Vacas S, Gallego C, Pons C, Arbona V, Rambla JL, Navarro-Llopis V, Granell A, Urbaneja A (2021) Plant exposure to herbivore-induced plant volatiles: a sustainable approach through eliciting plant defenses. J Pest Sci 94(4):1221–1235. https://doi.org/10.1007/S10340-021-01334-X

Perkins LE, Cribb BW, Brewer PB, Hanan J, Grant M, de Torres M, Zalucki MP (2013) Generalist insects behave in a jasmonate-dependent manner on their host plants, leaving induced areas quickly and staying longer on distant parts. Proc R Soc B Biol Sci 280(1756). https://doi.org/10.1098/rspb.2012.2646

Picket JA, Khan ZR (2016) Plant volatile-mediated signalling and its application in agriculture: successes and challenges. New Phytol 212:856–870. https://doi.org/10.1111/nph.14274

Piesik D, Aksoy J et al (2022) Relationships between volatile organic compounds released by wheat plants following artificial stress and their potential influence on natural pest management. Appl Sci 12:7762. https://doi.org/10.3390/app12157762

Pilet-Nayel ML, Moury B, Caffier V, Montarry J, Kerlan MC, Fournet S, Durel CE, Delourme R (2017) Quantitative resistance to plant pathogens in pyramiding strategies for durable crop protection. Front Plant Sci 8:1838. https://doi.org/10.3389/fpls.2017.01838

Pimentel D, Levitan L (1986) Pesticides: amounts applied and amounts reaching pests. Bioscience 36(2):86–91

Pizetta CSR, Ribeiro WR et al (2022) RNA interference-mediated tolerance to whitefly (*Bemisia tabaci*) in genetically engineered tomato. Plant Cell Tissue Org Cult 148:281–291. https://doi.org/10.1007/s11240-021-02185-1

Poelman EH, Dicke M (2018) Plant-mediated interactions among insects within a community ecological perspective. In: Annual plant reviews online. Wiley, pp 309–337

Quandahor P, Lin C, Gou YA, Coulter J, Liu C (2019) Leaf morphological and biochemical responses of three potato (*Solanum tuberosum* L.) cultivars to drought stress and aphid (*Myzus persicae* Sulzer) infestation. Insects 10(12):435. https://doi.org/10.3390/insects10120435

Quintana-Rodriguez E, Duran-Flores D, Heil M, Camacho-Coronel X (2018) Damage-associated molecular patterns (DAMPs) as future plant vaccines that protect crops from pests. Sci Hortic 237:207–220

Raeyat Z, Razmjou J, Naseri B, Ebadollahi A et al (2021) Evaluation of the susceptibility of some eggplant cultivars to green peach aphid, *Myzus persicae* (Sulzer) (Hemiptera: Aphididae). Agriculture 11:31. https://doi.org/10.3390/agriculture11010031

Rajadurai G, Kalaivani A, Varanavasiyappan S, Balakrishnan N, Udayasuriyan V, Sudhakar D, Natarajan N (2018) Generation of insect resistant marker-free transgenic rice with a novel cry2AX1 gene. Electron J Plant Breed 9(2):723–732

Rashid MH, Chung YR (2017) Induction of systemic resistance against insect herbivores in plants by beneficial soil microbes. Front Plant Sci 8:1816. https://doi.org/10.3389/fpls.2017.01816

Rivera-Vega LJ, Grunseich JM, Aguirre NM, Valencia CU, Sword GA, Helms AM (2022) A beneficial plant-associated fungus shifts the balance toward plant growth over resistance, increasing cucumber tolerance to root herbivory. Plants 11:282. https://doi.org/10.3390/plants11030282

Rivero J, Lidoy J, Llopis-Giménez Á, Herrero S, Flors V, Pozo MJ (2021) Mycorrhizal symbiosis primes the accumulation of antiherbivore compounds and enhances herbivore mortality in tomato. J Exp Bot 72(13):5038–5050. https://doi.org/10.1093/JXB/ERAB171

Rosen R, Kanakala S, Kliot A, Cathrin Pakkianathan B, Farich BA, Santana-Magal N, Elimelech M, Kontsedalov S, Lebedev G, Cilia M, Ghanim M (2015) Persistent, circulative transmission of begomoviruses by whitefly vectors. Curr Opin Virol 15:1–8. https://doi.org/10.1016/J.COVIRO.2015.06.008

Rossouw LT, Madala NE, Tugizimana F, Steenkamp PA, Esterhuizen LL, Dubery IA (2019) Deciphering the resistance mechanism of tomato plants against whitefly-mediated tomato curly stunt virus infection through ultra-high- performance liquid chromatography coupled to mass spectrometry (UHPLC-MS)-based metabolomics approaches. Metabolites 9:4. https://doi.org/10.3390/metabo9040060

Roumani M, Le Bot J, Boisbrun M, Magot F, Péré A, Robin C, Hilliou F, Larbat R (2022) Transcriptomics and metabolomics analyses reveal high induction of the phenolamide pathway in tomato plants attacked by the leafminer *Tuta absoluta*. Metabolites 12(6):484. https://doi.org/10.3390/metabo12060484

Russo A, Pollastri S et al (2022) Volatile organic compounds in the interaction between plants and beneficial microorganisms. J Plant Interact 17:840–852. https://doi.org/10.1080/17429145.2022.2107243

Sahu PK, Jayalakshmi K, Tilgam J et al (2022) ROS generated from biotic stress: effects on plants and alleviation by endophytic microbes. Front Plant Sci 13:1042936. https://doi.org/10.3389/fpls.2022.1042936

Samadia DK, Haldhar SM (2017) Breeding strategies and scope of improvement in arid zone fruit crop-plants under abiotic stressed agro-climate: an analysis. J Agric Ecol 4:1–13

Samadia DK, Haldhar SM (2019) Scope and strategies for genetic improvement in vegetable crop-plants under high temperature and abiotic stressed climate of Rajasthan: a gap analysis. J Agric Ecol 8:1–18

Santamaria ME, Arnaiz A, Gonzalez-Melendi P, Martinez M, Diaz I (2018) Plant perception and short-term responses to phytophagous insects and mites. Int J Mol Sci 19:5

Schettino M, Grasso DA, Weldegergis BT, Castracani C, Mori A, Dicke M, Van Lenteren JC, Van Loon JJA (2017) Response of a predatory ant to volatiles emitted by aphid and caterpillar-infested cucumber and potato plants. J Chem Ecol 43(10):1007–1022. https://doi.org/10.1007/S10886-017-0887-Z

Schuman MC, Baldwin IT (2016) The layers of plant responses to insect herbivores. Annu Rev Entomol 61:373–394. https://doi.org/10.1146/ANNUREV-ENTO-010715-023851

Scott JG, Michel K, Bartholomay LC, Siegfried BD, Hunter WB, Smagghe G et al (2013) Towards the elements of successful insect RNAi. J Insect Physiol 59(12):1212–1221

Selale H et al (2017) *Cry1Ac*-mediated resistance to tomato leaf miner (*Tuta absoluta*) in tomato. Plant Cell Tissue Org Cult 131:65–73. https://doi.org/10.1007/s11240-017-1262-z

Sharma HC, Ortiz R (2002) Host plant resistance to insects: an eco-friendly approach for pest management and environment conservation. J Environ Biol 23(2):111–135

Sharma HC, Jaba J, Vashisth S (2017) Distinguishing proof and utilization of resistance of insect pests in grain legumes: progress and limitations. In: Arora R, Sandhu S (eds) Breeding insect resistant crops for sustainable agriculture. https://doi.org/10.1007/978-981-10-6056-4_5

Shi X, Chen G, Tian L, Peng Z, Xie W, Wu Q et al (2016) The salicylic acid-mediated release of plant volatiles affects the host choice of *Bemisia tabaci*. Int J Mol Sci 17(7):1048

Silva DB, Weldegergis BT, Van Loon JJA, Bueno VHP (2017) Qualitative and quantitative differences in herbivore-induced plant volatile blends from tomato plants infested by either *Tuta absoluta* or *Bemisia tabaci*. J Chem Ecol 43(1):53–65. https://doi.org/10.1007/S10886-016-0807-7

Smith CM (ed) (2005) Plant resistance to arthropods: molecular and conventional approaches. Springer, Dordrecht

Sobhy IS, Woodcock CM, Powers SJ, Caulfield JC, Pickett JA, Birkett MA (2017) Cis-jasmone elicits aphid-induced stress signalling in potatoes. J Chem Ecol 43(1):39–52. https://doi.org/10.1007/S10886-016-0805-9

Sobhy IS, Caulfield JC, Pickett JA, Birkett MA (2020) Sensing the danger signals: cis-jasmone reduces aphid performance on potato and modulates the magnitude of released volatiles. Front Ecol Evol 7:499. https://doi.org/10.3389/FEVO.2019.00499

Soliman HI, Abo-El-Hasan FM, El-Seedy AS, Mabrouk YM (2021) Agrobacterium-mediated transformation of tomato (*Lycopersicon esculentum* Mill.) using a synthetic *cry1ab* gene for enhanced resistance against *Tuta absoluta* (Meyrick). J Microbiol Biotechnol Food Sci 7:67–74

Sperdouli I, Andreadis SS, Adamakis IDS, Moustaka J, Koutsogeorgiou EI, Moustakas M (2022) Reactive oxygen species initiate defence responses of potato photosystem II to sap-sucking insect feeding. Insects 13:409. https://doi.org/10.3390/insects13050409

Stahl E, Hilfiker O, Reymond P (2018) Plant-arthropod interactions: who is the winner? Plant J 93(4):703–728. https://doi.org/10.1111/TPJ.13773

Steglińska A, Pielech-Przybylska K, Janas R, Grzesik M, Borowski S, Kręgiel D, Gutarowska B (2022) Volatile organic compounds and physiological parameters as markers of potato (*Solanum tuberosum* L.) infection with phytopathogens. Molecules 27:3708. https://doi.org/10.3390/molecules27123708

Stenberg JA, Muola A (2017) How should plant resistance to herbivores be measured? Front Plant Sci 8:663. https://doi.org/10.3389/fpls.2017.00663

Stenberg JA, Heil M, Åhman I, Björkman C (2015) Optimizing crops for biocontrol of pests and disease. Trends Plant Sci 20:698–712. https://doi.org/10.1016/j.tplants.201508.007

Su Q, Chen G, Mescher MC, Peng Z, Xie W, Wang S, Wu Q, Liu J, Li C, Wang W, Zhang Y (2018) Whitefly aggregation on tomato is mediated by feeding-induced changes in plant metabolites that influence the behaviour and performance of conspecifics. Funct Ecol 32(5):1180–1193. https://doi.org/10.1111/1365-2435.13055

Su Q, Yang F, Zhang Q, Tong H, Hu Y, Zhang X, Xie W, Wang S, Wu Q, Zhang Y (2020) Defence priming in tomato by the green leaf volatile (Z)-3-hexenol reduces whitefly transmission of a plant virus. Plant Cell Environ 43(11):2797–2811. https://doi.org/10.1111/PCE.13885

Sugimoto K, Matsui K, Iijima Y, Akakabe Y, Muramoto S, Ozawa R, Uefune M, Sasaki R, Alamgir KM, Akitake S, Nobuke T, Galis I, Aoki K, Shibata D, Takabayashi J (2014) Intake and transformation to a glycoside of (Z)-3-hexenol from infested neighbors reveals a mode of plant odor reception and defense. PNAS USA 111:7144–7149. https://doi.org/10.1073/PNAS.1320660111

Sun D, Guo Z, Liu Y, Zhang Y (2017) Progress and prospects of CRISPR/Cas systems in insects and other arthropods. Front Physiol 8:608. https://doi.org/10.3389/fphys.2017.00608

Talakayala A, Katta S, Garladinne M (2020) Genetic engineering of crops for insect resistance: an overview. J Biosci 45(1):114

Tanpure RS, Barbole RS et al (2017) Improved tolerance against *Helicoverpa armigera* in transgenic tomato over-expressing multi-domain proteinase inhibitor gene from *Capsicum annum*. Physiol Mol Biol Plants 23(2):597–604. https://doi.org/10.1007/s12298-017-0456-5

Tiwari S, Kate A et al (2020) Volatile organic compounds (VOCs): biomarkers for quality management of horticultural commodities during storage through e-sensing. Trends Food Sci Technol 106:417–433. https://doi.org/10.1016/j.tifs.2020.10.039

Trung NP, Fitches E, Gatehouse JA (2006) A fusion protein containing a lepidopteran-specific toxin from the South Indian red scorpion (*Mesobuthus tamulus*) and snowdrop lectin shows oral toxicity to target insects. BMC Biotechnol 6(1):1–12

Turlings TCJ, Erb M (2018) Tritrophic interactions mediated by herbivore-induced plant volatiles: mechanisms, ecological relevance, and application potential. Annu Rev Entomol 63:433–452. https://doi.org/10.1146/ANNUREV-ENTO-020117-043507

Tyagi S, Kesiraju K, Saakre M, Rathinam M, Raman V, Pattanayak D, Sreevathsa R (2020) Genome editing for resistance to insect pests: an emerging tool for crop improvement. ACS Omega 5(33):20674–20683

Varshney RK, Bohra A, Yu J, Graner A, Zhang Q, Sorrells ME (2021) Designing future crops: genomics-assisted breeding comes of age. Trends Plant Sci 26:631–649. https://doi.org/10.1016/j.tplants.2021.03.010

Vázquez-González C, Quiroga V, Martín-Cacheda L, Rasmann S, Röder G, Abdala-Roberts L, Moreira X (2023) Effect of herbivore load on VOC-mediated plant communication in potato. Planta 257(2):1–6. https://doi.org/10.1007/S00425-023-04075-6

Venegas-Molina J, Proietti S et al (2020) Induced tolerance to abiotic and biotic stresses of broccoli and Arabidopsis after treatment with elicitor molecules. Sci Rep 10:10319. https://doi.org/10.1038/s41598-020-67074-7

Villarroel CA, Jonckheere W, Alba JM, Glas JJ, Dermauw W, Haring MA, Van Leeuwen T, Schuurink RC, Kant MR (2016) Salivary proteins of spider mites suppress defenses in

Nicotiana benthamiana and promote mite reproduction. Plant J 86(2):119–131. https://doi.org/10.1111/TPJ.13152

Vosman B, van't Westende WPC, Henken B, van Eekelen HDLM, de Vos RCH, Voorrips RE (2018) Broad spectrum insect resistance and metabolites in close relatives of the cultivated tomato. Euphytica 214(3):1–14

Wang G, Dong Y, Liu X, Yao G, Yu X, Yang M (2018) The current status and development of insect-resistant genetically engineered poplar in China. Front Plant Sci 9:1408. https://doi.org/10.3389/fpls.2018.01408

Wang A, Luca A, Edelenbos M (2019) Emission of volatile organic compounds from yellow onion (*Allium cepa* L.) bulbs during storage. J Food Sci Technol 56(6):2940–2948. https://doi.org/10.1007/s13197-019-03764-z

Wang D, Wang Q, Sun X, Gao Y, Ding J (2020) Potato tuberworm *Phthorimaea operculella* (Zeller) (Lepidoptera: Gelechioidea) leaf Infestation affects performance of conspecific larvae on harvested tubers by inducing chemical defenses. Insects 11(9):633. https://doi.org/10.3390/INSECTS11090633

Wang F, Park YL, Gutensohn M (2021a) Glandular trichome-derived mono- and sesquiterpenes of tomato have contrasting roles in the interaction with the potato aphid *Macrosiphum euphorbiae*. J Chem Ecol 47(2):204–214. https://doi.org/10.1007/S10886-021-01243-4

Wang Y, Mostafa S et al (2021b) Function and mechanism of JA in plant responses to abiotic and biotic stresses. Int J Mol Sci 22:8568. https://doi.org/10.3390/ijms22168568

Wani SH, Choudhary M et al (2022) Molecular mechanisms, genetic mapping, and genome editing for insect pest resistance in field crops. Theor Appl Genet 135:3875–3895. https://doi.org/10.1007/s00122-022-04060-9

War AR, Paulraj MG, Ahmad T, Buhroo AA, Hussain B, Ignacimuthu S, Sharma HC (2012) Mechanisms of plant defense against insect herbivores. Plant Signal Behav 7(10):1306–1320. https://doi.org/10.4161/psb.21663

War AR, Taggar GK, Hussain B, Taggar SM, Nair RM, Sharma HC (2018) Plant defense against herbivory and insect adaptations. AoB Plants. https://doi.org/10.1093/aobpla/ply037

Weintraub R, Garrido E, Poveda K (2018) Age-dependent potato tolerance to herbivory in different nutrient environment. Am J Potato Res 95:642–649. https://doi.org/10.1007/s12230-018-9670-0

Wheeler AG, Krimmel BA (2015) Mirid (Hemiptera: Heteroptera) specialists of sticky plants: adaptations, interactions and ecological implications. Annu Rev Entomol 60:393–414

Wouters FC, Blanchette B, Gershenzon J, Vassao DG (2016) Plant defense and herbivore counter-defense: benzoxazinoids and insect herbivores. Phytochem Rev 15:1127–1151

Xia J, Guo Z, Yang Z, Han H, Wang S, Xu H et al (2021) Whitefly hijacks a plant detoxification gene that neutralizes plant toxins. Cell 184(7):1693–1705

Xiao Y, Qian J, Hou X, Zeng L, Liu X, Mei G, Liao Y (2023) Diurnal emission of herbivore-induced (Z)-3-hexenyl acetate and allo-ocimene activates sweetpotato defense responses to sweetpotato weevils. J Integr Agric. https://doi.org/10.1016/J.JIA.2023.02.020

Xing Z, Liu Y, Cai W, Huang X, Wu S, Lei Z (2017) Efficiency of trichome-based plant defense in *Phaseolus vulgaris* depends on insect behavior, plant ontogeny, and structure. Front Plant Sci 8:2006. https://doi.org/10.3389/fpls.2017.02006

Xue S, Dong M, Liu X, Xu S, Pang J, Zhang W et al (2019) Classification of fruit trichomes in cucumber and effects of plant hormones on type II fruit trichome development. Planta 249:407–416

Yang F, Zhang Q, Yao Q, Chen G, Tong H, Zhang J, Li C, Su Q, Zhang Y (2020) Direct and indirect plant defenses induced by (Z)-3-hexenol in tomato against whitefly attack. J Pest Sci 93(4):1243–1254. https://doi.org/10.1007/S10340-020-01234-6

Yang JN, Wei JN, Kang L (2021) Feeding of pea leafminer larvae simultaneously activates jasmonic and salicylic acid pathways in plants to release a terpenoid for indirect defense. Insect Sci 28(3):811–824. https://doi.org/10.1111/1744-7917.12820

Yao Q, Peng Z, Tong H, Yang F, Xing G, Wang L, Zheng J, Zhang Y, Su Q (2019) Tomato plant flavonoids increase whitefly resistance and reduce spread of tomato yellow leaf curl virus. J Econ Entomol 112(6):2790–2796. https://doi.org/10.1093/jee/toz199

Yates-Stewart AD, Daron J, Wijeratne S, Shahid S, Edgington HA, Slotkin RK, Michel A (2020) Soybean aphids adapted to host-plant resistance by down regulating putative effectors and up regulating transposable elements. Insect Biochem Mol Biol 121:103363

Zalucki MP, Zalucki JM, Perkins LE, Schramm K, Vassão DG, Gershenzon J, Heckel DG (2017) A generalist herbivore copes with specialized plant defence: the effects of induction and feeding by *Helicoverpa armigera* (Lepidoptera: Noctuidae) larvae on intact *Arabidopsis thaliana* (Brassicales) plants. J Chem Ecol 43(6):608–616. https://doi.org/10.1007/S10886-017-0855-7

Zhang J, Khan SA, Hasse C, Ruf S, Heckel DG, Bock R (2015) Full crop protection from an insect pest by expression of long double-stranded RNAs in plastids. Science 347(6225):991–994

Zhang M, Demeshko Y et al (2019a) Elevated α-linolenic acid content in extra-plastidial membranes of tomato accelerates wound-induced jasmonate generation and improves tolerance to the herbivorous insects *Heliothis peltigera* and *Spodoptera littoralis*. J Plant Growth Regul 38:723–738. https://doi.org/10.1007/s00344-018-9885-9

Zhang PJ, Wei JN, Zhao C, Zhang YF, Li CY, Liu SS, Dicke M, Yu XP, Turlings TCJ (2019b) Airborne host-plant manipulation by whiteflies via an inducible blend of plant volatiles. PNAS USA 116(15):7387–7396

Zhang Y, Bouwmeester HJ, Kappers IF (2020) Combined transcriptome and metabolome analysis identifies defence responses in spider mite-infested pepper (*Capsicum annuum*). J Exp Bot 71(1):330–343

Zhang J, Gu X et al (2022a) Characterization of differences in the composition and content of volatile compounds in cucumber fruit. Foods 11:1101. https://doi.org/10.3390/foods11081101

Zhang Z, He H, Yan M, Zhao C, Lei C, Li J, Yan F (2022b) Widely targeted analysis of metabolomic changes of *Cucumis sativus* induced by cucurbit chlorotic yellows virus. BMC Plant Biol 22(1):1–12. https://doi.org/10.1186/S12870-022-03555-3

Zhong Y, Ahmed S, Deng G, Fan W, Zhang P, Wang H (2019) Improved insect resistance against *Spodoptera litura* in transgenic sweet potato by overexpressing Cry1Aa toxin. Plant Cell Rep 38:1439–1448

Zhou S, Jander G (2022) Molecular ecology of plant volatiles in interactions with insect herbivores. J Exp Bot 73:449–462. https://doi.org/10.1093/jxb/erab413

Zhu Z, Xu X et al (2015) Pyramiding of AtEDT1/HDG11 and Cry2Aa2 into pepper (*Capsicum annuum* L.) enhances drought tolerance and insect resistance without yield decrease. Plant Cell Tissue Org Cult 120:919–932. https://doi.org/10.1007/s11240-014-0600-7

Zsögön A, Čermák T, Naves ER, Notini MM, Edel KH, Weinl S, Freschi L, Voytas DF, Kudla J, Peres LEP (2018) *De novo* domestication of wild tomato using genome editing. Nat Biotechnol 36(12):1211–1216

Zunjarrao SS, Tellis MB, Joshi SN, Joshi RS (2020) Plant-insect interaction: the saga of molecular coevolution. In: Reference series in phytochemistry. Springer, pp 19–45. https://doi.org/10.1007/978-3-319-96397-642

Role of Induced Resistance in Insect-Pest Management

10

Liyaqat Ayoub, Munazah Yaqoob, Sobiya Zahoor, Fazil Fayaz Wani, Sheikh Salma Irshad, Audil Gull, and Mohmmad Aasif Sheikh

Abstract

Induced resistance is a process by which plants increase their resistance against pathogens and herbivores through the activation of their own defence mechanisms. This phenomenon has been extensively studied in the context of plant–pest interactions and has proven to be an effective strategy for pest management. Induced resistance can be triggered by various stimuli, including insect feeding, application of plant growth regulators or phytohormones, and exposure to certain chemicals. Once induced, plants can produce a range of defence compounds, such as alkaloids, terpenoids, and phenolics, which can deter or kill herbivores and pathogens. The use of induced resistance as a pest management strategy has several advantages over traditional chemical pesticides. Firstly, it is an environmentally friendly approach that does not harm non-target organisms or pollute the environment. Secondly, induced resistance is a long-term solution as it strengthens the plant's natural defence mechanisms, making it more resistant to future pest attacks. Finally, induced resistance can be integrated with other pest management strategies such as biological control, cultural control,

L. Ayoub (✉) · M. Yaqoob · S. Zahoor · M. A. Sheikh
Division of Entomology, Sher-e-Kashmir University of Agricultural Sciences and Technology of Kashmir, Kashmir, India

F. F. Wani
Division of Pathology, Sher-e-Kashmir University of Agricultural Sciences and Technology of Kashmir, Kashmir, India

S. S. Irshad
Department of Entomology, Punjab Agricultural University, Ludhiana, India

A. Gull
Division of Genetics and Plant Breeding, Sher-e-Kashmir University of Agricultural Sciences and Technology of Kashmir, Kashmir, India

S. Kumar, M. Furlong (eds.), *Plant Resistance to Insects in Major Field Crops*,
https://doi.org/10.1007/978-981-99-7520-4_10

and physical control. However, there are some challenges associated with the use of induced resistance. The effectiveness of induced resistance can vary depending on the plant species, the type of pest, and the environmental conditions. Additionally, the cost and time involved in inducing resistance can be higher than those of conventional pest control methods.

In conclusion, induced resistance is a promising strategy for pest management that can be used in combination with other pest control methods. Further research is needed to better understand the mechanisms underlying induced resistance and to optimize its application in agriculture.

Keywords

Biotic stress · Defence · Herbivore · Pest management · Plant secondary metabolites

10.1 Introduction

Plants and insects are highly diverse groups due to their ability to exploit a wide range of niches, from deserts to arctic zone. Since the Devonian Era, plants and insects have co-existed (Labandeira 2007). In the course of evolution, numerous interactions between insects and plants have evolved, including antagonistic interactions with insect herbivores and symbiotic interactions with insect carnivores (natural enemies) and pollinators (Dicke 2001). Both plants and insect herbivores have developed a variety of defence systems to combat one another throughout the course of these lengthy interactions and coevolution. Evidently, plants are able to survive and thrive despite significant tissue damage and occasionally full defoliation. Plants respond to insect herbivory through an intricate and dynamic defence system, which is mediated physiologically and biochemically. Host plant defence against herbivores is a complex array of structural, chemical, and physiological traits intended to perceive the attacking organisms and restrain them before they are able to cause extensive damage (Howe and Jander 2008; Sharma 2009; War et al. 2011a; Hanley et al. 2007; Karban 2011). To combat herbivore attack, plants react to them through a variety of morphological, biochemical, and molecular processes (War and Sharma 2014). Direct and indirect defence systems may both be present naturally or may be produced as a result of insect attack. Constitutive defence occurs in plants irrespective of the external stimuli such as insect damage and/or elicitor application, often providing a first line of defence. Induced defences are a second line of defence and are activated in plants in response to the external stimuli such as insect damage, pathogen infestation, abiotic stress, and/or elicitor application (Bown et al. 2002). Induced defence is key element of plant defensive strategy and has gained increased focus in insect control programmes (Sharma 2009; War et al. 2011b; Karban 2011). Induced defences make the plants phenotypically pliable and thus, less vulnerable to insect attack. However, the timing of induced resistance is crucial; the earlier the response, the greater the benefit to the plant and the lower the risk of subsequent herbivore and pathogen attack, in addition to improvement in overall plant fitness (Cipollini et al. 2004; War et al. 2011a). Application of induced resistance in pest management has the potential to reduce pesticide use on crops. By thorough understanding of the mechanism of induced resistance, we can forecast about its effect on the target species. Crop plants can be sprayed with elicitors of induced responses to strengthen their natural defences against herbivore harm. The genetically modified induced responses allow producing the defence chemicals naturally in plants after herbivory. Therefore, a thorough understanding of insect–plant interactions and defence mechanisms of plants will provide opportunities to develop new strategies, such as development of resistant crops or pest management strategics. This in turn will help reduce pesticide load on a crop and promote safer crop production. This chapter focuses on various aspects of induced direct and indirect defence strategies exhibited by plants against insect herbivores. Besides this, the role of phytohormones in defence mechanism, their cross talk, and plant defence priming, a chemical-free technique and its role in protecting the plants and boosting the sustainable agricultural production have been discussed.

10.2 Early Signalling Events at the Plant–Insect Interface

Successful implementation of an induced defence response requires quick and accurate plant response to herbivory. Early signalling events at the plant–insect interface, which occur well before changes in host plant gene expression and defence-related metabolism, are critical for the process of herbivore recognition (Maffei et al. 2007b). The presence of dislocated plant molecules and fragments of macromolecules from disrupted cells allows plants to recognize damage (Duran-Flores and Heil 2016). These so-called DAMPs (damage-associated molecular patterns) are sufficient in many systems to cause induced resistance following mechanical wounding. Plants also respond to molecules specifically from herbivores (herbivore-associated molecular patterns, or HAMPs) that the plants encounter during herbivore feeding or other activities (Acevedo et al. 2015). For example, legumes recognize inceptin, a catabolic plant product that is broken down by enzymes from a feeding caterpillar (Schmelz et al. 2006).

Plants most commonly respond to cues that are released by their own damaged tissues, as opposed to those of other individuals (Karban et al. 2006; Frost et al. 2007; Heil and Bueno 2007). These cues may travel from cell to cell, through the vascular system of the plant, or through the gaseous headspace that connects plant tissues. Volatile cues that are emitted during damage allow nearby individuals to eavesdrop on neighbours and adjust their defences (Karban et al. 2014).

10.3 Induced Responses to Herbivory

Damage from herbivores causes a variety of changes in plants, and some of these "induced responses" offer "induced resistance" to further animal exploitation. (Karban and Myers 1989). While induced resistance has been proven to be ubiquitous, “induced defences," which boost plant fitness under conditions of high risk of herbivory, have only been demonstrated infrequently. The connection between reductions in the performance of herbivores and benefits to plants has often been implicit (Hare 2011; Erb 2018) although a limited number of studies have found compelling evidence for this link, e.g. for direct defences (Agrawal 1998) and indirect defences that involve predators of herbivores (Schuman et al. 2012).

Herbivory causes large-scale changes in gene expression. For example, in hybrid poplar *Populus trichocarpa* × *deltoides*, it is estimated that 11% of the transcriptome is modified by forest tent caterpillar (Ralph et al. 2006). A common aspect of insect feeding (unlike pathogen attack) is some degree of mechanical damage, and many studies have focused on the regulatory signals generated as a result of artificial wounding. Plants may discriminate insect herbivory from experimental wounding by the different temporal and spatial patterns of natural and artificial injuries (Mithofer et al. 2005). Moreover, the type of feeding (e.g. chewing vs sucking) and blend of oral secretions (saliva or regurgitant of the attacker) may determine the specific response signature of the host plant. In particular, insect-derived elicitors produced during feeding have been shown to trigger direct and indirect defence responses, as detected by changes in gene and protein expression, and production of VOCs and other secondary metabolites (Felton 2008) (Fig. 10.1).

10.3.1 Direct Defences

There are many different herbivore resistance mechanisms in plants, which are typically divided into two main categories: tolerance and avoidance (Rosenthal and Kotanen 1994; Boege and Marquis 2005) that draw a distinction between a plant’s tolerance of herbivore attack and its ability to avoid herbivory through

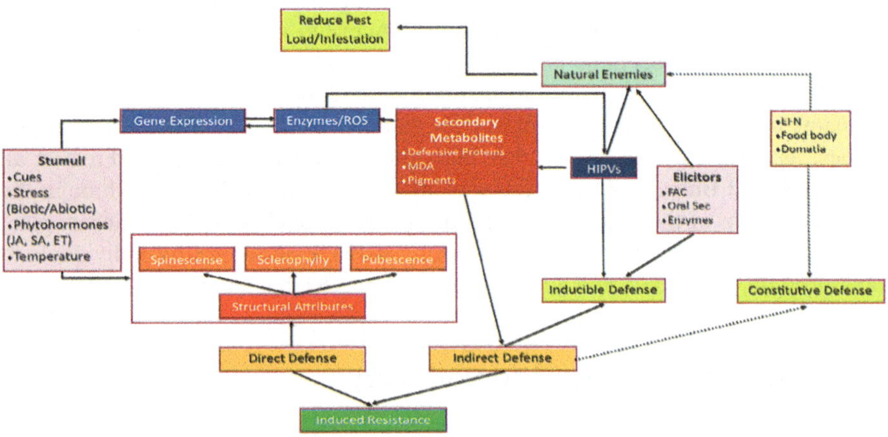

Fig. 10.1 Direct and indirect defence mechanism in induced resistance

defence. Some authors (Strauss and Agrawal 1999; Stowe et al. 2000) view "defence" as an umbrella term covering both avoidance and tolerance. Avoidance involves structural defence (leaves surrounded by thorns—Gowda 1996). In order to prevent herbivores from continuing to feed after taking a bite, plants produce chemicals called phenolics (Hanley and Lamont 2001). Phenological defences, such as quick turnover of vulnerable parts or timing of the life cycle, are also used to prevent herbivores from feeding (Saltz and Ward 2000). Structural defences are avoidance strategies based on structural traits, whether these be obtrusive substances kept by the plant or minute changes to cell wall thickness. Consequently, a good definition of structural defence may be any morphological or anatomical feature that gives the plant a fitness advantage by actively deterring herbivores from feeding on it (consistent with Rosenthal and Kotanen 1994; Boege and Marquis 2005). The initial physical defence against insect eating is provided by plant structural features like waxy leaf surfaces, trichomes, thick cell walls, and lignification. The second line of defence is provided by secondary metabolites that serve as poisons and inhibit growth, development, and digestibility. Furthermore, the defence mechanism of plants against insect invaders is strengthened by the synergistic effects of many defensive components. Trypsin proteinase inhibitors and nicotine expression worked together to strengthen wild tobacco plant (*Nicotiana attenuata*) defences against beet armyworm (*Spodoptera exigua*) (War et al. 2012).

10.3.1.1 Structural Defence

The first line of defence against insect pests is a plant's structure, which also contributes significantly to the host plant's insect resistance. The construction of a physical barrier, such as a waxy cuticle or the growth of spines, setae, and trichomes, is a plant's first line of defence against insect pests. Anatomical and morphological traits known as structural defences give a plant a fitness benefit by directly deterring insects or herbivores from feeding on it. These characteristics can range from conspicuous features on a plant to minute adjustments in cell wall thickness brought on by lignification and suberization. It is well known that induced lignification and suberization of cell walls strengthen cell walls and create a region of water-impervious tissue that isolates the wound from nearby healthy cells. (Eyles et al. 2003).

Plants are primarily protected against insect pests by structural characteristics like spines and thorns (spinescence), pubescence, toughened or hardened leaves (sclerophylly), incorporation of granular minerals into plant tissues, and divaricated branching (shoots with wiry stems produced at wide axillary angles).

10.3.1.1.1 Spinescence

The aggregate term "spinescence" is used to describe the plant parts with spines, thorns, and prickles. A spine is a woody, sharp-pointed branch, a thorn is a sharp-pointed branch, and a prickle is any sharp-pointed protrusion from the cortex or epidermis of an organ (Grubb 1992; Gutschick 1999). According to reports, spinescence protects plants from a variety of insects particularly vertebrates as compared to invertebrates, due to the size relations of the plant–herbivore interactions (Cooper and Owen-Smith 1986). Obeso (1997) demonstrated that holly shrubs with very spiny leaves were far less likely to suffer herbivory by large ungulates than nearby less spiny plants. The European holly (*Ilex aquifolium*) demonstrates great diversity in leaf spinescence.

10.3.1.1.2 Pubescence

The layer of hairs (trichomes) found on stems, leaves, and even fruits is referred to as pubescence. Trichomes are hair-like appendages that protrude from the epidermis of aerial tissues. They might be straight, spiral, stellate, hooked, or glandular (Levin 1973). According to Haberlandt (1914), leaf hairs play a part in herbivore defence as well. Trichomes can limit insect oviposition in addition to hindering herbivore mobility by changing how firmly the eggs are adhered to leaves (Haddad and Hicks 2000; Handley et al. 2005). For example, hooked trichomes can entrap or even puncture some insects, including the primary predator of Mentzelia, the beetle *Hippodamia convergens* (Quiring et al. 1992; Eisner et al. 1998).

10.3.1.1.3 Sclerophylly

Schimper (1903) coined the term "sclerophylly," which literally translates to "hard-leaved." Through longer leaf life, hard leaves may also boost total (but not immediate) absorption efficiency and defence compound buildup (Wright et al. 2004). Scleromorphic leaves and shoots make plant material less palatable and digestible (Grubb 1986; Robbins 1993), which eventually reduces herbivore fitness (Perez-Barberia and Gordon 1998). For instance, Bjorkman and Anderson (1990) demonstrated how butterfly larvae typically steer clear of feeding on the blackberry's (*Rubus bogotensis*) hardened leaves in South America.

10.3.1.2 Inducible Anatomical Defences

A well-recognized form of anatomical modification is the formation of traumatic resin ducts (TRDs) in the xylem and/or phloem of many conifer species. TRDs are associated with induction of terpene biosynthesis and increased resin flow within 2–3 weeks after attack (Luchi et al., 2005). The formation of the wound (necrophylactic) periderm, particularly its rate of formation, is considered to be a critical resistance mechanism to phloem-feeding borers (e.g. Buprestid beetles) and phloem-invading microorganisms (e.g. stem cankers; Eyles et al. 2003).

10.3.1.3 Secondary Metabolites for Chemical Defence of Plants

Plants produce a large and diverse array of organic compounds that appear to have no direct functions in growth and development but play a significant role in direct defence, impair herbivore performance by one of two general mechanisms: these chemicals may reduce the nutritional value of plant food, or they may act as feeding deterrents or toxins (Berenbaum 1995). In response to an insect or microbe attack, the defensive (secondary) metabolites may be either induced or constitutively stored in inactive forms. Both are referred to as phytoalexins and phytoanticipins, respectively (antimicrobial compounds synthesized by plants that accumulate rapidly at areas of pathogen infection). During herbivory, –glucosidase primarily activates the

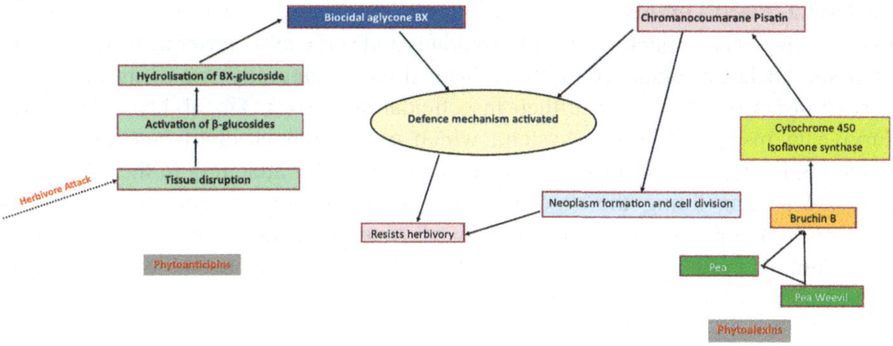

Fig. 10.2 Secondary metabolites as chemical defence in plants. (**a**) Phytoanticpins-Hydrolysation of BX-glucosides (Benzoxinoid) by plastid-targeted β-glucosidases during tissue damage leads to the production of biocidal aglycone BXs, which play an important role in plant defence against insects. (**b**) Phytoalexins-insect inducibility in peas, where the elicitor Bruchin B from the pea weevil was shown to induce a cytochrome P450 isoflavone synthase, resulting in successive accumulation of the chromanocoumaranepisatin, which stimulates the cell division and neoplasm formation and activates other plant defence responses

phytoanticipins, which then mediate the release of different biocidal aglycone metabolites. Glucosinolates, which are degraded by myrosinases (endogenous-thioglucoside glucohydrolases) during tissue rupture, are the standard illustrations of phytoanticipins. Benzoxazinoids (BXs), which are extensively dispersed throughout Gramineae, are among the other phytoanticipins. Isoflavonoids, terpenoids, alkaloids, and other phytoalexins affect how well and how long insects can survive (Fig. 10.2). The plants' secondary metabolites not only protect them from various challenges but also improve their fitness. According to reports, the presence of the secondary metabolites C-glycosyl flavone maysin and the phenylpropanoid derivative, chlorogenic acid, contributes significantly to maize's susceptibility to the corn earworm, *Helicoverpa zea*. Shoot fly resistance in sorghum has been linked to the compound 4,4-dimethyl cyclooctene (Chamarthi et al. 2011).

10.3.1.3.1 Plant Phenolic Compounds

Plant phenols are one of the most prevalent and extensive classes of defensive chemicals among secondary metabolites, and they play a significant role in the host plant's ability to resist insects. Phenols serve as a defence strategy for plants not just against insects but also against competitive plants and microbes.

A key component of plant defence against pests and pathogens is lignin, a phenolic heteropolymer. It restricts the entry of diseases by physically obstructing them or making the leaf tougher, which inhibits insect feeding and lowers the leaf's nutritional value. It has been discovered that insect or pathogen attack induces lignin synthesis, and that its fast deposition inhibits further disease or insect fecundity growth.

10.3.1.3.2 Plant Lectins

Lectins are proteins that bind to carbohydrates and serve as a barrier against a variety of pests. Different plant lectins have been used as naturally occurring pesticides against insect pests because of their insecticidal properties. The ability of lectins to survive in insects' digestive system, which gives them a significant insecticidal potential, is one of their most crucial characteristics. By attaching to the membrane glycosyl groups lining the digestive tract, they function as antinutritive and/or poisonous chemicals and cause a variety of systemic adverse effects. The damage to the luminal epithelial membranes caused by lectins, which are stable across a wide pH range, prevents the digestion and absorption of nutrients.

10.3.1.3.3 Proteinase Inhibitors

One of the most prevalent kinds of protective proteins in plants is known as a proteinase inhibitor (PI). One to ten per cent of the total proteins in storage organs like seeds and tubers contain PIs, which inhibit a variety of enzymes and are crucial for plant defence against insects. PIs may be found even in higher concentrations in these organs. Insects are starved or slowed down in their development as a result of the lack of amino acids caused by PIs' binding to the digestive enzymes in their stomach and inhibiting their action. Many PIs have been researched for their defensive properties against lepidopteran and hemipteran insects, either directly or by expression in transgenic plants to increase plant resistance to insects. The success of transgenic crops in expressing PIs against insect pests has accentuated the need to understand the mechanisms, interactions of multiple PIs with other defences, and the adaptive responses of the insects.

10.3.1.3.4 Enzymes

Enzymes have a significant role in the disruption of insect nutrition and are one of the key components of host plant resistance to insects. Peroxidases, polyphenol oxidases, ascorbate peroxidases, and other peroxidases all prevent insects from absorbing nutrients by creating electrophiles when they oxidize mono- or dihydroxyphenols. PPOs, in particular, appear to reduce the absorption of amino acids by oxidizing orthodiphenolic chemicals to quinones, which cross-link proteins in the stomach of insects and render them indigestible (Felton et al. 1992).

10.3.2 Indirect Defence Methods

A crucial part of defending plants from insect attack is the defensive response in plants that attracts natural enemies of insects. As a result of the combined impact of mechanical damage and elicitors from the attacking herbivore, indirect defences may be constitutive or induced. Plants interact with natural enemies of insect pests (i.e. parasitoids or predators) through the production of volatiles and the secretion of additional floral nectar, which actively lowers the quantity of feeding herbivores. Induced indirect defences have been examined at the genetic, biochemical, physiological, and ecological levels in recent years (Furlong et al. 2018).

10.3.2.1 Herbivore-Induced Plant Volatiles (HIPVs)

Plants emit a mixture of volatile and non-volatile chemicals to covertly protect themselves from insect feeding. Insect-induced plant volatiles (HIPVs) are essential for plant defence because they either draw an insect's natural predators or prevent an insect from laying eggs. In reaction to an insect attack, plants emit HIPVs, which are lipophilic chemicals with a higher vapour pressure, into the atmosphere through their leaves, flowers, and fruits as well as into the soil through their roots.

10.3.2.1.1 Biosynthesis of HIPVs

Different metabolic routes, including terpenoids, fatty acid derivatives, and phenylpropanoids or benzenoids, are used to synthesize HIPVs (Mumm and Dicke 2010).

Terpenoids

Monoterpenes (C10), sesquiterpenes (C15), and homoterpenes (C11 or C16) are the main terpenoid volatiles, and they all considerably contribute to any mixture of volatiles obtained from plants. The cytosolic mevalonate (MVA) and the methylerythritol 4-phosphate (MEP) pathways are the two pathways used to produce all terpenoids (Chappell 1995; Aharoni et al. 2005; Cheng et al. 2007). It has been suggested that JA, SA, and ET signalling pathways interact antagonistically or synergistically to govern the distinctive mixture of terpenoids in response to herbivory (Ozawa et al. 2000; Engelberth et al. 2001; Horiuchi et al. 2001; Schmelz et al. 2003). Terpenoids have also been proven to have the ability to attract natural enemies by themselves or in conjunction with other plant volatiles or herbivore pheromones (Dicke et al. 1990; Erbilgin and Raffa 2001; Pettersson 2001; De Boer et al. 2004; Mumm and Hilker 2005).

Volatile Fatty Acid Derivatives

Green leaf volatiles (GLVs), also known as volatile fatty acid derivatives, are frequently linked to the green leaf odour released following tissue damage. GLVs originate from C18 unsaturated fatty acids, such as linoleic acid and linolenic acid involving deoxygenation catalysed by lipoxygenases (Feussner and Wasternack 2002), and are synthesized via the octadecanoid pathway, which is also used to make the phytohormone JA. They assist in systemic induction or priming of changes in plant phenotype or serve as attractants for herbivores' natural enemies in indirect plant defence (Reddy 2002; Shiojiri et al. 2006; Frost et al. 2007; Heil and Bueno 2007; Van Wijk et al. 2008; Wei and Kang 2011).

Phenypropanoids/benzenoids

A sizable class of HIPVs known as phenypropanoids and benzenoids play a significant role in plant defence against herbivores. They are produced from L-phenylalanine, which is transformed into trans-cinnamic acid by L-phenylalanine ammonia-lyase. Then, methylation and hydroxylation are used to transform hydroxyl cinnamic acid into a range of hydroxycinnamic acids, aldehydes, and alcohols (Humphreys and Chapple 2002). Volatiles are released from some of these intermediary molecules. *Trans*-cinnamic acid can also be used to synthesize benzenoids. However, a C2 unit is added to the *trans*-cinnamic acid during their production. The CoA-dependent non-oxidative process, the CoA-independent non-oxidative pathway, and a combination of these two mechanisms have all been postulated as possible pathways for chain shortening (Boatright et al. 2004). The benzenoid ester methyl salicylate (MeSA), which is regularly released by plants with herbivore infestations and serves as a plant indirect defence, is the most researched of this class of substances (Dicke et al. 1990; Van Poecke and Dicke 2002; James 2003; Ament et al. 2004; De Boer et al. 2004).

10.3.2.1.2 Role of HIPVs in Plant Defence Against Herbivores

As long as the generation of volatiles continues, parasitoids will use these volatiles as cues to hunt their prey and will give the emitting plants an adaptive edge. These interactions are particular to an interaction between an insect and a plant. For example, *Zea mays* TPS10 is a herbivore-induced terpene synthase that forms (E)-β-farnesene, (E)-α-bergamotene, and other sesquiterpenes in *Arabidopsis thaliana*, which does not produce significant amounts of volatile terpenes, suggesting that a single herbivore-induced gene from *Z. mays* is sufficient to elicit this indirect defence (Schnee et al. 2006). Damage from the larvae of the corn rootworm *Diabrotica virgifera* causes the release of (E)-caryophyllene, which draws the nematode *Heterorhabditis megidis*, which feeds on the larvae of *D. virgifera* (Rasmann et al. 2005).

Due to increased volatile emissions from HIPV-exposed leaves, *Lymantria dispar* larval weight was reduced by 70% on branches exposed to HIPVs (Fig. 10.3). Many caterpillars are repulsed by many volatiles produced by Gipsy moth in *Vaccinium corymbosum*, including linalool and farnesenes (Rodriguez-Saona et al. 2009; Markovic et al. 1996).

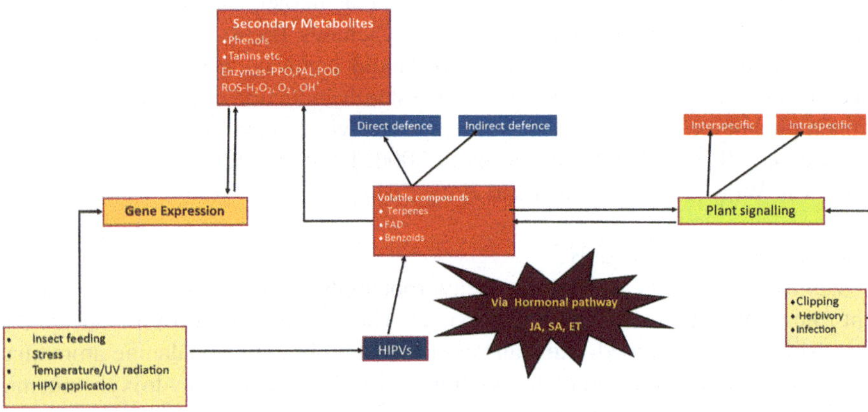

Fig. 10.3 Role of HIPVs in plant defence

10.3.2.2 Defence Elicitors (Insect Oral Secretion)

In response to physical and chemical signals caused by herbivores, such as chemicals found in oviposition fluids and oral secretions of insects, plants experience a dynamic alteration in their transcriptomes, proteomes, and metabolomes. The widespread consensus is that herbivore oral secretions and regurgitates mediate insect-induced plant responses. Depending on the type of elicitor and the biological processes involved, different elicitors produce different defences. The enzyme-glucosidase has been identified as a potential elicitor of herbivore-induced plant volatiles from the regurgitation of *Pieris brassicae* L. larvae. This enzyme causes mechanically injured cabbage leaves to emit a volatile mixture that attracts the parasitic wasp *Cotesia glomerata* (L.) (Steppuhn and Baldwin 2007).

10.3.2.3 Inducible Constitutive Traits in Indirect Plant Defence

Numerous plant features are expressed constitutively, but they can also be induced under specific circumstances. These inducible constitutive features frequently serve many purposes. Extrafloral nectar, food bodies, and domatia are a few instances of these inducible constitutive features. (Turlings and Wackers 2004).

10.3.2.3.1 Extrafloral Nectar

Extrafloral nectary organs secrete extrafloral nectar, an aqueous solution with sugars and amino acids (Koptur 1992). More than a thousand different plant species' shoots, leaves, and inflorescences have been seen to contain extrafloral nectar (Koptur 1992; Wackers et al. 2001; Turlings and Wackers 2004). Extrafloral nectar draws parasitoids and predators alike (Heil and McKey 2003). Regardless of whether there are herbivores present, most plants still exude some extrafloral nectar (Turlings and Wackers 2004). This constitutive nectar's baseline secretion enables plants to host and feed some natural enemies (Wackers et al. 2001). However, in the presence of herbivores, extrafloral nectar can be stimulated to a much higher level (Heil et al. 2001; Wackers et al. 2001; Huntzinger et al. 2004; Gonzalez-Teuber and Heil 2009), in response to mechanical damage, and by exogenous JA application (Heil et al. 2001; Heil et al. 2004; Kost and Heil 2005, 2008). Extrafloral nectar-fed parasites have an increased rate of reproduction and a much longer life span (Rose et al. 2006).

Extrafloral nectars are frequently combined with other defensive strategies (Arimura et al. 2005). When plants are exposed to exogenous application of JA, both volatile emission and extrafloral nectars secretion increase in the lima bean (Heil 2004). Lima bean plants treated with volatiles emit higher quantities of extrafloral nectars, such as (Z)-3-hexenyl acetate, (E)-3-hexenyl acetate, (E)-2-hexenyl acetate, 5-hexenyl acetate, (Z)-3-hexenyl isovalerate, and (Z)-3-hexenyl butyrate. (Kost and Heil 2008). Other plants from a variety of plant taxa frequently activate both extrafloral nectar release and volatile emission after herbivore attack (Kost and Heil 2006; Choh et al. 2006).

10.3.2.3.2 Food Body (Pearl Body)

Although the components of nutrients found in food bodies on various plants vary, polysaccharides, lipids, and proteins are frequently present (O'Dowd 1982; Silva and Machado 1999; Heil et al. 2004; Webber et al. 2007; Paiva et al. 2009). Similar to extrafloral nectar, food bodies can draw in and feed ants and other herbivores' natural enemies (Turlings and Wackers 2004).

10.3.2.3.3 Domatia

Small morphogenetic hollow structures called "domatia" can be seen on plants and act as homes for mites and other arthropods (Norton et al. 2000; Heil and McKey 2003; Romero and Benson 2004, 2005). In exchange, mites and other arthropods residing in domatia serve the plants by removing pathogenic fungus spores and hyphae as well as by consuming phytophagous mites and other undesirable arthropods (O'Dowd and Willson 1997; Norton et al. 2000; Romero and Benson 2004, 2005; Duso et al. 2005; Zemek 2005; English-Loeb et al. 2005; Monks et al. 2007; Pozzebon et al. 2009; O'Connell et al. 2015; Tempfli et al. 2015). Even though other arthropod species, including thrips, may also benefit from the protection provided by this structure, the majority of arthropod species found in domatia are mites and ants (Monks et al. 2007; O'Dowd and Willson 1997).

10.3.2.3.4 Inducible Civilian Defences

Plants can minimize the negative fitness consequences of tissue lost to herbivory by activating physiological processes that allow the plant to compensate for the reduction in total photosynthetic capacity. These are termed "civilian" defences. Unlike host resistance, civilian defences do not directly affect the biotic agent's performance. For example, in insect–host interactions, host tolerance allows plants to support herbivore populations similar to a susceptible host without a concomitant reduction in plant fitness (Karban and Baldwin 1997; Haukioja and Koricheva 2000). Mechanisms of tolerance appear common in cases of attack by both leaf-feeding herbivores and foliar pathogens. These include up-regulation of photosynthetic rates in remaining uninfected/undamaged leaves (Quentin et al. 2010), alteration in growth patterns to favour development of leaf area (Frost and Hunter 2008) and shifts in resource allocation patterns within and between the above- and below-ground organs of a tree. The latter can be achieved by remobilization of reserves from storage tissues or by mobilization of resources to temporary storage in organs that are less susceptible to damage (e.g. the root system) (Babst et al. 2008; Frost and Hunter 2008). Collectively, these changes enhance the plant's ability to tolerate subsequent pathogen and herbivore attack.

10.3.2.4 Role of Phytohormones in Induced Resistance in Plants

Numerous signal transduction pathways that are mediated by a network of phytohormones are involved in plant defence against insect attack. Plant hormones are essential for controlling a plant's development, growth, and defence systems. In plants harmed by insects, a number of plant hormones have been linked to intra- and inter-plant communication. Jasmonic acid, salicylic acid, and ethylene-mediated signal transduction pathways are responsible for activating the majority of plant defence responses against insects. These pathways during injury or insect feeding activate particular sets of defence-related genes. Depending on the attacker, these hormones may operate alone, synergistically, or antagonistically.

10.3.2.4.1 Jasmonic Acid in Plant Defence

Upon insect feeding, jasmonic acid (JA) synthesis is rapidly triggered, inducing massive defence-related genes, the production of diverse secondary metabolites (terpenoids, phenolics, as well as nitrogenous and sulphur-containing compounds), specific defence proteins (protease inhibitors, polyphenol oxidases, leucine aminopeptidase, lectins, and chitinases), and the formation of a physical barrier (e.g. trichomes) to suppress or prevent the feeding (Boter et al. 2004; Falk et al. 2014). JA is changed in plants into a variety of functional forms, including as JA-Ile, methyl jasmonate (MeJA), and the JA precursor 12-oxophytodienoic acid (OPDA) (Staswick and Tiryaki 2004; Wang and Jiang 2007; Woldemariam et al. 2012). Plants respond to MeJA by launching a number of JA-mediated defence mechanisms, increasing their resistance to herbivores (Farmer and Ryan 1992; Grimes et al. 1992; Melan et al. 1993; Kahl et al. 2000; Truitt et al. 2004). Comparable to plants aroused with herbivore elicitors, the MeJA-treated plants enhance volatile emissions and draw predators and parasitoids (Dicke et al. 1999; Thaler 1999; Meiners and Hilker 2000; Kessler and Baldwin 2001; Mumm et al. 2003; Heil 2004; Bruinsma et al. 2009).

Arabidopsis plants deficient in JA biosynthesis and signalling typically suffer more damage from molluscan herbivores (Falk et al. 2014). MeJA treatment induced the expression of glucosinolate (GS) synthesis genes, as well as GS accumulation (Mikkelsen et al. 2003; Reymond et al. 2004). MYC2/3/4 directly binds to the promoters of GS biosynthetic genes and interacts with GS-related MYBs, thereby promoting the JA-mediated synthesis of secondary metabolites and defence against external assaults (Schweizer et al. 2013). Consistent with this, the Arabidopsis myc2/3/4 triple mutant is completely devoid of GS and is extremely susceptible to the generalist herbivore *Spodoptera littoralis* (Schweizer et al. 2013) and spider mite herbivory (Zhurov et al. 2013). In rice, the concentrations of JAs were dramatically increased after a brown planthopper (BPH) attack, along with an increase in H_2O_2 level (Qi et al. 2015). BPH performed better on JA-deficient lines (AOC and MYC2 knockout) than on wild-type (WT) plants due to the attenuation of defensive secondary metabolites accumulation (Xu et al. 2021). Additionally, rice COI1 RNAi lines increase susceptibility to chewing insect *Cnaphalocrocis medinalis* as a result of impairing inducible defence by induction of trypsin protease inhibitor (TrypPI), peroxidase (POD), and polyphenol oxidase (PPO) (Fonseca et al. 2014). Tomato plants treated with JA showed reduced numbers of *Frankliniella occidentalis* (thrips), *Helicoverpa armigera*, flea beetles, and aphids due to an increase in the activities of PIs and polyphenol oxidase (Thaler et al. 2001). Furthermore, the development of glandular trichomes in tomato leaves is controlled, in part, by the JA pathway (Peiffer et al. 2009), providing an important anti-insect defence layer (Kang et al. 2014) (Fig. 10.4).

Fig. 10.4 Multiple roles of Jasmonic acid in plant immunity to insects. (**a**) Promotes resistance to a wide range of arthropod herbivores, including caterpillars (Lepidoptera), beetles (Coleoptera), thrips (Thysanoptera), leafhoppers (Homoptera), spider mites (Acari), fungal gnats (Diptera), (**b**) regulation of global changes in gene expression in response to both mechanical wounding and herbivory, (**c**) regulation of tritrophic interactions, (**d**) host plant resistance to phloem-feeding insects, (**e**) trichome-based defences (**f**) priming of direct and indirect defences, (**g**) systemic transmission of defence signals, (**h**) plant development

10.3.2.4.2 Salicylic Acid in Plant Defence

The monohydroxybenzoic acid salicylic acid (SA) is formed from cinnamate, which is created from phenylalanine by the action of phenylalanine ammonia lyases (Chen et al. 2009). SA has important role in controlling plant growth and development, just like other plant hormones (Rivas-San Vicente and Plasencia 2011), and is involved in plant defence against herbivores (Ryals et al. 1996; Potlakayala et al. 2007; Benedetti et al. 2015; Khan et al. 2015). More and more evidence points to the involvement of SA-related signalling pathways in indirect defence through the induction of volatile emission (Ozawa et al. 2000; Kessler and Baldwin 2002; de

Boer et al. 2004; Maffei et al. 2007a; Diezel et al. 2009). First, the bioactive SA derivative MeSA can be directly derived from SA. After spider mite infestation, lima bean and tomato plants showed elevated levels of MeSA and a strong attraction to natural herbivore foes (Dicke et al. 1990; Dicke et al. 1998; Ozawa et al. 2000), pear plants upon spider mite attack (de Boer et al. 2004); pear plants in responding to psyllid infestation (Scutareanu et al. 1997); and potato plants in responding to Colorado potato beetle infestation (Bolter et al. 1997). MeSA-treated poplar tree leaves show elevated expression of defence genes and increased volatile chemical emission (Arimura et al. 2004). The MeSA-treated lima bean plants also emit more of two homoterpenes, which can draw predatory mites that are out foraging (Dicke et al. 1999; Ozawa et al. 2000). These findings imply that MeSA can both directly attract natural predators of insect herbivores and indirectly boost indirect defence by inducing the emission of other volatiles.

However, not all insect species have the ability to cause MeSA to develop. Caterpillar injury increases levels of SA and activates genes that are SA-inducible, but it does not cause the emission of MeSA. (Turlings et al. 1993; Bi et al. 1997; Ozawa et al. 2000; Diezel et al. 2009). Similarly, not every plant species responds to MeSA treatments. Wild tobacco plants treated with MeSA are not attractive to predatory foragers (Kessler and Baldwin 2001). Arabidopsis plants treated with SA are not attractive to the parasitoid *Cotesia rubecula*, a parasite of caterpillar *P. rapae*, (van Poecke et al. 2001). These findings imply that while SA increases defence against piercing and sucking insects, JA is primarily engaged in defence against leaf-chewing herbivores (Walling 2000; van Poecke and Dicke 2004; Zhao et al. 2009).

10.3.2.4.3 Ethylene in Plant Defence

Ethylene (ET) plays a critical role in the activation of plant defences against different biotic stresses through its participation in a complex signalling network that includes jasmonic acid (JA), salicylic acid (SA), and abscisic acid (ABA). Pathogen attack, wounding, and herbivory trigger asymmetric activation of this defence signalling network, thereby affecting the final balance of interactions between its components and establishing a targeted response to the initial threat. Ethylene's contribution to the modulation of this defence network relies on the complexity of the regulation of multigene families involved in ET biosynthesis, signal transduction, and cross talk and enables the plant to fine-tune its response. It is generally accepted that ET cooperates with JA in the activation of defences against necrotrophic pathogens and antagonizes SA-dependent resistance against biotrophic pathogens (Adie et al. 2007).

Fine-tuning of local versus systemic defence responses by ET has been highlighted in *Nicotiana attenuata* by Kahl et al. (2000). They found that ET locally decreased nicotine accumulation in leaves following herbivory by larvae of the nicotine intolerant *Manduca sexta*. However, local volatile terpenoids and endogenous JA pools remained unaffected. It was hypothesized that this adaptive tailoring of defences would reduce nicotine uptake by the larvae, thereby making them more susceptible to their nicotine-sensitive parasitoids.

10.3.2.5 Hormone Cross Talk in Plant Defence

Cross talk between several hormones is how plant hormones control the regulation of plant growth, development, and differentiation (Dicke and van Poecke 2002; Kessler and Baldwin 2002; van Poecke and Dicke 2004; Felton and Tumlinson 2008; Wu and Baldwin 2010; Erb et al. 2012). A phenomenon called hormone cross talk occurs when various hormone signalling pathways interact either antagonistically or synergistically, offering a significant regulatory potential to nimbly adjust the plant's adaptive response to a variety of environmental inputs. As a key regulatory mechanism of plant immunity, cross talk between the SA, JA, and ET signalling pathways has developed (Spoel and Dong 2008; Grant and Jones 2009; Pieterse et al., 2009). SA-mediated defences are predominantly effective against biotrophic pathogens, such as *P. syringae*, whereas JA-mediated defences are primarily effective against herbivorous insects and necrotrophic pathogens. Numerous investigations have shown that endogenous SA accumulation counteracts JA-dependent defences, favouring SA-dependent resistance over JA-dependent defence (Koornneef and Pieterse 2008). ET frequently modulates this in this way (Leon-Reyes et al. 2009; Zander et al. 2009). However, JA can also obstruct the SA pathway, depending on the plant type and the attacker's scheme of attack. For instance, it has been shown that in wild tobacco (*Nicotiana attenuata*), JA and ET bursts caused by elicitors originating from herbivores block the SA burst, causing tuning of the SA-JA signal interaction and, consequently, the defence responses of the plant to herbivory (Diezel et al. 2009). Auxins, gibberellins, brassinosteroids, and abscisic acid (ABA) have all recently become powerful players on the battlefield. These hormones frequently have interactions with the SA-JA-ET core of the plant immune signalling network that are antagonistic or synergistic, which guide the output of the defence mechanism (Wang et al. 2007; Navarro et al. 2008; Yasuda et al. 2008; Campos et al. 2009; De Torres-Zabala et al. 2009; Ton et al. 2009; De Vleesschauwer et al. 2010; Jiang et al. 2010).

10.3.2.6 Plant Defence Priming

Plant defence priming is a strategy for crop protection which is operated by the exploitation of the immune capacity of the plant. Priming protects the plants by boosting their responsiveness to attackers based on prior experience. Plant defence priming is a new tool for sustainable crop improvement. This strategy could also be called green vaccination. As we know, in the case of humans, in vaccination, mild doses of inactivated or heat killed germs are introduced into the body to develop antibodies against that particular organism. In the same way, plants can be vaccinated by intentionally exposing them to a mild level of stress (biotic or abiotic) as a primary stimulus and letting the plants know about their immune capability (Tiwari and Singh 2021) In priming, the plant is challenged intentionally with a mild level of any biotic (pathogens, beneficial microbes, insects, chemical elicitors), or abiotic stress (primary stimulus) which brings the plant into a vigilant or alarmed state. This alarmed state is called the primed state (Tiwari and Singh 2021), which is marked by accumulation of calcium, tricarboxylic acids, reactive oxygen species, hormone conjugates, amino acids, sugars, and post-transcriptional modifications and activation of defence-related genes (Beneloujaephajri et al. 2013, Chassot et al. 2008). In the priming phase, the plant remains only in an alert or vigilant state but does not exhibit any defence response. Upon subsequent attack by attackers (secondary stimulus), the plant exhibits enhanced perception towards attackers and faster and stronger defence response against attackers, which is called the post challenge priming phase, which is marked by accumulation of glucosinolates, phytoalexins, callose, phenolics, hormones such as SA and JA, pathogenesis-related (PR) proteins, and histone modifications (Mauch-Mani et al. 2017). Seedlings of *Vigna radiata* primed with sodium nitroprusside showed defence response against salinity through enhancement in the level of protective metabolites, alteration in the level of anti-oxidative enzymes, and reduction in the extent of cell wall damage and chlorophyll loss (Roychoudhury et al. 2021). The priming with silicon dioxide nanoparticles enhanced the ability of wheat plants to cope with drought conditions and also improved the rate of seed germination and biomass production (Rai-Kalal et al. 2021).

10.3.2.6.1 Synthetic Chemical Inducers as Priming Agents

Synthetic chemical inducers, also known as plant strengtheners, are structurally different from the natural plant defence elicitors. They may activate or prime plant immunity by simply mimicking the structures of natural immune inducers. Alternatively, they can also be structurally unrelated to natural elicitors and target a subset of defence signalling components, and little is known about other mechanisms by which they may increase plant resistance (Wang et al. 2020). The exploitation of induced plant defences for pest control is a promising strategy to reduce the use of pesticides in agriculture.

JA Analogue

The phytotoxin, coronatine, is a natural structural and functional mimic of JA-Ile (Weiler et al. 1994; Fonseca et al. 2009). Coronatine can elicit similar responses as JA. In an effort to identify more potent mimics of coronatine, the synthetic JA mimic coronalon was synthesized (Schuler et al. 2001). Later research revealed that the coronon mediates stress responses in a variety of plant species (Schuler et al. 2004). It can activate defence proteins known to be activated by MeJA as well as MeJA-responsive genes (Pluskota et al. 2007). Several artificial JA mimics, including coronalon, have been investigated and demonstrated to trigger JA signalling and defence responses in lima bean, soybean, and coyote tobacco (Krumm et al. 1995; Fliegmann et al. 2003; Pluskota et al. 2007). Based on the co-receptor structure, a coronatine derivative, coronatine-O-methyloxime (COR-MO), was synthesized through direct chemical derivation and identified as a potent competitive antagonist of jasmonate perception (Monte et al. 2014).

10.3.2.6.2 β-Aminobutyric Acid (BABA)

BABA is a non-protein amino acid that has been known to induce plant resistance since 1963 (Papavizas and Davey 1963). It has been shown to protect about 40 different plant species against a diverse range of pathogen and pests including viruses, bacteria, oomycetes, fungi, nematodes, and arthropods (Cohen et al. 2016). BABA primes multiple defence mechanisms regulated by SA-dependent and SA-independent pathways (Zimmerli et al. 2000; Ton et al. 2005). The priming effects elicited by BABA can be maintained to the next generation, making BABA the first plant immune inducer with transgenerational efficacy (Slaughter et al. 2012). BABA is sensed by an aspartyl-tRNA synthetase, IBI1 (Luna et al. 2014). Binding of BABA to IBI1 primes it for alternative defence activity. However, the inhibition of BABA on the aspartyl-tRNA synthetase activity leads to toxicity in plants, which makes BABA unsuitable for agricultural use. While BABA has long been considered as a synthetic plant immune priming agent, a recent study has unequivocally identified BABA as an endogenous metabolite synthesized by various plant species including Arabidopsis, Chinese cabbage, maize, teosinte, and wheat (Thevenet et al. 2017).

10.3.2.6.3 Induction of Defence in Cereals by 4-Fluorophenoxyacetic Acid Suppresses Insect Pest Populations

Exogenous application of 4-fluorophenoxyacetic acid (4-FPA) protects cereals from piercing-sucking insects and thereby increases rice yield in the field. 4-FPA does not stimulate hormonal signalling, but modulates the production of peroxidases, H_2O_2, and flavonoids and directly triggers the formation of flavonoid polymers. The increased deposition of phenolic polymers in rice parenchyma cells of 4-FPA-treated plants is associated with a decreased capacity of the white-backed planthopper (WBPH) *Sogatella furcifera* to reach the plant phloem. Thus, the application of 4-PFA in the field enhances rice yield by reducing the abundance of, and damage caused by, insect pests. 4-FPA also increases the resistance of other major cereals such as wheat and barley to piercing-sucking insect pests (Wang et al. 2020).

10.4 Outlook and Future Challenges

In all natural habitats, plants are surrounded by an enormous number of potential enemies (biotic) and various kinds of abiotic environmental stresses. Nearly all ecosystems contain a wide variety of bacteria, viruses, fungi, nematodes, mites, insects, mammals, and other herbivorous animals, greatly responsible for heavy reduction in crop productivity. By their nature, plants protect themselves directly by developing different morphological structures and by producing some compounds called as secondary metabolites. Plant mechanical defences act negatively on herbivorous insects, affecting their larval and adult performance.

The ecological interactions between plants and herbivores can be understood through an understanding of induced resistance in plants, which can also be used to manage pests in crops. The elicitors of these pathways could be employed as inducers in various crops because the metabolic pathways that result in induced resistance are highly conserved among plants. The next task will be to identify the genes that encode proteins that are up- or down-regulated during a plant's reaction to a herbivore attack. These genes can then be used to genetically modify plants so that they can confer resistance to herbivores. However, its crucial to understand the chemical changes that elicitors cause in plants, how they affect herbivores, particularly in the field, and whether there have been any changes to plant development and yield before applying them effectively in agricultural systems. The Ecogenomic technique, which combines association and correlation studies, mapping of natural selection, and population genomics, allows for the assessment of variable selection at loci and distinguishes it from processes that act on the entire genome, such as migration and genetic drift.

From a biotechnological, food-developmental, and breeding point of view, understanding the defence systems of plants and learning how to apply the knowledge are of course of huge interest. For instance, modifications of the jasmonic acid pathway have been proposed (Grant et al. 2013). However, due to the extensive cross talk with other hormone signalling pathways, increased resistance against one certain insect herbivore might result in susceptibility towards another. Furthermore, some defence responses might have negative effects on the environment and humans as well, as they involve toxic bioactive natural products and proteins reducing digestibility of plant material. Still, reducing the need for synthetic insecticides, by developing crop plants resistant to insect herbivores, would be of significant gain for the food and production industry, both at an economical and environmental level.

Defence priming is very effective strategy to protect plants and enhance agricultural productivity to achieve the aim of sustainable global food security. Priming is a big solution to the problem of global hunger and application of chemicals in our food. There is urgent need of the chemical-free technique of plant protection. Priming can provide increased resistance to a wide range of biotic and abiotic stresses. This mechanism helps to avoid the unnecessary wastage of valuable resources in expression of defence during absence of challenge. This approach can brighten the future of agriculture because it has the potential to protect the crop without incurring yield penalty. Transgenerational potential of priming makes it more sustainable by increasing the profitability of farming.

In agriculture, the discovery of natural and synthetic inducers that mimic the action of the natural signals prompted a strong interest in IR as a strategy for crop protection (Vallad and Goodman 2004; Goellner and Conrath 2008). The great expansion of synthetic immune inducers has also provided opportunities to dissect the signalling networks of plant immune system that is not accessible to genetic screens due to the lethality and gene redundancy. With the discovery of the hidden drug-able targets in plant immune system, new synthetic immune inducers may be developed to target these hidden points. These new inducers can again enhance our ability to dissect plant immune system and keep this discovery cycle going on.

References

Acevedo FE, Rivera-Vega LJ, Chung SH, Ray S, Felton GW (2015) Cues from chewing insects – the intersection of DAMPs, HAMPs, MAMPs and effectors. Curr Opin Plant Biol 26:80–86

Adie B, Chico JM, Rubio-Somoza I, Solano R (2007) Modulation of plant defences by ethylene. J Plant Growth Regul 26:160–177

Agrawal AA (1998) Induced responses to herbivory and increased plant performance. Science 279: 1201–1202

Aharoni A, Jongsma MA, Bouwmeester HJ (2005) Volatile science? Metabolic engineering of terpenoids in plants. Trends Plant Sci 10:594–602

Ament K, Kant MR, Sabelis MW, Haring MA, Schuurink RC (2004) Jasmonic acid is a key regulator of spider mite-induced volatile terpenoid and methyl salicylate emission in tomato. Plant Physiol 135:2025–2037

Arimura G, Huber DPW, Bohlmann J (2004) Forest tent caterpillars (*Malacosoma disstria*) induce local and systemic diurnal emissions of terpenoid volatiles in hybrid poplar (*Populus trichocarpa×deltoides*): CDNA cloning, functional characterization, and patterns of gene expression of (−)-germacrene D synthase, PtdTPS1. Plant J 37:603–616

Arimura G, Kost C, Boland W (2005) Herbivore-induced, indirect plant defences. Biochem Biophys Acta Mol Cell Biol Lipids 1734:91–111

Babst BA, Ferrieri RA, Thorpe MR, Orians CM (2008) *Lymantria dispar* herbivory induces rapid changes in carbon transport and partitioning in *Populus nigra*. Entomol Exp Appl 128:117–125

Benedetti M, Pontiggia D, Raggi S, Cheng Z, Scaloni F, Ferrari S, Ausubel FM, Cervone F, De Lorenzo G (2015) Plant immunity triggered by engineered in vivo release of oligogalacturonides, damage-associated molecular patterns. Proc Natl Acad Sci U S A 112: 5533–5538

Beneloujaephajri E, Costa A, L'Haridon F, Metraux JP, Binda M (2013) Production of reactive oxygen species and wound-induced resistance in *Arabidopsis thaliana* against *Botrytis cinerea* are preceded and depend on a burst of calcium. BMC Plant Biol 13(1):1–10

Berenbaum MR (1995) Turnabout is fair play: secondary roles for primary compounds. J Chem Ecol 21(7):925–940

Bi JL, Murphy JB, Felton GW (1997) Does salicylic acid act as a signal in cotton for induced resistance to *Helicoverpa zea*? J Chem Ecol 23:1805–1818

Bjorkman C, Anderson DB (1990) Trade off among antiherbivore defences in a South American blackberry (*Rubus bogotensis*). Oecologia 85:247–249

Boatright J, Negre F, Chen XL, Kish CM, Wood B, Peel G et al (2004) Understanding in vivo benzenoid metabolism in Petunia petal tissue. Plant Physiol 135:1993–2011

Boege K, Marquis RJ (2005) Facing herbivory as you grow up: the ontogeny of resistance in plants. Trends Ecol Evol 20:441–448

Bolter CJ, Dicke M, van Loon JJ, Visser JH, Posthumus MA (1997) Attraction of Colorado potato beetle to herbivore-damaged plants during herbivory and after its termination. J Chem Ecol 23: 1003–1023

Boter M, Ruíz-Rivero O, Abdeen A, Prat S (2004) Conserved MYC transcription factors play a key role in jasmonate signaling both in tomato and *Arabidopsis*. Genes Dev 18:1577–1591

Bown AW, Hall DE, MacGregor KB (2002) Insect footsteps on leaves stimulate the accumulation of 4-aminobutyrate and can be visualized through increased chlorophyll fluorescence and superoxide production. Plant Physiol 129:1430–1434

Bruinsma M, Posthumus MA, Mumm R, Mueller MJ, van Loon JA, Dicke M (2009) Jasmonic acid-induced volatiles of *Brassica oleracea* attract parasitoids: effects of time and dose, and comparison with induction by herbivores. J Exp Bot 60:2575–2587

Campos ML, de Almeida M, Rossi ML, Martinelli AP, Litholdo CG, Figueira A, Rampelotti-Ferreira FT, Vendramim JD, Benedito VA, Peres LEP (2009) Brassinosteroids interact negatively with jasmonates in the formation of anti-herbivory traits in tomato. J Exp Bot 60:4346–4360

Chamarthi SK, Sharma HC, Vijay PM, Narasu LM (2011) Leaf surface chemistry of sorghum seedlings influencing expression of resistance to sorghum shoot fly (*Atherigona soccata*). J Plant Biochem Biotechnol 20(2):211–216

Chappell J (1995) Biochemistry and molecular biology of the isoprenoid biosynthetic pathway in plants. Rev Plant Physiol Plant Mol Biol 46:521–547

Chassot C, Buchala A, Schoonbeek HJ, Metraux JP, Lamotte O (2008) Wounding of *Arabidopsis* leaves causes a powerful but transient protection against *Botrytis* infection. Plant J 55(4): 555–567

Chen Z, Zheng Z, Huang J, Lai Z, Fan B (2009) Biosynthesis of salicylic acid in plants. Plant Signal Behav 4:493–496

Cheng AX, Lou YG, Mao YB, Lu S, Wang LJ, Chen XY (2007) Plant terpenoids: biosynthesis and ecological functions. J Integr Plant Biol 49:179–186

Choh Y, Kugimiya S, Takabayashi J (2006) Induced production of extrafloral nectar in intact lima bean plants in response to volatiles from spider mite-infested conspecific plants as a possible indirect defense against spider mites. Oecologia 147:455–460

Cipollini D, Enright S, Traw MB, Bergelson J (2004) Salicylic acid inhibits jasmonic acid-induced resistance of *Arabidopsis thaliana* to *Spodoptera exigua*. Mol Ecol 13:1643–1653

Cohen Y, Vaknin M, Mauch-Mani B (2016) BABA-induced resistance: milestones along a 55-year journey. Phytoparasitica 44:513–538

Cooper SM, Owen-Smith N (1986) Effects of plant spinescence on large mammalian herbivores. Oecologia 68:446–455

De Boer JG, Posthumus MA, Dicke M (2004) Identification of volatiles that are used in discrimination between plants infested with prey or nonprey herbivores by a predatory mite. J Chem Ecol 30:2215–2230

De Torres-Zabala M, Bennett MH, Truman WH, Grant MR (2009) Antagonism between salicylic and abscisic acid reflects early host-pathogen conflict and moulds plant defence responses. Plant J 59:375–386

De Vleesschauwer D, Yang Y, Cruz CV, Hofte M (2010) Abscisic acid induced resistance against the brown spot pathogen *Cochliobolus miyabeanus* in rice involves MAP kinase-mediated repression of ethylene signaling. Plant Physiol 152:2036–2052

Dicke M (2001) Inducible defence of plants against insects: from molecular to evolutionary biology. Appl Entomol Nev Amsterdam 12:9–18

Dicke M, van Poecke RMP (2002) Signalling in plant insect interactions: signal transduction in direct and indirect plant defence. In: Scheel D, Wasternack C (eds) Plant Signal Transduction. Oxford University Press, UK, pp 289–316

Dicke M, Van Beek TA, Posthumus MA, Ben Dom N, Van Bokhoven H, De Groot A (1990) Isolation and identification of volatile kairomone that affects acarine predator-prey interactions: involvement of host plant in its production. J Chem Ecol 16:381–396

Dicke M, Takabayashi J, Posthumus MA, Schutte C, Krips OE (1998) Plant-phytoseiid interactions mediated by herbivore-induced plant volatiles: variation in production of cues and in responses of predatory mites. Exp Appl Acarol 22:311–333

Dicke M, Gols R, Ludeking D, Posthumus MA (1999) Jasmonic acid and herbivory differentially induce carnivore attracting plant volatiles in lima bean plants. J Chem Ecol 25:1907–1922

Diezel C, Von Dahl CC, Gaquerel E, Baldwin IT (2009) Different lepidopteran elicitors account for cross-talk in herbivory-induced phytohormone signaling. Plant Physiol 150:1576–1586

Duran-Flores D, Heil M (2016) Sources of specificity in plant damaged-self recognition. Curr Opin Plant Biol 32:77–87

Duso C, Pozzebon A, Capuzzo C, Malagnini V, Otto S, Borgo M (2005) Grape downy mildew spread and mite seasonal abundance in vineyards: effects on *Tydeus caudatus* and its predators. Biol Control 32:143–154

Eisner T, Eisner M, Hoebeke ER (1998) When defence backfires: detrimental effect of a plant's protective trichomes on an insect beneficial to the plant. Proc Natl Acad Sci U S A 95:4410–4444

Engelberth J, Koch T, Schuler G, Bachmann N, Rechtenbach J, Boland W (2001) Ion channel-forming alamethicin is a potent elicitor of volatile biosynthesis and tendril coiling. Cross talk between jasmonate and salicylate signaling in Lima bean. Plant Physiol 125:369–377

English-Loeb G, Norton AP, Gadoury D, Seem R, Wilcox W (2005) Tri-trophic interactions among grapevines, a fungal pathogen, and a mycophagous mite. Ecol Appl 15:1679–1688

Erb M (2018) Plant defenses against herbivory: closing the fitness gap. Trends Plant Sci 20:187–194

Erb M, Meldau S, Howe GA (2012) Role of phytohormones in insect-specific plant reactions. Trends Plant Sci 17:250–259

Erbilgin N, Raffa KF (2001) Kairomonal range of generalist predators in specialized habitats: Responses to multiple phloeophagous species emitting pheromones vs. host odors. Entomol Exp Appl 99:205–210

Eyles A, Davies NW, Yuan ZQ, Mohammed C (2003) Host responses to natural infection by *Cytonaema* sp. in the aerial bark of *Eucalyptus globulus*. For Pathol 33:317–331

Falk KL, Kastner J, Bodenhausen N, Schramm K, Paetz C, Vassao DG, Reichelt M, von Knorre D, Bergelson J, Erb M et al (2014) The role of glucosinolates and the jasmonic acid pathway in resistance of *Arabidopsis thaliana* against molluscan herbivores. Mol Ecol 23:1188–1203

Farmer EE, Ryan CA (1992) Octadecanoid precursors of jasmonic acid activate the synthesis of wound-inducible proteinase inhibitors. Plant Cell 4:129–134

Felton GW (2008) Caterpillar secretions and induced plant responses. In: Schaller A (ed) Induced plant resistance to herbivory. Springer, New York, pp 369–389

Felton GW, Tumlinson JH (2008) Plant-insect dialogs: complex interactions at the plant-insect interface. Curr Opin Plant Biol 11:457–463

Felton GW, Donato KK, Broadway RM, Duffey SS (1992) Impact of oxidized plant phenolics on the nutritional quality of dietary protein to a noctuid herbivore, *Spodoptera exigua*. J Insect Physiol 38:277–285

Feussner I, Wasternack C (2002) The lipoxygenase pathway. Annu Rev Plant Biol 53:275–297

Fliegmann J, Schuler G, Boland W, Ebel J, Mithofer A (2003) The role of octadecanoids and functional mimics in soybean defense responses. Biol Chem 384:437–446

Fonseca S, Chini A, Hamberg M, Adie B, Porzel A, Kramell R et al (2009) (+)-7-iso-Jasmonoyl-L-isoleucine is the endogenous bioactive jasmonate. Nat Chem Biol 5:344–350

Fonseca S, Calvo PF, Fernandez GM, Diez-Diaz M, Gimenez-Ibanez S, Lopez-Vidriero I, Godoy M, Fernandez-Barbero G, Van Leene J, De Jaeger G et al (2014) bHLH003, bHLH013 and bHLH017 are new targets of JAZ repressors negatively regulating JA responses. PLoS One 9:e86182

Frost CJ, Hunter MD (2008) Herbivore-induced shifts in carbon and nitrogen allocation in red oak seedlings. New Phytol 178:835–845

Frost CJ, Appel HM, Carlson JE, De Moraes CM, Mescher MC, Schultz JC (2007) Within plant signaling via volatiles overcomes vascular constraints on systemic signaling and primes responses against herbivores. Ecol Lett 10:490–498

Furlong MJ, Ang Gurion CK, Rehan S, Zalucki MP (2018) Bringing ecology back: How can the chemistry of indirect plant defenses against herbivore be manipulated to improve pest management? Front Plant Sci 9:1436. https://doi.org/10.3389/fpls.2018.01436

Goellner K, Conrath U (2008) Priming: it's all the world to induced disease resistance. Eur J Plant Pathol 121:233–242

Gonzalez-Teuber M, Heil M (2009) The role of extrafloral nectar amino acids for the preferences of facultative and obligate ant mutualists. J Chem Ecol 35:459–468

Gowda JH (1996) Spines of *Acacia tortilis*: what do they defend and how? Oikos 77:279–284

Grant MR, Jones JDG (2009) Hormone (dis)harmony moulds plant health and disease. Science 324:750–752

Grant MR, Kazan K, Manners JM (2013) Exploiting pathogens' tricks of the trade for engineering of plant disease resistance: challenges and opportunities. Microb Biotechnol 6(3):212–222

Grimes HD, Koetje DS, Franceschi VR (1992) Expression, activity, and cellular accumulation of methyl jasmonate responsive lipoxygenase in soybean seedlings. Plant Physiol 100:433–443

Grubb PJ (1986) Sclerophyllys, pachyphylls and pycnophylls: the nature and significance of hard surfaces. In: Juniper B, Southwood TRE (eds) Insects and the plant surface. Arnold, London, pp 137–150

Grubb PJ (1992) A positive distrust in simplicity – lessons from plant defences and from competition among plants and among animals. J Ecol 80:585–610

Gutschick VP (1999) Biotic and abiotic consequences of differences in leaf structure. New Phytol 143:3–18

Haberlandt G (1914) Physiological Plant Anatomy. McMil-lan, London

Haddad NM, Hicks WM (2000) Host pubescence and the behavior and performance of the butterfly *Papilio troilus* (Lepidoptera: Papilionidae). Environ Entomol 29:299–303

Handley R, Ekbom B, Agren J (2005) Variation in trichome density and resistance against a specialist insect herbivore in natural populations of *Arabidopsis thaliana*. Ecol Entomol 30: 284–292

Hanley ME, Lamont BB (2001) Herbivory, serotiny and seedling defence in Western Australian Proteaceae species. Oecologia 126:409–417

Hanley ME, Lamont BB, Fairbanks MM, Rafferty CM (2007) Plant structural traits and their role in antiherbivore defense. Perspec Plant Ecol Evol Syst 8:157–178

Hare JD (2011) Ecological role of volatiles produced by plants in response to damage by herbivorous insects. Annu Rev Entomol 56:161–180

Haukioja E, Koricheva J (2000) Tolerance to herbivory in woody vs. herbaceous plants. Evol Ecol 14:551–562

Heil M (2004) Induction of two indirect defences benefits lima bean (*Phaseolus lunatus*, Fabaceae) in nature. J Ecol 92:527–536

Heil M, Bueno JCS (2007) Within-plant signaling by volatiles leads to induction and priming of an indirect plant defense in nature. Proc Natl Acad Sci U S A 104:5467–5472

Heil M, McKey D (2003) Protective ant-plant interactions as model systems in ecological and evolutionary research. Annu Rev Ecol Evol Syst 34:425–453

Heil M, Koch T, Hilpert A, Fiala B, Boland W, Linsenmair K (2001) Extrafloral nectar production of the ant associated plant, *Macaranga tanarius*, is an induced, indirect, defensive response elicited by jasmonic acid. Proc Natl Acad Sci U S A 98:1083–1088

Heil M, Greiner S, Meimberg H, Kruger R, Noyer J, Heubl G, Linsenmair KE, Boland W (2004) Evolutionary change from induced to constitutive expression of an indirect plant resistance. Nature 430:205–208

Horiuchi JI, Arimura GI, Ozawa R, Shimoda T, Takabayashi J, Nishioka T (2001) Exogenous ACC enhances volatiles production mediated by jasmonic acid in Lima bean leaves. FEBS Lett 509: 332–336

Howe GA, Jander G (2008) Plant immunity to herbivores. Annu Rev Plant Biol 59:41–66

Humphreys JM, Chapple C (2002) Rewriting the lignin roadmap. Curr Opin Plant Biol 5:224–229

Huntzinger M, Karban R, Young TP, Palmer TM (2004) Relaxation of induced indirect defenses of *acacias* following exclusion of mammalian herbivores. Ecology 85:609–614

James DG (2003) Field evaluation of herbivore-induced plant volatiles as attractants for beneficial insects: Methyl salicylate and the green lacewing, *Chrysopa nigricornis*. J Chem Ecol 29:1601–1609

Jiang CJ, Shimono M, Sugano S, Kojima M, Yazawa K, Yoshida R, Inoue H, Hayashi N, Sakakibara H, Takatsuji H (2010) Abscisic acid interacts antagonistically with salicylic acid signaling pathway in rice-*Magnaporthe grisea* interaction. Mol Plant-Microbe Interact 23:791–798

Kahl J, Siemens DH, Aerts RJ, Gabler R, Kuhnemann F et al (2000) Herbivore-induced ethylene suppresses a direct defense but not a putative indirect defense against an adapted herbivore. Planta 210:336–342

Kang JH, McRoberts J, Shi F, Moreno J, Jones AD, Howe GA (2014) The flavonoid biosynthetic enzyme chalcone isomerase modulates terpenoid production in glandular trichomes of tomato. Plant Physiol 164:1161–1174

Karban R (2011) The ecology and evolution of induced resistance against herbivores. Funct Ecol 25:339–347

Karban R, Baldwin IT (1997) Induced responses to herbivory. University of Chicago Press, Chicago, IL, p 320p
Karban R, Myers JH (1989) Induced plant responses to herbivory. Annu Rev Ecol Syst 20:331–348
Karban R, Shiojiri K, Huntzinger M, McCall AC (2006) Damage-induced resistance in sagebrush: volatiles are key to intra and interplant communication. Ecology 87:922–930
Karban R, Yang LH, Edwards KF (2014) Volatile communication between plants that affects herbivory: a meta-analysis. Ecol Lett 17:44–52
Kessler A, Baldwin IT (2001) Defensive function of herbivore-induced plant volatile emissions in nature. Science (New York, NY) 291:2141–2144
Kessler A, Baldwin IT (2002) Plant responses to insect herbivory: the emerging molecular analysis. Annu Rev Plant Biol 53:299–328
Khan MIR, Fatma M, Per TS, Anjum NA, Khan NA (2015) Salicylic acid-induced abiotic stress tolerance and underlying mechanisms in plants. Front Plant Sci 6(462). https://doi.org/10.3389/fpls.2015.00462
Koornneef A, Pieterse CMJ (2008) Cross-talk in defense signaling. Plant Physiol 146:839–844
Koptur S (1992) Extrafloral nectary-mediated interactions between insects and plants. Insect Plant Interactions 4:81–129
Kost C, Heil M (2005) Increased availability of extrafloral nectar reduces herbivory in lima bean plants (*Phaseolus lunatus*, Fabaceae). Basic Appl Ecol 6:237–248
Kost C, Heil M (2006) Herbivore-induced plant volatiles induce an indirect defence in neighbouring plants. J Ecol 94:619–628
Kost C, Heil M (2008) The defensive role of volatile emission and extrafloral nectar secretion for lima bean in nature. J Chem Ecol 34:2–13
Krumm T, Bandemer K, Boland W (1995) Induction of volatile biosynthesis in the Lima bean (*Phaseolus lunatus*) by leucine- and isoleucine conjugates of 1-oxo- and 1-hydroxyindan-4-carboxylic acid: evidence for amino acid conjugates of jasmonic acid as intermediates in the octadecanoid signalling pathway. FEBS Lett 377:523–529
Labandeira C (2007) The origin of herbivory on land: initial patterns of plant tissue consumption by arthropods. Insect Sci 14:259–275
Leon-Reyes A, Spoel SH, De Lange ES, Abe H, Kobayashi M, Tsuda S, Millenaar FF, Welschen RAM, Ritsema T, Pieterse CMJ (2009) Ethylene modulates the role of nonexpressor of pathogenesis-related genes1 in cross talk between salicylate and jasmonate signaling. Plant Physiol 149:1797–1809
Levin DA (1973) The role of trichomes in plant defence. Q Rev Biol 48:3–15
Luchi N, Ma R, Capretti P, Bonello P (2005) Systemic induction of traumatic resin ducts and resin flow in Austrian pine by wounding and inoculation with *Sphaeropsis sapinea* and *Diplodia scrobiculata*. Planta 221:75–84
Luna E, Van Hulten M, Zhang Y, Berkowitz O, Lopez A, Petriacq P et al (2014) Plant perception of β-aminobutyric acid is mediated by an aspartyl-tRNA synthetase. Nat Chem Biol 10:450–456
Maffei ME, Mithofer A, Boland W (2007a) Insects feeding on plants: rapid signals and responses preceding the induction of phytochemical release. Phytochemistry 68:2946–2959
Maffei ME, Mithofer A, Boland W (2007b) Before gene expression: early events in plant-insect interaction. Trends Plant Sci 12:310–316
Markovic I, Norris DM, Phillips JK, Webster FX (1996) Volatiles involved in the nonhost rejection of *Fraxinus pennsylvanica* by *Lymantria dispar* larvae. J Agric Food Chem 44:929–935
Mauch-Mani B, Baccelli I, Luna E, Flors V (2017) Defense priming: an adaptive part of induced resistance. Annu Rev Plant Biol 68:485–512
Meiners T, Hilker M (2000) Induction of plant synomones by oviposition of a phytophagous insect. J Chem Ecol 26:221–232
Melan MA, Dong X, Endara ME, Davis KR, Ausubel FM, Peterman TK (1993) An *Arabidopsis thaliana* lipoxygenase gene can be induced by pathogens, abscisic acid, and methyl jasmonate. Plant Physiol 101:441–450

Mikkelsen MD, Petersen BL, Glawischnig E, Jensen AB, Andreasson E, Halkier B (2003) Modulation of CYP79 genes and glucosinolate profiles in *Arabidopsis* by defense signaling pathways. Plant Physiol 131:298–308

Mithofer A, Wanner G, Boland W (2005) Effects of feeding *Spodoptera littoralis* on lima bean leaves. II. Continuous mechanical wounding resembling insect feeding is sufficient to elicit herbivory-related volatile emission. Plant Physiol 137:1160–1168

Monks A, O'Connell DM, Lee WG, Bannister JM, Dickinson KJ (2007) Benefits associated with the domatia mediated tritrophic mutualism in the shrub *Coprosma lucida*. Oikos 116:873–881

Monte I, Hamberg M, Chini A, Gimenez-Ibanez S, Garcia-Casado G, Porzel A et al (2014) Rational design of a ligand-based antagonist of jasmonate perception. Nat Chem Biol 10:671–676

Mumm R, Dicke M (2010) Variation in natural plant products and the attraction of bodyguards involved in indirect plant defense. Can J Zool 88:628–667

Mumm R, Hilker M (2005) The significance of background odour for an egg parasitoid to detect plants with host eggs. Chem Senses 30:337–343

Mumm R, Schrank K, Wegener R, Schulz S, Hilker M (2003) Chemical analysis of volatiles emitted by *Pinus sylvestris* after induction by insect oviposition. J Chem Ecol 29:1235–1252

Navarro L, Bari R, Achard P, Lison P, Nemri A, Harberd NP, Jones JDG (2008) DELLAs control plant immune responses by modulating the balance of jasmonic acid and salicylic acid signaling. Curr Biol 18:650–655

Norton AP, English-Loeb G, Gadoury D, Seem RC (2000) Mycophagous mites and foliar pathogens: leaf domatia mediate tritrophic interactions in grapes. Ecology 81:490–499

O'Connell DM, Monks A, Dickinson KJ, Lee WG (2015) Is domatia production in *Coprosma rotundifolia* (Rubiaceae) induced by mites or foliar pathogens? N Z J Ecol 39:214–220

O'Dowd DJ (1982) Pearl bodies as ant food: an ecological role for some leaf emergences of tropical plants. Biotropica 14:40–49

O'Dowd DJ, Willson MF (1997) Leaf domatia and the distribution and abundance of foliar mites in broadleaf deciduous forest in Wisconsin. Am Mid Nat 137:337–348

Obeso JR (1997) The induction of spinescence in European holly leaves by browsing ungulates. Plant Ecol 129:149–156

Ozawa R, Arimura GI, Takabayashi J, Shimoda T, Nishioka T (2000) Involvement of jasmonate- and salicylate-related signaling pathways for the production of specific herbivore induced volatiles in plants. Plant Cell Physiol 41:391–398

Paiva EA, Buono RA, Lombardi JA (2009) Food bodies in *Cissus verticillata* (Vitaceae): ontogenesis, structure and functional aspects. Ann Bot 103:517–524

Papavizas GC, Davey CB (1963) Effect of amino compounds and related substances lacking sulfur on aphanomyces root rot of peas. Phytopathology 53:116–122

Peiffer M, Tooker J, Luthe DS, Felton GW (2009) Plants on early alert: Glandular trichomes as sensors for insect herbivores. New Phytol 184:644–656

Perez-Barberia FJ, Gordon IJ (1998) Factors affecting food comminution during chewing in ruminants: a review. Biol J Linn Soc 63:233–256

Pettersson EM (2001) Volatile attractants for three Pteromalid parasitoids attacking concealed spruce bark beetles. Chemoecol 11:89–95

Pieterse CMJ, Leon-Reyes A, Van der Ent S, Van Wees SCM (2009) Networking by small-molecules hormones in plant immunity. Nat Chem Biol 5:308–316

Pluskota WE, Qu N, Maitrejean M, Boland W, Baldwin IT (2007) Jasmonates and its mimics differentially elicit systemic defence responses in *Nicotiana attenuata*. J Exp Bot 58:4071–4082

Potlakayala SD, Reed DW, Covello PS, Fobert PR (2007) Systemic acquired resistance in canola is linked with pathogenesis-related gene expression and requires salicylic acid. Phytopathology 97:794–802

Pozzebon A, Loeb G, Duso C (2009) Grape powdery mildew as a food source for generalist predatory mites occurring in vineyards: effects on life-history traits. Ann Appl Biol 155:81–89

Qi J, Li J, Han X, Li R, Wu J, Yu H, Hu L, Xiao Y, Lu J, Lou Y (2015) Jasmonic acid carboxyl methyltransferase regulates development and herbivory-induced defense response in rice. J Integr Plant Biol 58:564–576

Quentin A, Pinkard EA, Beadle C, Wardlaw TJ, O'Grady AP, Paterson S, Mohammed CL (2010) Do artificial and natural defoliation have similar effects on physiology of *Eucalyptus globules* Labill. seedlings? Ann For Sci 67:203–211

Quiring DT, Timmins PR, Park SJ (1992) Effect of variations in hooked trichome densities of *Phaseolus vulgaris* on longevity of *Liriomyza trifolii* (Diptera: Agromyzidae) adults. Environ Entomol 21:1357–1361

Rai-Kalal P, Tomar RS, Jajoo A (2021) Seed nanopriming by silicon oxide improves drought stress alleviation potential in wheat plants. Funct Plant Biol 48

Ralph S, Yueh H, Friedmann MF, Aeschliman D, Zeznik JA, Nelson CC, ButterWeld YSN, Kirkpatrick R, Liu J, Jones SJM et al (2006) Conifer defense against insects: microarray gene expression profiling of Sitka spruce (*Picea sitchensis*) induced by mechanical wounding or feeding by spruce budworm (*Choristoneura occidentalis*) or white pine weevil (*Pissodes strobi*) reveals large-scale changes of the host transcriptome. Plant Cell Environ 29:545–570

Rasmann S, Kollner TG, Degenhardt J, Hiltpold I, Toepfer S, Kuhlmann U et al (2005) Recruitment of entomopathogenic nematodes by insect-damaged maize roots. Nature 434:732–737

Reddy GVP (2002) Plant volatiles mediate orientation and plant preference by the predator *Chrysoperla carnea* Stephens (Neuroptera: Chrysopidae). Biol Control 25:49–55

Reymond P, Bodenhausen N, Van Poecke RMP, Krishnamurthy V, Dicke M, Farmer EE (2004) A conserved transcript pattern in response to a specialist and a generalist herbivore. Plant Cell 16: 3132–3147

Rivas-San Vicente M, Plasencia J (2011) Salicylic acid beyond defence: its role in plant growth and development. J Exp Bot 62:3321–3338

Robbins CT (1993) Wildlife feeding and nutrition. Academic Press, San Diego

Rodriguez-Saona CR, Rodriguez-Saona LE, Frost CJ (2009) Herbivore-induced volatiles in the perennial shrub, *Vaccinium corymbosum*, and their role in interbranch signaling. J Chem Ecol 35:163–175

Romero GQ, Benson WW (2004) Leaf domatia mediate mutualism between mites and a tropical tree. Oecologia 140:609–616

Romero GQ, Benson WW (2005) Biotic interactions of mites, plants and leaf domatia. Curr Opin Plant Biol 8:436–440

Rose U, Lewis J, Tumlinson J (2006) Extrafloral nectar from cotton (*Gossypium hirsutum*) as a food source for parasitic wasps. Funct Ecol 20:67–74

Rosenthal JP, Kotanen PM (1994) Terrestrial plant tolerance to herbivory. Trends Ecol Evol 9:145–148

Roychoudhury A, Singh A, Aftab T, Ghosal P, Banik N (2021) Seedling priming with sodium nitroprusside rescues *Vigna radiata* from salinity stress-induced oxidative damages. J Plant Growth Regul 40:2454–2464

Ryals JA, Neuenschwander UH, Willits MG, Molina A, Steiner HY, Hunt MD (1996) Systemic acquired resistance. Plant Cell 8:1809–1819

Saltz D, Ward D (2000) Responding to a three-pronged attack: desert lilies subject to herbivory by dorcas gazelles. Plant Ecol 148:127–138

Schimper AFW (1903) Plant geography upon a physiological basis. Fisher, WR (Transl.). Clarendon Press, Oxford

Schmelz EA, Alborn HT, Tumlinson JH (2003) Synergistic interactions between volicitin, jasmonic acid and ethylene mediated insect-induced volatile emission in *Zea mays*. Physiol Plant 117: 403–412

Schmelz EA, Carroll MJ, LeClere S, Phipps SM, Meredith J, Chourey PS et al (2006) Fragments of ATP synthase mediate plant perception of insect attack. Proc Natl Acad Sci U S A 103:8894–8899

Schnee C, Kollner TG, Held M, Turlings TCJ, Gershenzon J, Degenhardt J (2006) The products of a single maize sesquiterpene synthase form a volatile defense signal that attracts natural enemies of maize herbivores. Proc Natl Acad Sci U S A 103:1129–1134

Schuler G, Gorls H, Boland W (2001) 6-Substituted indanoyl isoleucine conjugates mimic the biological activity of coronatine. Eur J Organ Chem 2001:1663–1668

Schuler G, Mithofer A, Baldwin IT, Berger S, Ebel J, Santos JG et al (2004) Coronalon: a powerful tool in plant stress physiology. FEBS Lett 563:17–22

Schuman MC, Barthel K, Baldwin IT (2012) Herbivory-induced volatiles function as defenses increasing fitness of the native plant, *Nicotiana attenuata*, in nature. eLife 1:e7

Schweizer F, Fernandez-Calvo P, Zander M, Diez-Diaz M, Fonseca S, Glauser G, Lewsey MG, Ecker JR, Solano R, Reymond P (2013) *Arabidopsis* basic helix-loop-helix transcription factors MYC2, MYC3, and MYC4 regulate glucosinolate biosynthesis, insect performance, and feeding behavior. Plant Cell 25:3117–3132

Scutareanu P, Drukker B, Bruin J, Posthumus MA, Sabelis MW (1997) Volatiles from *Psylla*-infested pear trees and their possible involvement in attraction of anthocorid predators. J Chem Ecol 23:2241–2260

Sharma HC (2009) Biotechnological approaches for pest management and ecological sustainability. CRC Press/Taylor and Francis, New York, p 526p

Shiojiri K, Kishimoto K, Ozawa R, Kugimiya S, Urashimo S, Arimura G, Horiuchi J, Nishioka T, Matsui K, Takabayashi J (2006) Changing green leaf volatile biosynthesis in plants: an approach for improving plant resistance against both herbivores and pathogens. Proc Natl Acad Sci U S A 103:16672–16676

Silva EMJ, Machado SR (1999) Ultrastructure and cytochemistry of the pearl gland in *Piper regnellii* (Piperaceae). Nord J Bot 19:623–634

Slaughter A, Daniel X, Flors V, Luna E, Hohn B, Mauch-Mani B (2012) Descendants of primed Arabidopsis plants exhibit resistance to biotic stress. Plant Physiol 158:835–843

Spoel SH, Dong X (2008) Making sense of hormone crosstalk during plant immune responses. Cell Host Microbe 3:348–351

Staswick PE, Tiryaki I (2004) The oxylipin signal jasmonic acid is activated by an enzyme that conjugates it to isoleucine in *Arabidopsis*. Plant Cell 16:2117–2127

Steppuhn A, Baldwin IT (2007) Resistance management in a native plant: nicotine prevents herbivores from compensating for plant protease inhibitors. Ecol Lett 10(6):499–511

Stowe KA, Marquis RJ, Hochwender CG, Simms EL (2000) The evolutionary ecology of tolerance to consumer damage. Annu Rev Ecol Syst 31:565–595

Strauss SY, Agrawal AA (1999) The ecology and evolution of plant tolerance to herbivory. Trends Ecol Evol 14:179–185

Tempfli B, Penzes B, Fail J, Szab ´ o, ´ A (2015) The occurrence of tydeoid mites (Acari: Tydeoidea) in Hungarian vineyards. Syst Appl Acarol 20:937–954

Thaler JS (1999) Jasmonate-inducible plant defences cause increased parasitism of herbivores. Nature 399:686–688

Thaler JS, Stout MJ, Karban R, Duffey SS (2001) Jasmonate-mediated induced plant resistance affects a community of herbivores. Ecol Èntomol 26:312–324

Thevenet D, Pastor V, Baccelli I, Balmer A, Vallat A, Neier R et al (2017) The priming molecule β-aminobutyric acid is naturally present in plants and is induced by stress. New Phytol 213:552–559

Tiwari M, Singh P (2021) Plant defense priming: a new tool for sustainable global food security. In: Shekhawat GS (ed) Agric Innov Sustain pp, pp 133–153

Ton J, Jakab G, Toquin V, Flors V, Iavicoli A, Maeder MN et al (2005) Dissecting the beta-aminobutyric acid-induced priming phenomenon in *Arabidopsis*. Plant Cell 17:987–999

Ton J, Flors V, Mauch-Mani B (2009) The multifaceted role of ABA in disease resistance. Trends Plant Sci 14:310–317

Truitt CL, Wei HX, Pare PW (2004) A plasma membrane protein from *Zea mays* binds with the herbivore elicitor volicitin. Plant Cell 16:523–532

Turlings TC, Wackers F (2004) Recruitment of predators and parasitoids by herbivore-injured plants. Adv Insect Chem Ecol 2:21–75

Turlings TCJ, McCall PJ, Alborn HT, Tumlinson JH (1993) An elicitor in caterpillar oral secretions that induces corn seedlings to emit chemical signals attractive to parasitic wasps. J Chem Eco 19:411–425

Vallad GE, Goodman RM (2004) Systemic acquired resistance and induced systemic resistance in conventional agriculture. Crop Sci 44:1920–1934

Van Poecke RMP, Dicke M (2002) Induced parasitoid attraction by *Arabidopsis thaliana*: Involvement of the octadecanoid and the salicylic acid pathway. J Exp Bot 53:1793–1799

Van Poecke RMP, Dicke M (2004) Indirect defence of plants against herbivores: using *Arabidopsis thaliana* as a model plant. Plant Biol 6:387–401

Van Poecke RM, Posthumus MA, Dicke M (2001) Herbivore-induced volatile production by *Arabidopsis thaliana* leads to attraction of the parasitoid *Cotesia rubecula*: chemical, behavioral, and gene-expression analysis. J Chem Ecol 27:1911–1928

Van Wijk M, De Bruijn PJA, Sabelis MW (2008) Predatory mite attraction to herbivore induced plant odors is not a consequence of attraction to individual herbivore-induced plant volatiles. J Chem Ecol 34:791–803

Wackers F, Zuber D, Wunderlin R, Keller F (2001) The effect of herbivory on temporal and spatial dynamics of foliar nectar production in cotton and castor. Ann Bot 87:365–370

Walling LL (2000) The myriad plant responses to herbivores. J Plant Growth Regul 19:195–216

Wang Y, Jiang H (2007) Reconstitution of a branch of the *Manduca sexta* prophenol oxidase activation cascade *in vitro*: snake-like hemolymph proteinase 21 (HP21) cleaved by HP14 activates prophenoloxidase-activating proteinase-2 precursor. Insect Biochem Mol Biol 37: 1015–1025

Wang D, Pajerowska-Mukhtar K, Hendrickson Culler A, Dong X (2007) Salicylic acid inhibits pathogen growth in plants through repression of the auxin signaling pathway. Curr Biol 17: 1784–1790

Wang W, Zhoua P, Mo X, Hu L, Jin N, Chen X, Yu Z, Meng J, Erb M, Shang Z, Gatehouse AMR, Wu J, Lou Y (2020) Induction of defense in cereals by 4-fluorophenoxyacetic acid suppresses insect pest populations and increases crop yields in the field. Proc Natl Acad Sci U S A 117(22): 12017–12028

War AR, Sharma HC (2014) Induced resistance in plants and counter-adaptation by insect pests. Short Views Insect. Biochem Mol Biol:1–16

War AR, Paulraj MG, War MY, Ignacimuthu S (2011a) Jasmonic acid-mediated induced resistance in groundnut (*Arachis hypogaea* L.) against *Helicoverpa armigera* (Hubner) (Lepidoptera: Noctuidae). J Plant Growth Regul 30:512–523

War AR, Paulraj MG, War MY, Ignacimuthu S (2011b) Herbivore- and elicitor-induced resistance in groundnut to Asian armyworm, *Spodoptera litura* (Fab.) (Lepidoptera: Noctuidae). Plant Signal Behav 6(11):1769–1777

War AR, Paulraj MG, Ahmad T, Buhroo AA, Hussain B et al (2012) Mechanisms of plant defense against insect herbivores. Plant Signal Behav 7(10):1306–1320

Webber BL, Abaloz BA, Woodrow IE (2007) Myrmecophilic food body production in the understorey tree, *Ryparosa kurrangii* (Achariaceae), a rare Australian rainforest taxon. New Phytol 173:250–263

Wei J, Kang L (2011) Roles of (Z)-3-hexenol in plant-insect interactions. Plant Signal Behav 6: 369–371

Weiler EW, Kutchan TM, Gorba T, Brodschelm W, Niesel U, Bublitz F (1994) The *Pseudomonas* phytotoxin coronatine mimics octadecanoid signalling molecules of higher plants. FEBS Lett 345:9–13

Woldemariam MG, Onkokesung N, Baldwin IT, Galis I (2012) Jasmonoyl-l-isoleucine hydrolase 1 (JIH1) regulates jasmonoyl-l-isoleucine levels and attenuates plant defenses against herbivores. Plant J 72:758–767

Wright IJ, Reich PB, Westoby M, Ackerly DD, Baruch Z, Bongers F, Cavender-Bares J, Chapin T, Cornelissen JHC, Diemer M, Flexas J, Garnier E, Groom PK, Gulias J, Hikosaka K, Lamont BB, Lee T, Lee W, Lusk C, Midgley JJ, Navas ML, Niinemets U, Oleksyn J, Osada N, Poorter H, Poot P, Prior L, Pyankov VI, Roumet C, Thomas SC, Tjoelker MG, Veneklaas EJ, Villar R (2004) The worldwide leaf economics spectrum. Nature 428:821–827

Wu J, Baldwin IT (2010) New insights into plant responses to the attack from insect herbivores. Annu Rev Genet 44:1–24

Xu J, Wang X, Zu H, Zeng X, Baldwin IT, Lou Y, Li R (2021) Molecular dissection of rice phytohormone signaling involved in resistance to a piercing-sucking herbivore. New Phytol 230:1639–1652

Yasuda M, Ishikawa A, Jikumaru Y, Seki M, Umezawa T, Asami T, Maruyama-Nakashita A, Kudo T, Shinozaki K, Yoshida S et al (2008) Antagonistic interaction between systemic acquired resistance and thc abscisic acid-mediated abiotic stress response in *Arabidopsis*. Plant Cell 20:1678–1692

Zander M, La Camera S, Lamotte O, Metraux JP, Gatz C (2009) *Arabidopsis thaliana* class-II TGA transcription factors are essential activators of jasmonic acid/ethylene-induced defense responses. Plant J 61:200–210

Zemek R (2005) The effect of powdery mildew on the number of prey consumed by *Typhlodromus pyri* (Acari: Phytoseiidae). J Appl Entomol 129:211–216

Zhao LY, Chen JL, Cheng DF, Sun JR, Liu Y, Tian Z (2009) Biochemical and molecular characterizations of *Sitobion avenae*-induced wheat defense responses. Crop Prot 28:435–442

Zhurov V, Navarro M, Bruinsma KA, Arbona V, Santamaria ME, Cazaux M, Wybouw N, Osborne EJ, Ens C, Rioja C et al (2013) Reciprocal responses in the interaction between *Arabidopsis* and the cell-content-feeding chelicerate herbivore spider mite. Plant Physiol 164:384–399

Zimmerli L, Jakab G, Metraux JP, Mauch-Mani B (2000) Potentiation of pathogen-specific defense mechanisms in Arabidopsis by beta-aminobutyric acid. Proc Natl Acad Sci U S A 97:12920–12925

Different Generations of Genetically Modified Crops for Insect Resistance

11

Saber Delpasand Khabbazi, Afsaneh Delpasand Khabbazi, and Cennet Yaman

Abstract

Crops are exposed to a variety of insect pests throughout their lifetime. Insect pests cause significant damage to crop plants by feeding on their tissues or sap. Besides the conventional methods which are based on using chemicals, the genetic transformation of plants with insecticidal toxin genes such as *Bt* has been widely applied to control insect pests. In addition to *Bt* genes, other toxin genes from different sources were also transferred to plants. Transgenic plants have been on the market for over two decades and have had remarkable achievements so far. However, current restrictions on these products, as well as public concern make scientists explore new approaches. The advent of RNA interference technology and later the CRISPR/Cas genome editing tool has opened up a promising new avenue in the development of next-generation biotech crops. These new approaches allow scientists to introduce new plant genotypes resistant to pests and diseases without transferring toxin genes, and all it takes is to edit target regions in the genome or apply modifications to the host transcriptome content. In this chapter, we will review different generations of biotech crops developed for insect resistance.

S. D. Khabbazi (✉)
Department of Agriculture and Food, Institute of Hemp Research, Yozgat Bozok University, Yozgat, Turkey

A. D. Khabbazi
Department of Biology, Faculty of Science, Karadeniz Technical University, Trabzon, Turkey

C. Yaman
Department of Field Crops, Faculty of Agriculture, Yozgat Bozok University, Yozgat, Turkey

S. Kumar, M. Furlong (eds.), *Plant Resistance to Insects in Major Field Crops*,
https://doi.org/10.1007/978-981-99-7520-4_11

Keywords

Bt crops · CRISPR/Cas system · Insect resistance · RNA interference · Transgenic crops

11.1 Introduction

Sustainable crop production is one of the biggest challenges we face to ensure the availability of adequate nutrition for the world's growing population. According to the United Nations report, the world population which is currently around 8 billion in 2023, will reach 9.7 billion in 2050. The increase in food demand by 59–98% during this time renders traditional agricultural practices insufficient to secure the food supply (Valin et al. 2014). In addition, unfortunately, global climate change, biotic and abiotic stress factors cause serious problems in agricultural production. According to a research report, global crop losses caused by pests and diseases have been calculated as up to 37%, with 13% of losses due to insects (Gatehouse et al. 1992). Insect pests damage different parts of plants including roots, stems, leaves, and fruits either by chewing these parts or sucking the plant sap. Moreover, these pests may cause indirect damage to the host plant as a vector for viral, bacterial, and fungal pathogens (Mahmood-ur-Rahman et al. 2021). In order to protect the crops against biotic factors, farmers traditionally adopt a variety of synthetic insecticides, however, the increasing use of chemicals has been proven to be harmful to the environment and public health (Curry et al. 2002). The advent of recombinant DNA technology has opened up a horizon of promise to reduce chemical use concerns. The successful transformation of tobacco plants with *Cry* gene from *Bacillus thuringiensis* (Vaeck et al. 1987) made the genetic modification of crops a novel approach to reducing insecticide use. Together with tobacco, transgenic cotton plant was also produced in 1987 (Umbeck et al. 1987). Since the introduction of the first biotech crop to the market in 1996, the production ratio of these crops has increased more than 100-fold, with 190.4 million hectares grown in 29 countries in 2019 (ISAAA 2019).

So far, numerous plant transformation studies with *Bt* genes have taken important steps forward. However, based on laboratory selection and data collected from the field, the resistance conferred by *Bt* genes proved fragile as some species developed resistance to Cry toxins (Tabashnik 1994; Ferré et al. 1995). Moreover, the genetic modification of crops has been questioned and criticized by the public and scientists (Godfrey 2000). Thus, despite the successes achieved by Bt crops, they turned out to be insufficient on their own to be considered as a guaranteed long-term alternative approach to agricultural production. Using resistance genes isolated from plants such as agglutinin lectin genes (*GNA*, *ASAL*, *ACA*, *WGA*), Potato inhibitor II genes, and the gene stacking strategy was then carried out to improve the utility of these crops (Bakhsh et al. 2015). Besides, the employment of genetic modification technologies such as RNAi and CRISPR/Cas system pave the way to novel insect pest management studies. This chapter reviewsdifferent generations of genetically modified crops including Bt-, RNAi- and CRISPR/Cas-based developed crops resistant to insect pests.

11.2 Transformation of Crop Plants with Resistance Genes

11.2.1 Bt Crops

The genetic transformation of plants with insecticidal toxin genes such as Cry toxins has been widely used to control insect pests. *Bacillus thuringiensis* is the source of different insecticidal agents, including *Cry* toxins, and has been deployed in pest management strategies. *B.thuringiensis* is a gram-positive soil-dwelling spore-producing bacterium that has been used as a biological control agent for nearly a century. *B. thuringiensis* is safe for humans and is the most environmentally compatible microbial insecticide worldwide (Ibrahim et al. 2010). This unique bacterium is the source of insecticidal toxin genes, mainly *Vip*, *Sip*, *Cry* (Crystal), and *Cyt* (Cytolytic) genes, which are produced throughout the bacterium's life cycle (Santos et al. 2022). *B. thuringiensis* can colonize inside the insect gut, therefore, it is an appropriate insecticidal agent for pest management strategies (Deist et al. 2014).

According to the classification by Crickmore et al. (1998), *Cry* genes are divided into 51 groups and subgroups, and Cry toxins based on the insect host specifications are classified into six main groups including group 1 lepidoptera (*Cry1*, *Cry9* and *Cry15*); group 2 lepidoptera and diptera (*Cry2*); group 3coleoptera (*Cry3*, *Cry7* and *Cry8*); group 4 diptera (*Cry4*, *Cry10*, *Cry11*, *Cry16*, *Cry17*, *Cry19* and *Cry20*); group 5 lepidoptera and coleoptera (*Cry1I*); and group 6 nematodes (*Cry6*) (reviewed in Ibrahim et al. 2010).

Using the advantage of recombinant DNA technology in the late 1980s, the first *Bt* gene was transferred to tobacco and cotton plants (Vaeck et al. 1987; Umbeck et al. 1987) and commercialization of transgenic crops expressing the *Bt* gene started in the mid-1990s and by 1999 different transgenic Bt crops such as potato, cotton, and corn were also introduced (Tabashnik et al. 2013). To date, different *Cry* genes have been transferred to agricultural crops to confer resistance to different pest species of lepidoptera, coleoptera, diptera (reviewed in Bakhsh et al. 2015). The introduction of Bt crops has reduced the use of chemical pesticides in the fields and their subsequent harmful side effects. Most Bt strains are harmful to lepidopterans; however, some are also lethal to coleopterans (McPherson et al. 1988) or dipterans (Yamamoto and McLaughlin 1981). It has been determined that Bt proteins do not show any toxicity to beneficial insects, other animals, or humans (Klausner 1984). Modification of *Bt* genes for improved expression in plants was a critical step toward achieving insect resistance in plants (Perlak et al. 1991). Codon-optimized genes conferring protection against insects of coleoptera and lepidoptera were respectively transferred to potato and cotton at first (Perlak et al. 1991). After the first reports of insect resistance, many successful studies were carried out to confer resistance against insect pests (Table 11.1).

Table 11.1 List of the toxin genes transferred to some of the crop plants

Crop	Toxin genes	Targeted insect order	Reference
Alfalfa	*Cry3a*	Coleoptera	Tohidfar et al. (2013)
Canola	*Cry1Ac*	Lepidoptera	Tabashnik et al. (1993) Stewart Jr et al. (1996) Ramachandran et al. (1998) Halfhill et al. (2001) Wang et al. (2014a, b) Rahnama and Sheykhhasan (2016)
Chickpea	*Cry1A (c)* *Cry2Aa* *Cry1Ac + Cry1Ab* *ASAL* *Vip3Aa*	Lepidoptera Hemiptera	Sanyal et al. (2005) Indurker et al. 2010 Chakraborti et al. (2009) Acharjee et al. (2010) Mehrotra et al. (2011) Singh et al. (2022)
Cotton	*Cry1Aa* *Cry1Ab* *Cry1Ac* *Cry2A* *Cry1EC* *Cry2Ab* *Cry3Bb1* *Cry3* *Cry11* *Cry1h* *Cry1Ia12* *potato proteinase inhibitor* *GNA* *ACA* *ASAL*	Lepidoptera Hemiptera	Perlak et al. (1990) Majeed (2005) Wu et al. (2006) Tohidfar et al. (2008) Khan et al. (2011) Pushpa et al. (2013) Vajhala et al. (2013) Anayol et al. (2016) Bakhsh et al. (2016) Khabbazi et al. (2018) Siddiqui et al. (2019) Zafar et al. (2022) Razzaq et al. (2023) Tariq et al. (2022)
Maize	*Cry3Bb1* *Cry1Ab* *Cry1Ab* (MON810) *Cry19c* *GNA*	Lepidoptera Hemiptera	Koziel et al. (1993) Vaughn et al. (2005) Wang et al. (2005) Gassmann et al. (2011)
Potato/sweet potato	*Cry3A* *Cry3Aa* *Cry1Ac* *Cry1Ab* Cowpea trypsin inhibitor *Cry1Ba1* *Cry1Ca5* *Cry9Aa2* *GNA* *ConA*	Coleoptera Lepidoptera Hemiptera	Peferoen et al. (1990) Cheng et al. (1992) Adang et al. (1993) Perlak et al. (1993) Newell et al. (1995) Morán et al. (1998) Gatehouse et al. (1999) Meiyalaghan et al. (2006) Jacobs et al. (2009) Mi et al. (2015) Salehian et al. (2021)
Rice	*Cry1A(b)* *Cry1A(c)*	Lepidoptera Hemiptera	Fujimoto et al. (1993) Wünn et al. (1996)

(continued)

Table 11.1 (continued)

Crop	Toxin genes	Targeted insect order	Reference
	PinII *Cry1C* *Cry2AX1* SBK + SCK *GNA* *ASAL* *DB1/ G95A-mALS*		Cheng et al. (1998) Bashir et al. (2005) Tang et al. (2006) Zhang et al. (2013) Ramesh et al. (2004) Yoshimura et al. (2012) Chandrasekhar et al. (2014) Chakraborty et al. (2016) Boddupally et al. (2018) Liu et al. (2022)
Soybean	*Cry1Ab* *Cry1Ac* *Cry8-like* *eCry1Gb.1Ig*	Lepidoptera	Parrott et al. (1994) Dufourmantel et al. (2005) Dang and Wei (2007) Qin et al. (2019) Je et al. (2022) Chae et al. (2022)
Tomato	*Cry1Ac* *Cry1Ab*	Lepidoptera	Mandaokar et al. (2000) Kumar and Kumar (2004) Koul et al. (2014)

Most of the transgenic crops are harboring constitutive promoters in particular the 35S CaMV promoter driving the foreign genes which providethe strong production of toxin protein in whole plant tissues and organs including root, stem, flowers, pollens, etc. Hence to restrict the unnecessary production of toxins in plants, foreign genes can be expressed by inducible promoters. For instance, to restrict the production of Crytoxin to insect-biting sites in plants toxin genes were expressed under the control of the wound-inducible promoter (AoPR1) isolated from *Asparagus officinalis* (Özcan et al. 1993; Bakhsh et al. 2016; Anayol et al. 2016; Khabbazi et al. 2018). The use of AoPR1 promoter confines the accumulation of Bt toxin to the wounding part of the plant, therefore, it is a valuable approach in insect pest management considering the public concerns regarding transgenic Bt crops.

11.2.2 Transgenics Harboring Plant-Derived Insect Resistance Genes

Cultivation of Bt crops increased crop production and reduced the use of chemical insecticides in the field (Toenniessen et al. 2003). Therefore, it has had an important contribution to global food security and poverty reduction. Reports indicate that this technology is beneficial for farming communities and consumers (Qaim 2009). To date, many *Cry* resistance genes have been transferred to crops to cope with damaging insects. While most of these have had a satisfactory outcome at first, the efficacy of resistance genes has been compromised by the widespread cultivation of transgenic crops. Based on reports some of the pest species have evolved resistance against Cry proteins which indicates that the toxic effect of these genes has diminished (Calles-Torrez et al. 2019; Smith et al. 2019; Tabashnik and Carrière 2019). For example, the excessive use of *Cry1Ac* has led to the development of resistance in insect pests. This resistance is due to mutations occurring in the midgut receptors like cadherin. Development of crops harboring the codon-optimized *Bt* genes is an efficient method to combat the field-evolved resistance to Bt toxins (Tabashnik and Carrière 2017; Benowitz et al. 2022; Siddiqui et al. 2023). In addition, the investigation of new insecticidal genes and approaches is a necessity for sustainable pest management strategies. Using plant-derived toxin genes alone or in combination with *Cry* genes could be another approach to this goal (Khabbazi et al. 2018; Boddupally et al. 2018). Different lectin genes are toxic to members of coleoptera, lepidoptera (Czapla and Lang 1990), and diptera (Eisemann et al. 1994). Lectins stimulate endocytosis and possibly other toxic metabolites in the midgut, resulting in the inhibition of nutrient absorption or disruption of midgut cells (Czapla and Lang 1990). Plant lectin genes are toxic to sap-sucking insects of hemiptera and have an inhibitory effect on their growth and fecundity (Wang et al. 2005; Chakraborti et al. 2009; Khabbazi et al. 2018).

Transformation of crops with *Galanthus nivalis* agglutinin lectin gene (*GNA*) isolated from the snowdrop plant confers resistance to *Aphis gossypii, Rhopalosiphum maidis, Sitobion avenae* and other sap-sucking members of hemiptera (Khabbazi et al. 2018; Wang et al. 2005; Stoger et al. 1999). This gene as well as other agglutinin lectin genes derived from garlic (*ASAL*), onion (*ACA*), wheat (*WGA*) etc. has no harmful effect on the mammalian oral system (Peumans and van Damme 1996; Khabbazi et al. 2016) and have been transferred to some of the important crops such as cotton, maize, chickpea and rice (Table 11.1) and resulted in increased resistance to different sap-sucking insects including aphids, jassids, planthoppers and whiteflies (Bakhsh et al. 2015). Along with the transformation of plants with resistance genes, RNAi and CRISPR/Cas-based methods are other relatively new approaches contributing to sustainable pest management strategies in agriculture.

11.3 RNA Interference-Mediated Modifications in Plants

11.3.1 What Is RNAi and How Does It Work?

RNA silencing is a process that causes the downregulation of a target gene expression.This technology is a worthy reverse genetics tool to study gene function (Harmon et al. 2000). It is divided into transcriptional gene silencing (TGS) and post-transcriptional gene silencing (PTGS). PTGS also known as RNA interference (RNAi) is a highly specific homology-based gene silencing tool that is frequently used to downregulate the expression of target genes via mRNA degradation and hence is also called a knockdown process (Small 2007; Tang and Galili 2004). RNAi is triggered by the introduction of double-stranded RNA (dsRNA) molecules microRNAs (miRNAs) and small interfering RNAs (siRNAs) are the two main

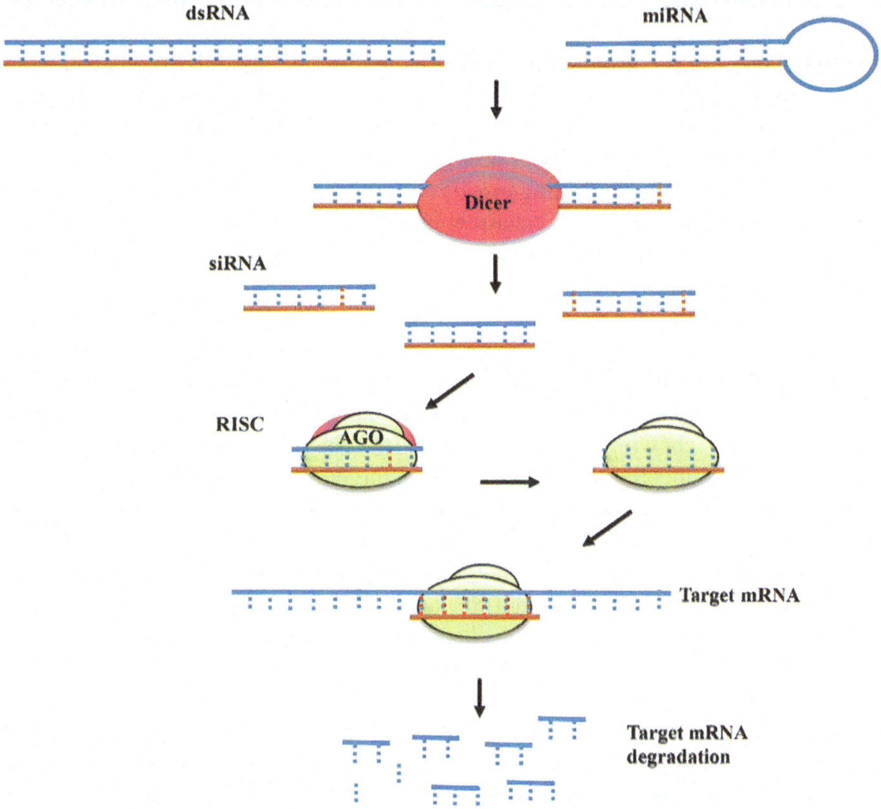

Fig. 11.1 Mechanism of the RNAi pathway. Long dsRNA or miRNA molecules are cleaved by the Dicer enzyme into short duplexes of 21–25 nucleotide RNAs. Small RNAs bind to the RISC complex and single-stranded short RNAs are produced and directed to the target mRNA, causing degradation of the mRNA

classes of small non-coding RNAs that initiate gene silencing in plants (Axtell 2013). Small interfering RNA constructs are short duplexes of 21 to 25 nucleotides produced after long dsRNA precursors are cleaved by the ribonuclease III Dicer-like (DCL) enzyme (Zamore et al. 2000; Bernstein et al. 2001). Afterward, miRNAs or siRNAs incorporated into RNA-induced silencing complex (RISC), argonaute, and other effector proteins bind to complementary mRNA molecules and subsequently degrade the mRNA and causing downregulation of the target gene (Fig. 11.1) (Bosher and Labouesse 2000; Kim and Rossi 2007; Mittal et al. 2011).

11.3.2 RNAi-Based Gene Regulation for Insect Resistance in Plants

The RNAi process is conserved in higher eukaryotes and naturally protects the host from viruses in plants, but is currently used in a variety of ways for different purposes, including insect-plant interaction studies (Khabbazi et al. 2020). RNAi has opened a new avenue in insect pest management strategies. This technology is particularly effective in controlling insects of the order coleoptera, whereas insects of lepidoptera and hemiptera are recalcitrant in response to RNAi which may be due to the biological barriers limiting the use of RNAi in these species (Terenius et al. 2011; Baum and Roberts 2014).

dsRNAs are either expressed by host plants or applied by methods like microinjection, feeding and spraying to control the insect pest damage on plants. Host-induced gene silencing (HIGS)-mediated RNAi has been successfully used in a variety of crop species to manage different agricultural insect pests including sap-sucking and chewing species (Table 11.1). In this approach, plant genetic background is engineered to produce dsRNAs targeting the essential genes in insect pests. After insects are fed with transgenic plants, dsRNAs are transported to the insect salivary glands or gut, and adsorbing cells subsequently activate the insect RNAi machinery and silence the targeted genes that interfere with insect vital metabolism. The utilization of plant-mediated RNAi provides a promising tool in crop protection without the use of chemicals and has the potential to target an unlimited number of genes in insects (Zhang et al. 2017).

Aphids are the members of the order hemiptera that damage crop plants by phloem-feeding and transmitting viral diseases. In HIGS-mediated RNAi studies in aphids, the focus has been on studying the management of *Myzus persicae* and *Sitobion avenae* aphid species in transgenic host plants including *Arabidopsis thaliana*, *Nicotiana tabacum*, *N. benthamiana*, and *Solanum lycopersicum* and *Triticum aestivum* (reviewed in Zhang et al. 2022). RNAi-mediated knocking down of the salivary effectors (MpC002, MpPIntO1, MpPIntO2, Mp55), Receptor of Activated Kinase C (Rack1), CuticularproteinMyCP, Acetylcholinesterase 1 (Ace1), Dynein heavy chain 64C (MpDhc64C), Chitin synthase 1(CHS1), Zinc finger protein (SaZFP), Carboxylesterase(CbE E4) and Lipase maturation factor 2-like gene adversely affected aphid fecundity and survival.

dsRNA-mediated downregulation of the *Sucrose non-fermenting 7* (*DvSnf7*) gene coding for an essential protein in vacuolar sorting in transgenic maize plant (*Zea mays*) conferred resistance to the western corn rootworm, *Diabrotica virgifera* (Coleoptera: Chrysomelidae) (Baum et al. 2007). *Snf7* dsRNA alone takes a long time to kill WCR larvae, so the RNAi pathway is accompanied by *Cry* genes from *B. thuringiensis* to accelerate the killing action. Further, combining the *Bt* and RNAi mechanisms reduces the occurrence of insect resistance to Bt crops. Maize plant expressing three different *Cry* genes plus dsRNA constructs for the *DvSnf7* gene, event MON87411, was approved for commercialization and release by The Canadian Food Inspection Agency (CFIA) in 2016 (Head et al. 2017). Later, in 2017, The United States Environmental Protection Agency (US-EPA) also granted permission for the commercial planting of MON87411 (Zotti et al. 2018).

Cotton bollworm, *Helicoverpa armigera,* is another devastating agricultural insect pest belonging to the order Lepidoptera. This pest has a wide host range and causes millions of dollars in losses each year (Sharma 2001). Cotton contains a polyphenolic compound called gossypol to protect itself from herbivorous insects, however, *H. armigera* can tolerate its moderate concentrations owing to the P450 monooxygenase gene, *CYP6AE14,* as this enzyme detoxifies the gossypol content. Feeding *H. armigera* larvae on leaves of transgenic Arabidopsis, tobacco, and cotton plants expressing dsRNA for *CYP6AE14*, resulted in suppression of the P450 monooxygenase gene in *H. armigera* and retarded larvae growth and enhanced host resistance to cotton bollworms (Mao et al. 2007, 2011).

Later Kumar et al. (2014) described how *Manduca sexta* larvae feeding on native *Nicotiana attenuata* can tolerate high concentrations of nicotine, a neurotoxin produced by tobacco species. Wolf spiders (*Camptocosa parallela*) avoid nicotine-fed larvae, therefore, *M. sexta* larvae deter its predator by exhaling nicotine through the spiracles as an anti-spider signal. Transformation of *N. attenuata* with constructs producing dsRNA to target the *M. sexta CYP6B46* gene and feeding the larvae with transgenic plants silenced the *CYP6B46* gene. Subsequently, insect vulnerability to spider predation was increased because of the less nicotine exhaled.

Another approach to managing insect pest damage is to interfere with chitin metabolism. Chitinase hydrolyzes chitin and, therefore, its function is vital for insect molting andmetamorphosis (Agrawal et al. 2013). Transgenic tobacco and tomato plants expressing RNAi constructs for the chitinase (*HaCHI*) gene significantly reduced chitinase production and adversely affected the overall growth and survival of *H. armigera* after continuous feeding with leaves of transgenic HaCHI-RNAi lines (Mamta and Rajam 2016).

In another study, tobacco plants were transformed to produce dsRNA targeting the *Sl102* gene in *Spodoptera littoralis*. *Sl102* is a gene involved in the immune cellular responses of *S.littoralis*, which was knocked down to increase the susceptibility of the insect pest against the pathogenicity of *B. thuringiensis*-based insecticides. Experimental larvae reared on transgenic leaves showed low transcript levels for the *Sl102* genewhich was positively associated with food consumption in the larvae (Di Lelio et al. 2022).

11.4 CRISPR/Cas System

11.4.1 Origin, Classification, and Efficiency

Genome editing of plants has achieved remarkable success since the advent of sequence-specific nucleases (Shelake et al. 2019). Zinc finger nucleases (ZNFs) and transcription activator-like effector nucleases (TALENs) were the pioneer editing tools. However, these tools are technically complex and cumbersome, with low efficiency, and therefore are not used any further (Kumar et al. 2018). In contrast, the discovery of the Clustered Regularly Interspaced Short Palindromic Repeats (CRISPR)/Cas system paved the way for a simple and precise method to

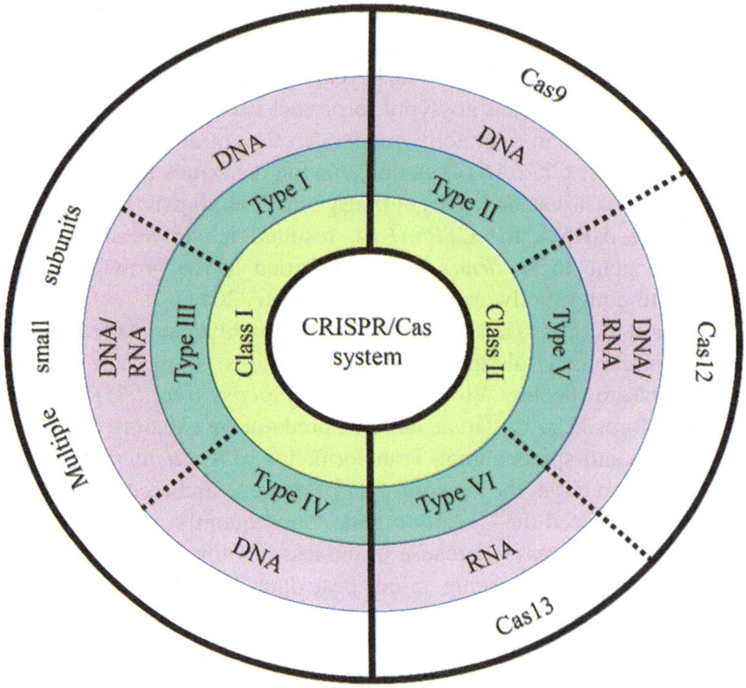

Fig. 11.2 The conventional classification of CRISPR/Cas system

modify several targets in the genome at the same time. CRISPR/Cas system was first discovered in the sequences of DNA from *Escherichia coli* (Ishino et al. 1987). Archaea and bacteria naturally use this system to protect themselves against viral invasions (Bhaya et al. 2011). After genetic elements such as phages invade the host cell, small nucleic acid fragments of invading pathogens are inserted into the host's CRISPR loci (spacers) and stored there for later encounters (Amitai and Sorek 2016). When the host cell faces a new invasion, spacer sequences are transcribed, and individual CRISPR RNAs (crRNAs) lead the Cas nuclease to the cognate nucleic acid sequences of the pathogen and cleave them (Barrangou et al. 2007). Depending on the nature of the interfering molecules, CRISPR/Cas system is divided into two classes (Fig. 11.2). Class 1 includes types I, III, and IV multiprotein effector modules that target DNA, DNA/RNA, and DNA molecules, respectively. Class 2 includes types II, V, and VI effector modules that associate with DNA, DNA/RNA, and RNA molecules, respectively. Unlike class 1, members of class 2 are single protein effector modules and the most notable examples of this class are Cas9, Cas12, and Cas13 (reviewed in Gostimskaya 2022). Type II CRISPR/Cas9 has been isolated from *Streptococcus pyogenes* and is based on RNA-guided interference with DNA and has the most contribution to genome editing studies in plants (Khabbazi et al. 2021). This system consists of a Cas9 nuclease and a single guide RNA (sgRNA) molecule. A twenty-nucleotide at the 5'end of the sgRNA

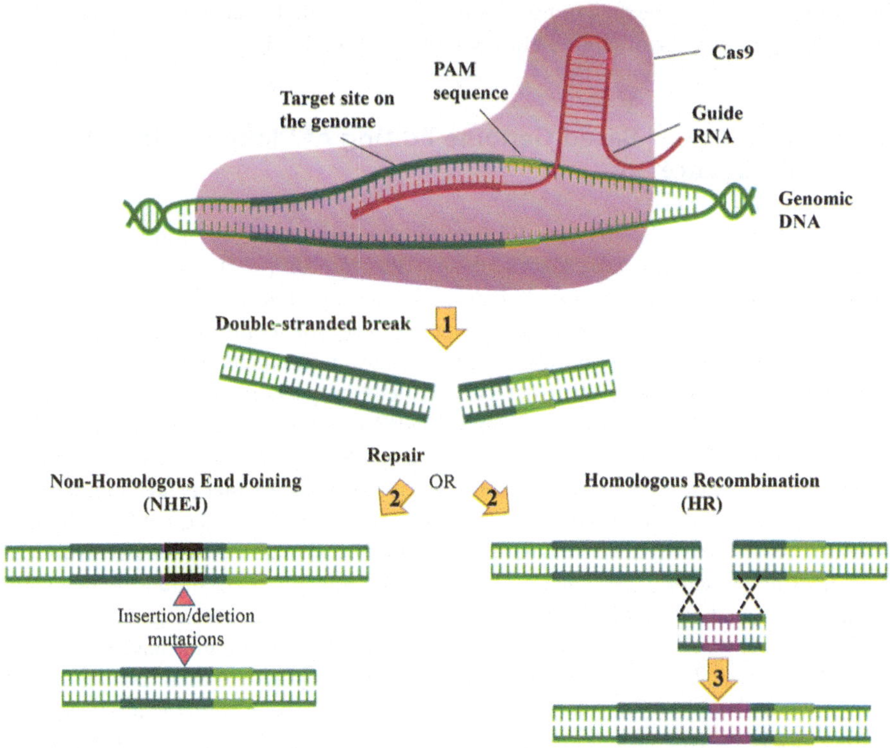

Fig. 11.3 The mechanism of CRISPR/Cas9 genome editing. Guide RNA molecule directs the Cas9 nuclease to the target site in the genome. The presence of PAM near the matching sequence in the genome is critical in finding the target site.After double-strand breaks are made, the broken ends are repaired via the NHEJ or HR processes

directs the CRISPR/Cas9 complex to the complementary sequence in the genome. The presence of conserved protospacer-adjacent motifs (PAM) near target sites in the genome plays a critical role in the in-target function of this complex. Following double-stranded DNA breaks by the Cas9 enzyme, it is subjected to cell repair machinery, which can be error-prone non-homologous end joining (NHEJ) repair or precise homology-directed repair (HDR) (Fig. 11.3). The targeted insertion or modification of desired sequences into the genome makes the HDR approach an outstanding tool for the genetic engineering of plants (Voytas and Gao 2014).

The importance of the breakthrough CRISPR/Cas technology is particularly emphasized as it causes heritable targeted modifications and also contributes to the development of transgene-free plants (Wang et al. 2014a, b; Pan et al. 2016). The first genome editing study in plants was reported by Feng et al. (2013), however, the same year there were other works that reported the successful use of the CRISPR system in genome modification of plants such as Arabidopsis, tobacco, wheat, and rice (Upadhyay et al. 2013; Jiang et al. 2013; Feng et al. 2013). Afterward, numerous studies have been conducted to apply desired modifications to a variety of plants including maize, soybean, potato, cotton, grapes, tomato, cucumber, Cacao tree, sweet orange, Grapefruit, apple, etc. (Khabbazi et al. 2021).

11.4.2 CRISPR/Cas-Based Genome Editing of Plants for Insect Resistance

Insects can damage crops by directly feeding on plant tissues or indirectly transmitting various diseases, thereby significantly reducing crop production and yield. The application of extensive chemicals has caused serious harm to human and animal health as well as the environment. After the successful contribution of transgenic crops for example Bt crops in reducing the usage of chemicals yet the existence of political, ethical, and societal resistance to these crops is a serious issue in many countries. The possibility of employing CRISPR/Cas technology in genome editing of plants towards insect resistance has already been discussed (Douglas 2018). Employing the CRISPR tool provides the opportunity to tackle the concerns in two ways; creating de novo resistance in case there is no convenient R-gene available, and controlling the insect pest population dynamics by breaking insecticide resistance, killing or causing sterility in insects. In such situations, CRISPR technology has the potential to develop designer plants for generating superior traits or to initiate a gene drive to selectively propagate mutations that lead to reduced fecundity or female death in the target insect population (Bisht et al. 2019).

Elucidation of molecular mechanisms of plant defense systems is a prerequisite for developing a new strategy to generate insect-resistant crops. Plants have developed a complex defense mechanism under millions of years of selection pressure from insects (Erb and Reymond 2019). Species of different orders show a strong spatio temporal variation in the expression of metabolites involved in defense against insects (Barton and Boege 2017). The expression level of immunity-associated genes in Arabidopsis plants is correlated with the duration of the vegetative stage (Davila Olivas et al. 2017; Glander et al. 2018), illustrating the relationship between flowering and resistance to insects.

Plant Calcium ion (Ca^{2+}) signals are involved in a wide variety of signaling pathways in the cell. Ca^{2+}enacts an important role in the circadian regulation of photoperiod-controlled flowering in the common morning glory (*Ipomoea purpurea*) (Dodd et al. 2010). Calcium-dependent protein kinase (*CDPK*) is one of the main receptors in the calcium signaling pathway and transduces the signal by phosphorylation (Harmon et al. 2000). In Arabidopsis, the loss of function of the *CPK33* causes late flowering (Kawamoto et al. 2015). $Ca^{2+}$is also involved in early defense signaling in plants (Yan et al. 2018), after insect feeding, there is a striking Ca^{2+}influx limited to a few cell layers lining the injured site (Maffei et al. 2007). In Arabidopsis plants, *CPK3* and *CPK13* activate the herbivore-induced network by increasing the transcription levels of plant defensin gene *PDF1.2* (Kanchiswamy et al. 2010). In another study, the knockdown of *NaCDPK4* and *NaCDPK5* genes in Coyote Tobacco (*Nicotiana attenuata*) up-regulated jasmonic acid accumulation and increased resistance to *Manduca sexta* (Yang et al. 2012).

CRISPR/Cas9 mediated knockout of *CPK* gene (*GmCDPK38*) in soybean resulted in late flowering time in *gmcdpk38* mutants regardless of the photoperiodic conditions. In addition to delayed flowering time, *gmcdpk38* mutants also exhibited enhanced resistance to *Spodoptera litura* (Li et al. 2022). This revealed the dual role of *GmCDPK38* in regulating photoperiod-induced flowering in soybean and resistance to *Spodoptera litura*, suggesting a possible link between flowering and insect resistance.

Resistance to insects is multifaceted with highly complex regulation in both insects and the host plant itself. Phytohormones such as jasmonic acid (JA), salicylic acid (SA), abscisic acid (ABA) and ethylene can affect plant response to insect pests. Deficiency of ABA in plants increases their susceptibility to insect pests (Thaler and Bostock 2004; Dinh et al. 2013) whereas exogenous application of ABA can increase plant resistance to brown planthopper (BPH) by promoting callose formation (Liu et al. 2017). In a recent study, overexpression of the 9-cis-epoxycarotenoid dioxygenase-3 (NCED3) enzyme in rice plants increased ABA biosynthesis and subsequent resistance to BPH (Sun et al. 2022).

Responses of plants to insect pests are also correlated with the feeding manner and the degree of damage at the feeding site. Therefore, the molecular response of plants induced by sap-sucking insects is different from chewing pests. One study demonstrated the role of serotonin regulation as part of the defense mechanism against insect pests in plants (Lu et al. 2018). In rice plants, the cytochrome P450 gene *CYP71A1* encodes tryptamine 5-hydroxylase, which catalyzes the conversion of tryptamine to serotonin (Fujiwara et al. 2010). Sap-sucking insects cause only a little damage to plant tissue therefore salicylic acid (SA) signaling pathway has the main role in insect infestation (Li et al. 2017). Serotonin biosynthesis is induced by insect infestation in rice, and its suppression confers resistance to BPH and striped stem borers (SSB). CRISPR-mediated *CYP71A1* gene knockout inhibits serotonin production resulting in higher salicylic acid levels and thus resistance to BPH and SSB in rice (Lu et al. 2018). However, *cyp71a1* mutant individuals showed increased resistance to rice blast, *Magnaporthe grisea* (Ueno et al. 2008) and susceptibility to rice brown spot disease *Bipolaris oryzae* (Ishihara et al. 2008).

11.5 Conclusion

Until the beginning of the current century, different approaches such as classical plant breeding methods and the application of chemicals in the field contributed to enhancing crop yield and production. The requirement of sufficient agricultural production for the increasing world population and ensuring global food security have led plant scientists to explore more efficient strategies, especially in terms of pest management. Existing criticism of traditional pest control approaches, such as environmental and health concerns and the development of resistance to insecticides, has rendered these methods inadequate on their own. The advent of recombinant DNA technology and the introduction of genetically modified crops expressing the *Bt* toxin gene was a new era in agriculture. Later advances in molecular biology discoveries such as RNAi and CRISPR/Cas technologies soon opened up a new avenue in the production of biotech crops. The next generation of GM crops has the potential to address concerns about transgenic crops and is of great importance for developing sustainable and environmentally friendly methods for crop improvements.

References

Acharjee S, Sarmah BK, Kumar PA, Olsen K, Mahon R, Moar WJ, Higgins TJV (2010) Expression of a sequence-modified cry 2Aa gene for resistance to *Helicoverpaarmigera* in chickpea (*Cicer arietinum* L.). Plant Sci 178(3):333–339

Adang MJ, Brody MS, Cardineau G, Eagan N, Roush RT, Shewmaker CK, McBride KE (1993) The reconstruction and expression of a *bacillus thuringiensis*cry IIIA gene in protoplasts and potato plants. Plant Mol Biol 21(6):1131–1145

Agrawal N, Sachdev B, Rodrigues J, Sree KS, Bhatnagar RK (2013) Development associated profiling of chitinase and micro RNA of *Helicoverpa armigera* identified chitinase repressive micro RNA. Sci Rep 3(1):1–6

Amitai G, Sorek R (2016) CRISPR–Cas adaptation: insights into the mechanism of action. Nat Rev Microbiol 14(2):67–76

Anayol, E., Bakhsh, A., Karakoç, Ö.C., Onarıcı, S., Köm, D., Aasim, M., ` and Özcan, S. (2016). Towards better insect management strategy: restriction of insecticidal gene expression to biting sites in transgenic cotton. Plant Biotechnol Rep 10:83–94

Axtell MJ (2013) Classification and comparison of small RNAs from plants. Annu Rev Plant Biol 64:137–159

Bakhsh A, Khabbazi SD, Baloch FS, Demirel U, Çalişkan ME, Hatipoğlu R, Özkan H (2015) Insect-resistant transgenic crops: retrospect and challenges. Turkish J Agric Forestr 39(4): 531–548

Bakhsh A, Anayol E, Khabbazi SD, Karakoç ÖC, Sancak C, Özcan S (2016) Development of insect-resistant cotton lines with targeted expression of insecticidal gene. Arch Biol Sci 68(4): 773–780

Barrangou R, Fremaux C, Deveau H, Richards M, Boyaval P, Moineau S, Horvath P (2007) CRISPR provides acquired resistance against viruses in prokaryotes. Science 315(5819): 1709–1712

Barton KE, Boege K (2017) Future directions in the ontogeny of plant defence: understanding the evolutionary causes and consequences. Ecol Lett 20(4):403–411

Bashir K, Husnain T, Fatima T, Riaz N, Makhdoom R, Riazuddin S (2005) Novel indica basmati line (B-370) expressing two unrelated genes of *bacillus thuringiensis* is highly resistant to two lepidopteran insects in the field. Crop Prot 24(10):870–879

Baum JA, Roberts JK (2014) Progress towards RNAi-mediated insect pest management. In: Advances in insect physiology, vol 47. Academic Press, pp 249–295

Baum JA, Bogaert T, Clinton W, Heck GR, Feldmann P, Ilagan O, Roberts J (2007) Control of coleopteran insect pests through RNA interference. Nat Biotechnol 25(11):1322–1326

Benowitz KM, Allan CW, Degain BA, Li X, Fabrick JA, Tabashnik BE, Matzkin LM (2022) Novel genetic basis of resistance to Bt toxin cry 1Ac in *Helicoverpa zea*. Genetics 221(1):iyac 037

Bernstein E, Caudy AA, Hammond SM, Hannon GJ (2001) Role for a bidentate ribonuclease in the initiation step of RNA interference. Nature 409(6818):363–366

Bhaya D, Davison M, Barrangou R (2011) CRISPR-Cas systems in bacteria and archaea: versatile small RNAs for adaptive defense and regulation. Annu Rev Genet 45:273–297

Bisht DS, Bhatia V, Bhattacharya R (2019) Improving plant-resistance to insect-pests and pathogens: the new opportunities through targeted genome editing. In: Seminars in Cell & Developmental Biology, vol 96. Academic Press, pp 65–76

Boddupally D, Tamirisa S, Gundra SR, Vudem DR, Khareedu VR (2018) Expression of hybrid fusion protein (cry 1Ac: ASAL) in transgenic rice plants imparts resistance against multiple insect pests. Sci Rep 8(1):8458

Bosher JM, Labouesse M (2000) RNA interference: genetic wand and genetic watchdog. Nat Cell Biol 2(2):E31–E36

Calles-Torrez V, Knodel JJ, Boetel MA, French BW, Fuller BW, Ransom JK (2019) Field-evolved resistance of northern and western corn rootworm (Coleoptera: Chrysomelidae) populations to corn hybrids expressing single and pyramided cry 3Bb1 and cry 34/35Ab1 Bt proteins in North Dakota. J Econ Entomol 112(4):1875–1886

Chae H, Wen Z, Hootman T, Himes J, Duan Q, McMath J, Bramlett M (2022) eCry1Gb. 1Ig, A novel chimeric cry protein with high efficacy against multiple fall armyworm (*Spodoptera frugiperda*) strains resistant to different GM traits. Toxins 14(12):852

Chakraborti D, Sarkar A, Mondal HA, Das S (2009) Tissue specific expression of potent insecticidal, *Allium sativum* leaf agglutinin (ASAL) in important pulse crop, chickpea (*Cicer arietinum* L.) to resist the phloem feeding *Aphis craccivora*. Transgenic Res 18:529–544

Chakraborty M, Reddy PS, Mustafa G, Rajesh G, Narasu VL, Udayasuriyan V, Rana D (2016) Transgenic rice expressing the cry 2AX1 gene confers resistance to multiple lepidopteran pests. Transgenic Res 25:665–678

Chandrasekhar K, Reddy GM, Singh J, Vani K, Vijayalakshmi M, Kaul T, Reddy MK (2014) Development of transgenic rice harbouring mutated *rice 5-enolpyruvylshikimate 3-phosphate synthase* (*Os-mEPSPS*) and *Allium sativum leaf agglutinin* (*ASAL*) genes conferring tolerance to herbicides and sap-sucking insects. Plant Mol Biol Report 32:1146–1157

Cheng J, Bolyard MG, Saxena RC, Sticklen MB (1992) Production of insect resistant potato by genetic transformation with a δ-endotoxin gene from *bacillus thuringiensis* var. *kurstaki*. Plant Sci 81(1):83–91

Cheng X, Sardana R, Kaplan H, Altosaar I (1998) Agrobacterium-transformed rice plants expressing synthetic cryIA (b) and cryIA (c) genes are highly toxic to striped stem borer and yellow stem borer. PNAS 95(6):2767–2772

Crickmore N, Zeigler DR, Feitelson J, Schnepf ESCHERICHIA, Van Rie J, Lereclus D, Dean D (1998) Revision of the nomenclature for the *bacillus thuringiensis*pesticidal crystal proteins. Microbiol Mol Biol Rev 62(3):807–813

Curry D, Browing H, Davis P, Ferguson I, Hutton D, Julius D, Wynne G (2002) Farming and food: a sustainable future. In: Report of the policy commission on the future of farming and food. Defra, London

Czapla TH, Lang BA (1990) Effect of plant lectins on the larval development of European corn borer (Lepidoptera: Pyralidae) and southern corn rootworm (Coleoptera: Chrysomelidae). J Econ Entomol 83(6):2480–2485

Dang W, Wei ZM (2007) Efficient agrobacterium-mediated transformation of soybean. J Mol Cell Biol 40(3):185–195

Davila Olivas NH, Frago E, Thoen MP, Kloth KJ, Becker FF, van Loon JJ, Dicke M (2017) Natural variation in life history strategy of *Arabidopsis thaliana* determines stress responses to drought and insects of different feeding guilds. Mol Ecol 26(11):2959–2977

Deist BR, Rausch MA, Fernandez-Luna MT, Adang MJ, Bonning BC (2014) Bt toxin modification for enhanced efficacy. Toxins 6(10):3005–3027

Di Lelio I, Barra E, Coppola M, Corrado G, Rao R, Caccia S (2022) Transgenic plants expressing immunosuppressive dsRNA improve entomopathogen efficacy against *Spodopteralittoralis* larvae. J Pest Sci:1–16

Dinh ST, Baldwin IT, Galis I (2013) The HERBIVORE ELICITOR-REGULATED1 gene enhances abscisic acid levels and defenses against herbivores in *Nicotiana attenuata* plants. Plant Physiol 162(4):2106–2124

Dodd AN, Kudla J, Sanders D (2010) The language of calcium signaling. Annu Rev Plant Biol 61: 593–620

Douglas AE (2018) Strategies for enhanced crop resistance to insect pests. Annu Rev Plant Biol 69: 637–660

Dufourmantel N, Tissot G, Goutorbe F, Garcon F, Muhr C, Jansens S, Dubald M (2005) Generation and analysis of soybean plastid transformants expressing *bacillus thuringiensis* Cry1Ab protoxin. Plant Mol Biol 58:659–668

Eisemann CH, Donaldson RA, Pearson RD, Cadogan LC, Vuocolo T, Tellam RL (1994) Larvicidal activity of lectins on *Luciliacuprina*: mechanism of action. Entomol Exp Appl 72(1):1–10

Erb M, Reymond P (2019) Molecular interactions between plants and insect herbivores. Annu Rev Plant Biol 70:527–557

Feng Z, Zhang B, Ding W, Liu X, Yang DL, Wei P, Zhu JK (2013) Efficient genome editing in plants using a CRISPR/Cas system. Cell Res 23(10):1229–1232

Ferré J, Escriche B, Bel Y, andvan Rie, J. (1995) Biochemistry and genetics of insect resistance to *bacillus thuringiensis* insecticidal crystal proteins. FEMS Microbiol Lett 132(1–2):1–7

Fujimoto H, Itoh K, Yamamoto M, Kyozuka J, Shimamoto KO (1993) Insect resistant rice generated by introduction of a modified δ-endotoxin gene of *bacillus thuringiensis*. Biotechnology 11(10):1151–1155

Fujiwara T, Maisonneuve S, Isshiki M, Mizutani M, Chen L, Wong HL, Shimamoto K (2010) Sekiguchi lesion gene encodes a cytochrome P450 monooxygenase that catalyzes conversion of tryptamine to serotonin in rice. J Biol Chem 285(15):11308–11313

Gassmann AJ, Petzold-Maxwell JL, Keweshan RS, Dunbar MW (2011) Field-evolved resistance to Bt maize by western corn rootworm. PLoS One 6(7):e22629

Gatehouse AM, Hilder VA, Powell K, Boulter D, Gatehouse JA (1992) Potential of plant-derived genes in the genetic manipulation of crops for insect resistance. In proceedings of the 8th international symposium on insect-plant relationships. Springer, Netherlands, pp 221–234

Gatehouse AM, Davison GM, Stewart JN, Gatehouse LN, Kumar A, Geoghegan IE, Gatehouse JA (1999) Concanavalin A inhibits development of tomato moth (*Lacanobia oleracea*) and peach-potato aphid (*Myzus persicae*) when expressed in transgenic potato plants. Mol Breed 5:153–165

Glander S, He F, Schmitz G, Witten A, Telschow A, de Meaux J (2018) Assortment of flowering time and immunity alleles in natural *Arabidopsis thaliana* populations suggests immunity and vegetative lifespan strategies coevolve. Gen Biol Evol 10(9):2278–2291

Godfrey J (2000) Do genetically modified foods affect human health? Lancet 355(9201):414

Gostimskaya I (2022) CRISPR–Cas9: A history of its discovery and ethical considerations of its use in genome editing. Biochem Mosc 87(8):777–788

Halfhill MD, Richards HA, Mabon SA, Stewart CN (2001) Expression of GFP and Bt transgenes in *Brassica napus* and hybridization with *Brassica rapa*. Theor Appl Genet 103:659–667

Harmon AC, Gribskov M, Harper JF (2000) CDPKs–a kinase for every Ca^{2+} signal? Trends Plant Sci 5(4):154–159

Head GP, Carroll MW, Evans SP, Rule DM, Willse AR, Clark TL, Meinke LJ (2017) Evaluation of SmartStax and SmartStax PRO maize against western corn rootworm and northern corn rootworm: efficacy and resistance management. Pest Manag Sci 73(9):1883–1899

Ibrahim MA, Griko N, Junker M, Bulla LA (2010) *Bacillus thuringiensis*: a genomics and proteomics perspective. Bioeng Bugs 1(1):31–50

Indurker S, Misra HS, Eapen S (2010) Agrobacterium-mediated transformation in chickpea (*Cicer arietinum* L.) with an insecticidal protein gene: optimisation of different factors. Physiol Mol Biol Plants 16:273–284

ISAAA (2019) Global status of commercialized biotech/GM crops in 2019: biotech crops drive socio-economic development and sustainable environment in the new frontier. In: ISAAA brief no, 55. Ithaca, NY, ISAAA

Ishihara A, Hashimoto Y, Tanaka C, Dubouzet JG, Nakao T, Matsuda F, Wakasa K (2008) The tryptophan pathway is involved in the defense responses of rice against pathogenic infection via serotonin production. The Plant J 54(3):481–495

Ishino Y, Shinagawa H, Makino K, Amemura M, Nakata A (1987) Nucleotide sequence of the iap gene, responsible for alkaline phosphatase isozyme conversion in *Escherichia coli*, and identification of the gene product. J Bacteriol 169(12):5429–5433

Jacobs JM, Takla MF, Docherty LC, Frater CM, Markwick NP, Meiyalaghan S, Conner AJ (2009) Potato transformation with modified nucleotide sequences of the cry 9Aa2 gene improves resistance to potato tuber moth. Potato Res 52:367–378

Je YH, Chung YS, Ngo XB (2022) Convergence of Bar and Cry1Ac mutant genes in soybean confers synergistic resistance to herbicide and lepidopteran insects. Key Advances and Future Perspectives, Inducing Plant Resistance Against Insects Using Exogenous Bioactive Chemicals

Jiang W, Zhou H, Bi H, Fromm M, Yang B, Weeks DP (2013) Demonstration of CRISPR/Cas9/sgRNA-mediated targeted gene modification in Arabidopsis, tobacco, sorghum and rice. Nucleic Acids Res 41(20):e188–e188

Kanchiswamy CN, Takahashi H, Quadro S, Maffei ME, Bossi S, Bertea C, Arimura GI (2010) Regulation of Arabidopsis defense responses against *Spodopteralittoralis* by CPK-mediated calcium signaling. BMC Plant Biol 10:1–10

Kawamoto N, Endo M, andAraki, T. (2015) Expression of a kinase-dead form of CPK33 involved in florigen complex formation causes delayed flowering. Plant Signal Behav 10(12):e1086856

Khabbazi SD, Bakhsh A, Sancak C, Özcan S (2016) Molecular characterization of snowdrop lectin (GNA) and its comparison with reported lectin sequences of Amaryllidaceae. Czech J Genet Plant Breed 52(3):94–100

Khabbazi SD, Khabbazi AD, Özcan SF, Bakhsh A, Başalma D, Özcan S (2018) Expression of GNA and biting site-restricted cry1Ac in cotton; an efficient attribution to insect pest management strategies. Plant Biotechnol Rep 12:273–282

Khabbazi SD, Khabbazi AD, Cevik V, Ergül A (2020) Genetic engineering of horticultural crops contributes to the improvement of crop nutritional quality and shelf life. In: Transgenic Technology Based Value Addition in Plant Biotechnology, pp 247–272

Khabbazi SD, Khabbazi AD, Cevik V, Ergül A (2021) CRISPR/Cas9 system, an efficient approach to genome editing of plants for crop improvement. In: RNA-Based Technologies for Functional Genomics in Plants, pp 369–391

Khan GA, Bakhsh A, Riazuddin S, Husnain T (2011) Introduction of cry1Ab gene into cotton (*Gossypiumhirsutum*) enhances resistance against lepidopteran pest (*Helicoverpaarmigera*). Span J Agric Res 9(1):296–302

Kim DH, Rossi JJ (2007) Strategies for silencing human disease using RNA interference. Nature Rev Genet 8(3):173–184

Klausner A (1984) Microbial insect control: using bugs to kill bugs. Biotechnology 2(5):408–419

Koul B, Srivastava S, Sanyal I, Tripathi B, Sharma V, Amla DV (2014) Transgenic tomato line expressing modified *bacillus thuringiensis* cry1Ab gene showing complete resistance to two lepidopteran pests. Springer Plus 3(1):1–13

Koziel MG, Beland GL, Bowman C, Carozzi NB, Crenshaw R, Crossland L, Evola SV (1993) Field performance of elite transgenic maize plants expressing an insecticidal protein derived from *bacillus thuringiensis*. Biotechnology 11(2):194–200

Kumar H, Kumar V (2004) Tomato expressing Cry1A (b) insecticidal protein from *bacillus thuringiensis* protected against tomato fruit borer, *Helicoverpaarmigera* (Hübner)(Lepidoptera: Noctuidae) damage in the laboratory, greenhouse and field. Crop Prot 23(2):135–139

Kumar P, Pandit SS, Steppuhn A, Baldwin IT (2014) Natural history-driven, plant-mediated RNAi-based study reveals *CYP6B46*'s role in a nicotine-mediated antipredator herbivore defense. PNAS 111(4):1245–1252

Kumar N, Stanford W, De Solis C, Abraham ND, Dao TMJ, Thaseen S, Ploski JE (2018) The development of an AAV-based CRISPR SaCas9 genome editing system that can be delivered to neurons *in vivo* and regulated via doxycycline and Cre-recombinase. Front Mol Neurosci:413

Li C, Luo C, Zhou Z, Wang R, Ling F, Xiao L, andChen, H. (2017) Gene expression and plant hormone levels in two contrasting rice genotypes responding to brown planthopper infestation. BMC Plant Biol 17(1):1–14

Li X, Hu D, Cai L, Wang H, Liu X, Du H, andWang, H. (2022) Calcium-dependent protein KINASE38 regulates flowering time and common cutworm resistance in soybean. Plant Physiol 190(1):480–499

Liu J, Du H, Ding X, Zhou Y, Xie P, Wu J (2017) Mechanisms of callose deposition in rice regulated by exogenous abscisic acid and its involvement in rice resistance to *Nilaparvata lugens* Stål (Hemiptera: Delphacidae). Pest Manag Sci 73(12):2559–2568

Liu Y, Han S, Yang S, Chen Z, Yin Y, Xi J, Hao D (2022) Engineered chimeric insecticidal crystalline protein improves resistance to lepidopteran insects in rice (*Oryza sativa* L.) and maize (*Zea mays* L.). Sci Rep 12(1):12529

Lu HP, Luo T, Fu HW, Wang L, Tan YY, Huang JZ, Shu QY (2018) Resistance of rice to insect pests mediated by suppression of serotonin biosynthesis. Nat Plants 4(6):338–344

Maffei ME, Mithöfer A, Boland W (2007) Insects feeding on plants: rapid signals and responses preceding the induction of phytochemical release. Phytochemistry 68(22–24):2946–2959

Majeed A (2005) Expression of proteinase inhibitor gene in cotton. University of the Punjab, Lahore, Pakistan

Mamta RKRK, Rajam MV (2016) Targeting chitinase gene of *Helicoverpaarmigera* by host-induced RNA interference confers insect resistance in tobacco and tomato. Plant Mol Biol 90: 281–292

Mandaokar AD, Goyal RK, Shukla A, Bisaria S, Bhalla R, Reddy VS, Kumar PA (2000) Transgenic tomato plants resistant to fruit borer (*Helicoverpaarmigera*Hubner). Crop Prot 19(5):307–312

Mao YB, Cai WJ, Wang JW, Hong GJ, Tao XY, Wang LJ, Chen XY (2007) Silencing a cotton bollworm P450 monooxygenase gene by plant-mediated RNAi impairs larval tolerance of gossypol. Nat Biotechnol 25(11):1307–1313

Mao YB, Tao XY, Xue XY, Wang LJ, Chen XY (2011) Cotton plants expressing CYP6AE14 double-stranded RNA show enhanced resistance to bollworms. Transgenic Res 20:665–673

McPherson SA, Perlak FJ, Fuchs RL, Marrone PG, Lavrik PB, Fischhoff DA (1988) Characterization of the coleopteran–specific protein gene of *bacillus thuringiensis* var. *tenebrionis*. Biotechnology 6(1):61–66

Mehrotra M, Singh AK, Sanyal I, Altosaar I, Amla DV (2011) Pyramiding of modified cry1Ab and cry1Ac genes of *bacillus thuringiensis* in transgenic chickpea (*Cicer arietinum* L.) for improved resistance to pod borer insect *Helicoverpaarmigera*. Euphytica 182:87–102

Meiyalaghan S, Jacobs JME, Butler RC, Wratten SD, Conner AJ (2006) Transgenic potato lines expressing cry 1Ba1 or cry 1Ca5 genes are resistant to potato tuber moth. Potato Res 49:203–216

Mi X, Ji X, Yang J, Liang L, Si H, Wu J, Wang D (2015) Transgenic potato plants expressing cry3A gene confer resistance to Colorado potato beetle. Comp Ren Biol 338(7):443–450

Mittal P, Yadav R, Devi R, Tiwari A, Upadhye SP, Gosal SS (2011) Wondrous RNAi-gene silencing. Biotechnology 10(1):41–50

Morán R, Garcıa R, López A, Zaldúa Z, Mena J, Garcıa M, Pimentel E (1998) Transgenic sweet potato plants carrying the delta-endotoxin gene from *bacillus thuringiensis* var. *tenebrionis*. Plant Sci 139(2):175–184

Newell CA, Lowe JM, Merryweather A, Rooke LM, Hamilton WDO (1995) Transformation of sweet potato (*Ipomoea batatas* (L.) lam.) with *agrobacterium tumefaciens* and regeneration of plants expressing cowpea trypsin inhibitor and snowdrop lectin. Plant Sci 107(2):215–227

Özcan S, Firek S, Draper J (1993) Selectable marker genesengineered for specificexpression in target cells for planttransformation. Nat Biotechnol 11:218–221

Pan C, Ye L, Qin L, Liu X, He Y, Wang J, Lu G (2016) CRISPR/Cas9-mediated efficient and heritable targeted mutagenesis in tomato plants in the first and later generations. Sci Rep 6(1): 24765

Parrott WA, All JN, Adang MJ, Bailey MA, Boerma HR, Stewart CN (1994) Recovery and evaluation of soybean plants transgenic for a *bacillus thuringiensis* var. *kurstaki* insecticidal gene. In Vitro Cell Dev Biol Plant 30:144–149

Peferoen M, Jansens S, Reynaerts A, Leemans J (1990) Potato plants with engineered resistance against insect attack. Mol Cel Biol Potato:193–204

Perlak FJ, Deaton RW, Armstrong TA, Fuchs RL, Sims SR, Greenplate JT, Fischhoff DA (1990) Insect resistant cotton plants. Biotechnology 8(10):939–943

Perlak FJ, Fuchs RL, Dean DA, McPherson SL, Fischhoff DA (1991) Modification of the coding sequence enhances plant expression of insect control protein genes. PNAS 88(8):3324–3328

Perlak FJ, Stone TB, Muskopf YM, Petersen LJ, Parker GB, McPherson SA, Fischhoff DA (1993) Genetically improved potatoes: protection from damage by Colorado potato beetles. Plant Mol Biol 22:313–321

Peumans WJ, van Damme EJM (1996) Prevalence, biological activity and genetic manipulation of lectins in foods. Trends Food Sci Technol 7(4):132–138

Pushpa R, Raveenderan TS, Rajeswari S, Amalabalu P, Punitha D (2013) Genetic transformation of cry1EC gene into cotton (*Gossypiumhirsutum* L.) for resistance against *Spodopteralitura*. Afr J Biotechnol 12(15)

Qaim M (2009) The economics of genetically modified crops. Ann Rev Resour Econ 1(1):665–694

Qin D, Liu XY, Miceli C, Zhang Q, Wang PW (2019) Soybean plants expressing the *bacillus thuringiensis* cry8-like gene show resistance to *Holotrichia parallela*. BMC Biotechnol 19:1–12

Mahmood-ur-Rahman KH, Khan MA, Bakhsh A, Rao AQ (2021) 01. An insight of cotton leaf curl virus: a devastating plant pathogenic begomovirus. Pure Appl Biol 1(3):52–58

Rahnama H, Sheykhhasan M (2016) Transformation and light inducible expression of cry1Ab gene in oilseed rape (*Brassica napus* L.). J Sci Islamic Rep Iran 27(4):313–319

Ramachandran S, Buntin GD, All JN, Tabashnik BE, Raymer PL, Adang MJ, Stewart CN Jr (1998) Survival, development, and oviposition of resistant diamondback moth (Lepidoptera: Plutellidae) on transgenic canola producing a *bacillus thuringiensis* toxin. J Econ Entomol 91(6):1239–1244

Ramesh S, Nagadhara D, Reddy VD, andRao, K.V. (2004) Production of transgenic indica rice resistant to yellow stem borer and sap-sucking insects, using super-binary vectors of *agrobacterium tumefaciens*. Plant Sci 166(4):1077–1085

Razzaq A, Ali A, Zahid S, Malik A, Pengtao L, Gong W, Zafar MM (2023) Engineering of cry genes "Cry11 and Cry1h" in cotton (*Gossypiumhirsutum* L.) for protection against insect pest attack. Arch Phytopathol plant Prot:1–13

Salehian H, Rahnama H, Dezhsetan S, Babaei S (2021) Constitutive expression of a synthetic cry1Ab gene confers resistance to potato tuber moth (*Phthorimaeaoperculella* Zeller) larva. Crop breed. Appl Biotechnol 21

Santos EN, Menezes LP, Dolabella SS, Santini A, Severino P, Capasso R, Jain S (2022) *Bacillus thuringiensis*: from biopesticides to anticancer agents. Biochimie 192:83–90

Sanyal I, Singh AK, Kaushik M, Amla DV (2005) Agrobacterium-mediated transformation of chickpea (*Cicer arietinum* L.) with *bacillus thuringiensis* cry1Ac gene for resistance against pod borer insect *Helicoverpaarmigera*. Plant Sci 168(4):1135–1146

Sharma HC (2001) Cotton bollworm/legume pod borer, *Helicoverpaarmigera* (Hubner) (Noctuidae: Lepidoptera): biology and management. Crop Prot Comp CABI, Oxon

Shelake RM, Pramanik D, Kim JY (2019) Evolution of plant mutagenesis tools: a shifting paradigm from random to targeted genome editing. Plant Biotechnol Rep 13:423–445

Siddiqui HA, Asif M, Asad S, Naqvi RZ, Ajaz S, Umer N, Mansoor S (2019) Development and evaluation of double gene transgenic cotton lines expressing cry toxins for protection against chewing insect pests. Sci Rep 9(1):11774

Siddiqui HA, Asif M, Naqvi RZ, Shehzad A, Sarwar M, Amin I, Mansoor S (2023) Development of modified Cry1Ac for the control of resistant insect pest of cotton, *Pectinophora gossypiella*. Gene 856:147113

Singh P, GK S, Thakur S, Rathore M, Verma OP, Singh NP, Das A (2022) Evaluation of transgenic chickpea harboring codon-modified Vip3Aa against gram pod borer (*Helicoverpaarmigera* H.). PLoS One 17(6):e0270011

Small I (2007) RNAi for revealing and engineering plant gene functions. Curr Opin Biotechnol 18(2):148–153

Smith JL, Farhan Y, Schaafsma AW (2019) Practical resistance of *Ostrinianubilalis* (Lepidoptera: Crambidae) to Cry1F *bacillus thuringiensis* maize discovered in Nova Scotia. Canada Sci Rep 9(1):18247

Stewart CN Jr, Adang MJ, All JN, Raymer PL, Ramachandran S, Parrott WA (1996) Insect control and dosage effects in transgenic canola containing a synthetic *bacillus thuringiensis*cryIAc gene. Plant Physiol 112(1):115–120

Stoger E, Williams S, Christou P, Down RE, Gatehouse JA (1999) Expression of the insecticidal lectin from snowdrop (*Galanthusnivalis* agglutinin; GNA) in transgenic wheat plants: effects on predation by the grain aphid *Sitobionavenae*. Mol Breed 5:65–73

Sun L, Li J, Liu Y, Noman A, Chen L, Liu J (2022) Transcriptome profiling in rice reveals a positive role for OsNCED3 in defense against the brown planthopper, *Nilaparvatalugens*. BMC Genomics 23(1):634

Tabashnik BE, Finson N, Johnson MW, Moar WJ (1993) Resistance to toxins from *bacillus thuringiensis* subsp. *kurstaki* causes minimal cross-resistance to *B. Thuringiensis* subsp. *aizawai* in the diamondback moth (Lepidoptera: Plutellidae). Appl Environ Microbiol 59(5):1332–1335

Tabashnik BE (1994) Evolution of resistance to *bacillus thuringiensis*. Annu Rev Entomol 39(1): 47–79

Tabashnik BE, Brévault T, Carrière Y (2013) Insect resistance to Bt crops: lessons from the first billion acres. Nat Biotechnol 31(6):510–521

Tabashnik BE, Carrière Y (2017) Surge in insect resistance to transgenic crops and prospects for sustainability. Nat Biotechnol 35(10):926–935

Tabashnik BE, Carrière Y (2019) Global patterns of resistance to Bt crops highlighting pink bollworm in the United States, China, and India. J Econ Entomol 112(6):2513–2523

Tang G, Galili G (2004) Using RNAi to improve plant nutritional value: from mechanism to application. Trends Biotechnol 22(9):463–469

Tang W, Chen H, Xu C, Li X, Lin Y, Zhang Q (2006) Development of insect-resistant transgenic indica rice with a synthetic cry1C* gene. Mol Breed 18:1–10

Tariq M, Tabassum B, Bakhsh A, Farooq AM, Qamar Z, Akram F, Nasir IA (2022) Heterologous expression of cry1Ia12 insecticidal gene in cotton encodes resistance against pink bollworm, *Pectinophoragossypiella* (Lepidoptera: Gelechiidae); an alternate insecticidal gene for insect pest management. Mol Biol Rep 49(11):10557–10564

Terenius O, Papanicolaou A, Garbutt JS, Eleftherianos I, Huvenne H, Kanginakudru S, Smagghe G (2011) RNA interference in Lepidoptera: an overview of successful and unsuccessful studies and implications for experimental design. J Insect Physiol 57(2):231–245

Thaler JS, Bostock RM (2004) Interactions between abscisic-acid-mediated responses and plant resistance to pathogens and insects. Ecology 85(1):48–58

Toenniessen GH, O'Toole JC, DeVries J (2003) Advances in plant biotechnology and its adoption in developing countries. Current Opin. Plant Biol 6(2):191–198

Tohidfar M, Zare N, Jouzani GS, Eftekhari SM (2013) Agrobacterium-mediated transformation of alfalfa (*Medicago sativa*) using a synthetic cry3a gene to enhance resistance against alfalfa weevil. Plant Cell Tissue Organ Cult 113:227–235

Tohidfar M, Gharahyazi B, Mousavi M, Yazdani S, Golabchian R (2008) Agrobacterium-mediated transformation of cotton (*Gossypiumhirsutum*) using a synthetic cry1Ab gene for enhanced resistance against *Heliothisarmigera*. Iran J Biotechnol 6(3):164–173

Ueno M, Imaoka A, Kihara J, Arase S (2008) Tryptamine pathway-mediated DNA fragmentation is involved in sekiguchi lesion formation for light-enhanced resistance in lesion mimic mutant of rice to *Magnaporthe grisea* infection. J Phytopathol 156(11–12):715–724

Umbeck P, Johnson G, Barton K, Swain W (1987) Genetically transformed cotton (*Gossypium hirsutum* L.) plants. Biotechnology 5(3):263–266

Upadhyay SK, Kumar J, Alok A, Tuli R (2013) RNA-guided genome editing for target gene mutations in wheat. G3: Genes Genom Genet 3(12):2233–2238

Vaeck M, Reynaerts A, Höfte H, Jansens S, De Beuckeleer M, Dean C, Leemans J (1987) Transgenic plants protected from insect attack. Nature 328(6125):33–37
Vajhala CS, Sadumpati VK, Nunna HR, Puligundla SK, Vudem DR, Khareedu VR (2013) Development of transgenic cotton lines expressing Allium sativum agglutinin (ASAL) for enhanced resistance against major sap-sucking pests. PLoS One 8(9):e72542
Valin H, Sands RD, Van der Mensbrugghe D, Nelson GC, Ahammad H, Blanc E, andWillenbockel, D. (2014) The future of food demand: understanding differences in global economic models. Agric Econ 45(1):51–67
Vaughn T, Cavato T, Brar G, Coombe T, DeGooyer T, Ford S, Pershing J (2005) A method of controlling corn rootworm feeding using a *bacillus thuringiensis* protein expressed in transgenic maize. Crop Sci 45(3):931–938
Voytas DF, Gao C (2014) Precision genome engineering and agriculture: opportunities andregulatory challenges. PLoS Biol 12:e1001877
Wang Y, Cheng X, Shan Q, Zhang Y, Liu J, Gao C, Qiu JL (2014a) Simultaneous editing of three homoeoalleles in hexaploid bread wheat confers heritable resistance to powdery mildew. Nat Biotechnol 32(9):947–951
Wang Y, Zhang Y, Wang F, Liu C, Liu K (2014b) Development of transgenic *Brassica napus* with an optimized cry1C* gene for resistance to diamondback moth (*Plutellaxylostella*). Can J Plant Sci 94(8):1501–1506
Wang Z, Zhang K, Sun X, Tang K, Zhang J (2005) Enhancement of resistance to aphids by introducing the snowdrop lectin gene gna into maize plants. J Biosci 30:627–638
Wu J, Luo X, Guo H, Xiao J, Tian Y (2006) Transgenic cotton, expressing *Amaranthuscaudatus* agglutinin, confers enhanced resistance to aphids. Plant Breed 125(4):390–394
Wünn J, Klöti A, Burkhardt PK, Biswas GCG, Launis K, Iglesias VA, Potrykus I (1996) Transgenic indica rice breeding line IR58 expressing a synthetic crylA (b) gene from *bacillus thuringiensis* provides effective insect pest control. Biotechnology 14(2):171–176
Yamamoto T, McLaughlin RE (1981) Isolation of a protein from the parasporal crystal of *bacillus thuringiensis* var.*kurstaki* toxic to the mosquito larva, *Aedes taeniorhynchus*. Biochemi Biophys Res Comm 103(2):414–421
Yan C, Fan M, Yang M, Zhao J, Zhang W, Su Y, Xie D (2018) Injury activates Ca2+/calmodulin-dependent phosphorylation of JAV1-JAZ8-WRKY51 complex for jasmonate biosynthesis. Mol Cell 70(1):136–149
Yang DH, Hettenhausen C, Baldwin IT, andWu, J. (2012) Silencing *Nicotianaattenuata* calcium-dependent protein kinases, CDPK4 and CDPK5, strongly up-regulates wound-and herbivory-induced jasmonic acid accumulations. Plant Physiol 159(4):1591–1607
Yoshimura S, Komatsu M, Kaku K, Hori M, Ogawa T, Muramoto K, Toriyama K (2012) Production of transgenic rice plants expressing *Dioscorea batatas* tuber lectin 1 to confer resistance against brown planthopper. Plant Biotechnol 29(5):501–504
Zafar MM, Mustafa G, Shoukat F, Idrees A, Ali A, Sharif F, Li F (2022) Heterologous expression of cry3Bb1 and cry3 genes for enhanced resistance against insect pests in cotton. Sci Rep 12(1): 10878
Zamore PD, Tuschl T, Sharp PA, Bartel DP (2000) RNAi: double-stranded RNA directs the ATP-dependent cleavage of mRNA at 21 to 23 nucleotide intervals. Cell 101(1):25–33
Zhang J, Khan SA, Heckel DG, Bock R (2017) Next-generation insect-resistant plants: RNAi-mediated crop protection. Trends Biotechnol 35(9):871–882
Zhang J, Li H, Zhong X, Tian J, Segers A, Xia L, Francis F (2022) RNA-interference-mediated aphid control in crop plants: A review. Agriculture 12(12):2108
Zhang QJ, Cong LI, Liu SK, Dong LAI, Qi QM, Lu CG (2013) Breeding and identification of insect-resistant rice by transferring two insecticidal genes, sbk and sck. Rice Sci 20(1):19–24
Zotti M, Dos Santos EA, Cagliari D, Christiaens O, Taning CNT, Smagghe G (2018) RNA interference technology in crop protection against arthropod pests, pathogens and nematodes. Pest Manag Sci 74(6):1239–1250

High-Throughput Phenotyping and Its Importance in Host Plant Resistance

12

Pritha Kundu, Heena Puri, Devasantosh Mohanty, Sanket Shinde, and Sajjan Grover

Abstract

Plant phenotyping has been the topic of research for several decades across the globe, which has made significant contributions to the advancements in plant cultivation and breeding. However, the implication of high-throughput phenotyping (HTP) in evaluating the study made in the field of plant-insect interaction is in its infancy. Here, we take the opportunity to revisit and highlight the different scopes of employing HTP for large-scale analysis of the insect damage signatures across different plant parts and the respective insect behavior. Screening for insect-resistant host plants and breeding for insect-resistant varieties require phenotyping data to identify and ensure the level of resistance in plants. HTP allows a nondestructive way of quantifying plant damage, insect feeding behavior, choice, and no-choice assays and generates a tremendous amount of data through image capturing, image processing, and video recordings. Different automated image capturing platforms are available to determine the plant traits and insect behavior, and by utilizing the HTP, a large number of plant accessions can be screened in a shorter time while maintaining the data accuracy and sensitivity. This chapter, thus, highlights the various mechanisms to employ HTP in large-scale plant-herbivore interaction studies to make the future better for the scientists of tomorrow.

P. Kundu · H. Puri · S. Shinde · S. Grover (✉)
Department of Entomology, University of Nebraska-Lincoln, Lincoln, NE, USA

D. Mohanty
Department of Biological Sciences, University of North Texas, Denton, TX, USA

S. Kumar, M. Furlong (eds.), *Plant Resistance to Insects in Major Field Crops*,
https://doi.org/10.1007/978-981-99-7520-4_12

Keywords

High-throughput phenotyping · Plant imaging · Plant-insect interactions · Insect behavior · Feeding guilds · Machine learning · Automation

12.1 Introduction

Plant pests can cause 20–40% crop yield losses globally with an annual loss of USD $290 billion (Kundu and Vadassery 2021). Insects can account for 23% of crop losses among all pests (https://www.agrivi.com/blog/yield-losses-due-to-pests/). It is believed that plants and insects have been coevolving for 400 million years (Labandeira 2013). In this coevolutionary process, plants have developed sophisticated defense mechanisms to protect against insect attacks and vice versa (War et al. 2012). Understanding the plant defense mechanisms and development of insect pest-resistant/tolerant varieties is a very promising, effective, and sustainable way of pest management (Panda and Khush 1995; Sharma and Ortiz 2002).

Host plant resistance to insects encompasses plant traits via three categories—antixenosis, antibiosis, and tolerance (Panda and Khush 1995; Stout 2014). First, antixenosis/non-preference refers to the morphological characteristics of plants that deter/repel insects to feed, reproduce, and/or grow. Second, antibiosis refers to the chemical characteristics of plants that can adversely affect insect growth and reproduction. Third, tolerance is the most durable plant resistance category that does not affect insect fitness but recovers/compensates for plant growth losses caused by insect feeding via enhanced plant physiological processes; hence, it does not allow faster development of insect biotypes (Peterson et al. 2017). Conventional plant breeding methods for insect resistance require extensive plant phenotyping in heterogeneous plant populations to understand the genetic basis of insect resistance. Furthermore, modern genomic approaches also require plant phenotyping to evaluate insect feeding-based plant damage and other agronomic growth drifts because of foreign gene incorporation. Looking at the farmers' interest/ideal scenario, insect-resistant plants should not compensate for plant yield. For instance, genetic engineering for insect resistance in plants could lead to off-types or yield drift, or any other undesirable characteristics. Therefore, plant phenotyping plays a crucial role in screening insect-resistant and -tolerant crop varieties (Goggin et al. 2015).

Several insect species, among lepidopterans, dipterans, homopterans, heteropterans, orthopterans, and coleopterans, chew leaves or burrow through fruits, stems, or roots during their larvae and/or adult stages of the life cycle (Satpathy et al. 2020; Fuentes-Jacques et al. 2021; Bhuvaneswari et al. 2022). Insects belonging to different feeding guilds can cause different types of damage such as defoliation, chlorosis, leaf mining, stunted plant growth, and development of galls. Insects with chewing type of mouthparts could result in visible feeding scars, whereas piercing-sucking-type insects keep the plant cell alive for their nourishment and result in chlorophyll loss (van Emden and Harrington 2017). Attack of insects on plants is a multistep process such as host finding, host selection, host acceptance, sustained feeding, and oviposition (Powell et al. 2006; Follett 2017; Finch and Collier 2000). Plants with different defense traits could impact different steps of the abovementioned process. The insect damage phenotyping usually directly relies on visual parameters such as percent insect-fed areas, chlorosis, or damage rating. Especially, the damage rating could be very subjective and not a very accurate measure. Inconsistencies and inaccuracies for insect damage ratings could hinder the process of identifying the genetic loci responsible for insect resistance. Furthermore, conventional phenotyping does not capture the physiological and biochemical phenotypes of plants. Detailed plant phenotyping beyond measuring direct yield can help to distinguish different host plant resistance mechanisms and plays a crucial role in genome-wide association, nested association mapping, and other genetic studies (Kloth et al. 2012; Araus et al. 2018).

The interaction between environment and genotype leads to phenotype. Evaluating different genotypes in different environmental conditions ends up in a very high number of combinations to be tested. Crop phenotyping involves the use of specific protocols and methodologies to measure plant morphological/structural traits, physiological traits, and component content traits of different parts of the plants (Dhondt et al. 2013). These methodologies are very time-consuming, resource-heavy, and labor-intensive. Despite considerable progress in plant genotyping and molecular breeding core facilities, plant phenotyping still remains a bottleneck in breeding efforts (Dhondt et al. 2013; Goggin et al. 2015; Walter et al. 2015; Ubbens and Stavness 2017).

To accelerate the development of insect-proof plants, it becomes imperative to overcome the plant phenotyping bottleneck by identifying the plant traits which are good measures of insect resistance and tolerance. For instance, SPAD meters have been used to measure chlorophyll loss in response to aphid attacks in a nondestructive way (Lage et al. 2003). Using the metabolomics approach, the abundance of secondary metabolites can be estimated to know the plant resistance status (Zogli et al. 2020; Grover et al. 2022a). Furthermore, methods like ELISA can be used to test for the presence of viruses in plants (Mondal et al. 2016). The phenotypic traits such as root and shoot biomass, fruit and/or seed yield, plant height, number of leaves, leaf area, and photosynthetic efficiency are heavily impacted by insect attack (Kessler and Baldwin 2002). The precise and accurate assessment of these phenotypic parameters is crucial to unravel and quantify the potential of different plant genotypes in the insect control space. High-throughput phenotyping strategies are gaining enormous attention these days in plant science studies due to (1) precise, accurate, and objective measurement of phenotypic variables in automated and nondestructive way; (2) time series assessment of plant performance under insect attack at high resolution; and (3) reduction in time, labor, and several other human resources through automation and digitization of data analyses (Dhondt et al. 2013; Goggin et al. 2015; Ubbens and Stavness 2017).

12.2 High-Throughput Phenotyping (HTP) for Plants

Phenome refers to a set of physical and biochemical traits that develop in an organism in response to varied environmental stimuli, and a measurement of these phenotypes is referred to as phenomics. The study of plant phenomics, thus, plays a crucial role in the detection of the first emergence of insects in millions of hectares of agro-economically important crop fields, thereby rendering us with a wide range of downstream datasets that give insight into future directions. HTP, thus, results in the production of huge datasets that leads to successful curation by the application of deep learning, which is one of the budding fields of research in plant phenomics (Arya et al. 2022). Despite modern digital techniques such as imaging, there are few non-imaging-based HTP methods available. For example, Xing et al. (2017) developed a dynamic and efficient HTP method for the evaluation of antixenosis against common cutworms in soybeans. Though this method shortened evaluation time to 4X and increased efficiency to 12X, these methods highly depend on the type of insects and crops. Therefore, digital imaging techniques could fill those gaps for plant HTP to insect attack. An improved high-throughput phenotyping method was recently developed to screen rice germplasm resistance against rice gall midge (*Orseolia oryzae*), which allowed to screen 60 lines/varieties of rice (Cheng et al. 2021).

12.2.1 Different Feeding Patterns of Insects

Pests employ diverse feeding strategies to obtain nutrition from their host plants. Insects of major feeding guilds include leaf-chewing insects, plant sap feeders, and root feeders (Carvalho et al. 2014). Chewing insects are voracious leaf eaters that completely vanish the whole aboveground part of the plants leaving the midveins and the vascular regions untouched (Meena et al. 2019). Piercing-sucking insects, however, directly inject their salivary components into the vascular tissues, thereby suppressing plant defense and slowly sucking the nutrients from the plant sap. Moreover, insect species from the hemipteran order indirectly affect crops and vegetables by transmitting bacterial, viral, or fungal diseases from plant to plant (Yi 2020). A wide range of root feeders feed on plant roots dwelling in the soil, thereby causing hydraulic changes to the entire plant body. As feeding patterns differ, the phenotypic characteristics of the damage caused by these wide range of insects also differ considerably (Fig. 12.1). The aboveground visible effects include leaf mining, defoliation, oviposition scars, cell content feeding, and stem boring. Herbivores also cause chlorophyll loss, discoloration of leaves with premature senescence, reduced water content, and developmental growth retardation (Campbell et al. 2018).

The acceleration in the utilization of plant phenotyping is highly based on the availability of different imaging systems, and the foundation for these imaging systems relies on the insect damage types, which could affect different plant characteristics like absorption, transmission, or reflectance of light. Measurement

Fig. 12.1 Insects of different feeding guilds incur diverse plant damage phenotypes

of variation in plant reflectance can be employed to study pest attacks and the inflicted damage by using high-throughput cameras.

12.2.2 Imaging Techniques for Insect Damage Plant Phenotyping

Eastman Kodak's invention of the digital camera in 1975 led to a huge implication of visible light imaging technology that was employed to analyze the shoot biomass, leaf area index (LAI), yield-related traits, panicle traits, and architecture traits of plant roots (Kundu et al. 2022; Kundu and Sahu 2021; Yang et al. 2013). The present-day imaging technologies have developed from red-green-blue:visible (RGB), infrared, hyperspectral, and thermal to fluorescence including 3D laser scanning and tomographic imaging by employing magnetic resonance imaging (MRI) or X-ray computed tomography (CT) (Song et al. 2021).

Though there are several commonalities among plant traits in response to biotic and abiotic stresses, HTP methods for screening plants against insects are limited. Recently, digital imaging has been shown to detect differences in the yellowing of rice leaves in response to brown plant hopper and white-backed plant hopper feeding using the differences in red and blue reflectance (Horgan et al. 2020). Whereas visible imaging (red-green-blue; RGB) is primarily used to capture the morphological traits of plants, several recent techniques incorporate the characteristic infrared radiation released from the molecules that can be screened using the near-infrared (NIR) and far-infrared (far-IR) imaging devices. Infrared thermography has recently been reported to be highly applicable for detecting the time and abundance of the emergence of spotted lanternfly, an invasive plant pest in the fields of North America (Liu et al. 2021). Visible/near-infrared hyperspectral imaging was used to find the early infestation of striped stem borer in rice and the degrees of pest infestation (Fan et al. 2017). Another recent study demonstrates that robotic technology facilitated precision agriculture to automatize the selection of 18 different traits from the aboveground parts of the plant, flower, and fruit traits relevant for strawberry breeding programs (James et al. 2022). Similar methodologies could be used in understanding the various aspects of insect-resistant crop breeding.

Another growing area of imaging, hyperspectral imaging, further divides images into a huge portion of electromagnetic spectrums that are used to understand plant growth and insect-mediated damage characteristics. This technique also provides the opportunity to discriminate between the healthy kernels from the insect-damaged ones of wheat or other important crops (Singh et al. 2010) and to detect the soil nematode, *Heterodera schachtii*, in the sugar beet field (Hillnhütter et al. 2011). Similarly, the potential of hyperspectral imaging was also used to accurately detect the attack of the aphid, *Myzus persicae* on Chinese cabbage and *Pieris rapae* caterpillar on cabbage plants (Wu et al. 2016; Zhao et al. 2017). The use of hyperspectral cameras helped to distinguish soybean plants infested with stink bugs and caterpillars where the caterpillar-infested plants displayed varied amounts of reflectance as compared to noninfested plants (Filho 2022).

Fluorescence imaging measures the physiological changes in the plants like photosynthetic efficiency, plant respiration, and chlorophyll loss during biotic or abiotic stresses (Pérez-Bueno et al. 2019). It allows the screening of plants with normal and impaired metabolism, and the degree of changes in metabolism in response to herbivory can potentially help to distinguish the insect-resistant and -susceptible plants. Moustaka et al. (2021) used chlorophyll fluorescence to measure the photosynthetic efficiency of *Spodoptera exigua*-fed tomato plants. Similarly, the chlorophyll fluorescence technique revealed that photosynthesis was inhibited in tobacco plants upon *Manduca sexta* feeding due to jasmonic acid signaling (Nabity et al. 2013). Active IR operates at the lower wavelengths; however, thermal imaging functions in the mid- to longer IR wavelength energies, which are passive and sense the difference in heat energy. These sensors then convert the invisible radiation pattern (temperature) of any object into a visible image that allows to record the surface temperature of any object at a high resolution in two dimensions: black for cold and white for hot. These thermal data can then be used directly or indirectly in multiple ways (Ali et al. 2022; Mahanti et al. 2022). For example, infrared thermal screening has been deployed since several decades to detect infestation by six different developmental stages of the rusty grain beetle, *Cryptolestes ferrugineus*, under the seed coat on the germ layer of the wheat kernels by undergoing temperature screening (Manickavasagan et al. 2008). Insect herbivory has been shown to elevate leaf temperature (Havko et al. 2020; Zavala et al. 2013). Thermal imaging can help in monitoring the pest attack by measuring the leaf temperature and examining the stomatal opening and transpiration. Occasionally, stress conditions result in the closing of stomata followed by a decrease in transpiration and changes in the temperature of plants (Zhang and Zhang 2018). A great advantage for thermal imaging sensors being on the longer wavelengths is that the sensors do not absorb the reflected light and thus remain unaffected by dust, smoke, headlights, and other foreign objects.

12.2.3 Advancements in Plant Phenotyping Through Machine Learning to Develop HTP Platforms and Commercial Products

Though HTP can help screen hundreds of plant genotypes, the data generated with HTP is another bottleneck in handling enormous datasets and downstream data analysis pipelines. Machine learning provides us with tools to develop imaging and video-based products for plant phenotyping. Machine learning has been developed over the years to provide us with a wide range of pest screening techniques that help us to not only detect their presence but also in several instances provide information about the stage of the infesting insects. Support vector machine (SVM) is another such powerful method that is widely employed for pattern recognition (Banerjee and Madhumathy 2022). For the ease of farmers, the health conditions of crops can be analyzed easily by image processing and SVM, which is further incorporated into the Internet of Things (IoT) and sensor networks. SVM has been employed for the detection of leaf damage types in cotton (fungal/bacterial/pests), which shows up to 94% accuracy. SVM has also been incorporated to analyze and confirm the damage degrees caused by the leaf miners by recording the spectral reflectance at several sensitive wavelengths (Dake and Chengwei 2006). The leaf near-infrared spectral reflectance is generally measured by the UV-VIS-NIR spectrophotometer by the Shimadzu Corp. LiDAR is another actively used sensor technique that employs the visible to NIR wavelengths to measure the LAI, alongside the canopy temperature, aboveground biomass, and plant heights (Araus et al. 2018; Campbell et al. 2018; Singh et al. 2016a, 2016b). LiDAR sensors with 360° scanning facility have been used for the detection of flying insects in the range of 50 m. However, the identification of the detected insect/object is not possible for such poor resolution. Scientists are thus working on developing a convolutional neural network (CNN)-based grayscale image detection that would further facilitate identifying the type of object and with the help of a camera recording the image post-detection (Hosseini et al. 2020). Recent advances in the implication of the high-throughput imaging system for the detailed analysis of the type of damage caused by insects include a remote pheromone trap system that continuously monitors the arrival of bugs in large fields using wireless sensor networks (Song et al. 2021).

Insects of any feeding guilds upon entry into the plant cell alter several parameters in the host that includes leaf surface temperature, photosynthetic efficiency, CO_2 gas exchange, stress-induced pigmentation, and other physiological features that are reflected upon spectral imaging. Scientists are thus proposing the launch of a UAV phenotyping platform as the best choice for monitoring insect-mediated plant damage in large fields (Kloth et al. 2012).

12.2.4 Plant Phenotyping for Photosynthetic Gas Exchange Parameters

Insect pests have been reported to affect plant photosynthesis, growth, and yield negatively (Rossato Jr et al. 2019). For instance, aphid infestation can heavily impact the photosynthesis, growth, and yield of Brassicaceae and wheat cultivars (Hussain et al. 2015; Shahzad et al. 2019). Chlorophyll fluorescence imaging has been immensely employed for evaluating the photosynthetic efficiency upon stress in plants. Furthermore, a handheld spectrophotometer (SPAD meter) has been employed for large-scale screening of sorghum plants against greenbugs that measures the levels of chlorosis in infested leaves as compared to the control ones (Girma et al. 1998; Nagaraj et al. 2002). In plant biology, there was an immense need of advanced instruments, which could measure photosynthetic parameters in a nondestructive way.

An advanced high-throughput phenotyping platform has been developed by LI-COR Biosciences that serves the research community by providing cutting-edge scientific instruments to collect and analyze large-scale scientific data with maximum accuracy and ease. Poljaković-Pajnik et al. (2016) determined the aphid feeding effects on multiple poplar cultivars by measuring some of the key physiological parameters such as stomatal conductance and relative chlorophyll concentration using LI-600. LI-600 enables researchers to integrate additional data such as GPS location and the angle of the leaf relative to sunlight. They reported a significant decrease in photosynthesis rate up to 25–75% and an increased respiration rate ranging from 15 to 56.77%. They concluded that aphid colonization reduced stomatal conductance that adversely affects other physiological processes. Similarly, determining leaf area index (LAI) of fruit canopy plays an important role in pest management such as amount of pesticide to be applied per unit area, developing efficient spraying strategies, improved pesticide delivery, and cost reduction (Zhai et al. 2018; Wu et al. 2017; Wu 2007; Verrelst et al. 2015). Liu et al. (2021) took the advantage of LAI-2200C plant canopy analyzer to nondestructively measure LAI with maximum accuracy and ease in parallel with multispectral data acquisition by an unmanned aerial vehicle to propose machine learning models to enable efficient management of large-scale apple orchards.

Similarly, Chen et al. (2017) measured the net photosynthesis rate (PN) broadleaf evergreen canopy, *Schima superb*, in response to the attack of the moth, *Thalassodes quadraria*, using Li-6400XT portable photosynthesis system and concluded that insect outbreaks negatively impact *S. superba* by retarding the radial growth. Furthermore, rice tungro bacilliform virus (RTBV) and rice tungro spherical virus (RTSV) cause a complex disease in rice named tungro that affects photosynthetic machinery, chlorophyll biosynthesis, Fe/Zn homeostasis, and gene regulation associated with this biological process (Hull 1996; Sailaja et al. 2013; Mangrauthia et al. 2012; Srilatha et al. 2019). Symptoms include stunting, twisting of leaf tips, leaf discoloration, and reduction in ear-bearing rice tillers (Azzam and Chancellor 2002; Hull 1996). Insect vector green leafhopper facilitates the disease transfer (Azzam and Chancellor 2002; Hibino et al. 1978). Srilatha et al. (2019) analyzed the effect of disease on rice plant photosynthesis, biosynthesis of chlorophyll, and ion homeostasis. In this study, they used a high-throughput platform, Li-Cor 6400 to determine stomatal conductance (gs), net photosynthetic rate (PN), intercellular CO_2 concentration (Ci), and transpiration rate (E). Their finding suggests the involvement of RTBV ORF1 protein in regulating host cell photosynthesis and symptom development. Though there are several reports available on the effects of insect infestation on plant photosynthesis, the exact underlying mechanisms are largely unknown. Recently, a breakthrough study revealed that thrips can ingest chloroplasts, which are the core of plant photosynthesis (Wu et al. 2022). These mechanisms would help us to understand the commonalities between biotic and abiotic stress responses and, consequently, help to develop common HTP platforms for multiple stress agents.

12.2.5 Connecting the Effects of Biotic and Abiotic Stresses on Plant Gas Exchange Parameters

Plant responses to biotic and abiotic stress agents may complement/antagonize to each other. It has been reported that plant defenses to insects may get enhanced/suppressed in interaction with abiotic stress conditions (Table 12.1). For instance, in a study conducted by Xu et al. (2018), authors used LI-190R equipped with an LI-1500 radiation illuminance measuring instrument from LI-COR biosciences to measure PAR data to study maize diurnal responses to different water stresses at the canopy scale and concluded that both leaf folding and physiological changes are deployed by maize to cope with water stress. Similarly, Gago et al. (2017) used Li-Cor 6400 to measure leaf gas exchange, photosynthesis, and stomatal conductance to study metabolic components of water stress within a vineyard. From this study, they proposed an integrative methodology that combines UAV-based remote sensing, metabolomics, and organ-level physiology of the vineyard canopy responses to water stress. Liu et al. (2022) took advantage of a fast-response infrared gas analyzer (LI-500A), leaf area index analyzer (LAI-2200C), and canopy surface transpiration analyzer (LI-6400XT) from Li-Cor Inc. to measure the water stress index for monitoring water status in an oak plantation. Above-cited examples suggest the efficacy of the Li-Cor platform to assess water-related stress by measuring related physiological parameters in a high-throughput manner. In a recent study, Kansman et al. (2022) reported that plant water stress negatively impacts aphid performance. They correlated poor aphid performance with water stress-induced plant defense and reduced nutritional quality. Drought stress promotes resistance to herbivory by induction of phytohormone-mediated defense response (Nguyen et al. 2016; Blundell et al. 2020; Lin et al. 2021). Similarly, drought stress also induces the level of sucrose in plant phloem (Marček et al. 2019). Previously, it has been reported that high concentrations of some sugars such as trehalose, fructose, and

Table 12.1 High-throughput phenotyping used in different plant-insect study systems

Crop(s)	Insect(s)	Tool(s)/technique (s) used	Parameter(s) recorded	Conclusions	References
Arabidopsis (*Arabidopsis thaliana*) and lettuce (*Lactuca sativa*)	Aphid (*Myzus persicae*)	Video tracking	Aphid behavior	Video tracking can be used to screen large plant populations for resistance to aphids	Kloth et al. (2015)
Soybean (*Glycine max*)	Stink bugs (*Nezara viridula* and *Piezodorus guildinii*)	LI-COR	Levels of photosynthetically active radiation	Isoflavonoid induction by UV-B radiation provides resistance to stink bugs in soybean	Zavala et al. (2015)
Tomato (*Solanum lycopersicum*)	Whitefly (*Bemisia tabaci*)	LI-COR	Net photosynthetic rate	Elevated CO_2 conditions increased plant photosynthetic rate and confer resistance to whitefly	Guo et al. (2016)
Tomato (*Solanum lycopersicum*)	Tomato leaf miner (*Tuta absoluta*) and silverleaf whitefly (*Bemisia tabaci*)	LI-COR	Photosynthetic rate, transpiration rate, and stomatal conductance	Plants under nitrogen or water-limited conditions significantly affected *T. absoluta* survival and development, while survival of *B. tabaci* was affected only by nitrogen limitation	Han et al. (2016)
Arabidopsis (*Arabidopsis thaliana*)	Green peach aphids (*Myzus persicae*)	Video tracking/ EthoVision	Aphid behavior	Feeding behavior on knockout mutants and overexpression lines showed that *WRKY22* increases susceptibility to aphids	Kloth et al. (2016)

Tomato (*Solanum lycopersicum*)	Glasshouse whitefly (*Trialeurodes vaporariorum*)	EPG	Feeding behavior	In wild tomato species, whiteflies exhibited a shorter duration of second feeding bout, reduced pathway phase probing, longer salivation in phloem, and more non-probing activities in the early stages of the EPG compared to the commercial tomato. It confirms resistance factors' presence in wild tomato	McDaniel et al. (2016)
Soybean (*Glycine max*)	Soybean looper (*Chrysodeixis includens*), corn earworm (*Helicoverpa zea*), fall armyworm (*Spodoptera frugiperda*), and velvet bean caterpillar (*Anticarsia gemmatalis*)	ImageJ	Defoliation	Soybean resistance to insects was enhanced by QTL-M and QTL-E. Pyramiding these QTLs with cry1Ac increases protection against Bt-tolerant pests	Ortega et al. (2016)
Maize (*Zea mays*)	Western corn rootworm (WCR; *Diabrotica virgifera virgifera*)	Positron-emission tomography, root autoradiography, and radiometabolite flux analysis	Maize tolerance to WCR by auxin biosynthesis	WCR feeding induces local changes in roots, which leads to increased auxin biosynthesis flux and provides tolerance to WCR	Qu et al. (2016)
Arabidopsis (*Arabidopsis thaliana*)	Western flower thrips (*Frankliniella occidentalis*)	Video tracking	Thrip behavior	Video tracking is a promising phenotyping tool to assess host-plant resistance to thrips in large plant populations	Thoen et al. (2016)

(continued)

Table 12.1 (continued)

Crop(s)	Insect(s)	Tool(s)/technique (s) used	Parameter(s) recorded	Conclusions	References
Tomato (*Solanum lycopersicum*)	Whitefly (*Bemisia tabaci*)	EPG	Stylet penetration activities and inoculation of tomato chlorosis virus (ToCV) in tomato	ToCV inoculation is mainly associated with stylet activities in phloem sieve elements	Prado Maluta et al. (2017)
Wheat, *Triticum aestivum* (L.), and *T. monococcum* lines	English grain aphid (*Sitobion avenae*)	EPG	Feeding behavior	Mycorrhizal colonization increased aphid phloem feeding on wheat varieties	Simon et al. (2017)
Chinese cabbage (*Brassica rapa*)	Fungal pathogen *Pandora neoaphidis*-infected and uninfected green peach aphid (*Myzus persicae*)	EPG	Feeding behavior	Fungal infection has an effect on the feeding behavior of aphids, which potentially decreases the feeding capacity	Chen et al. (2018)
Olive (*Olea europaea*)	Meadow spittlebug (*Philaenus spumarius*)	EPG, video recordings, microcomputer tomography (micro-CT)	Feeding behavior	Correlation was observed between EPG waveforms and specific biological behaviors of *P. spumarius* captured by video recordings	Cornara et al. (2018)
Cabbage (*Brassica oleracea* L.)	Diamondback moth (*Plutella xylostella*) and aphids (*Myzus persicae*)	ImageJ	Photosynthetic area of leaves	Seed coating with jasmonic acid and chitosan leads to long-term defense priming in chewing and sap-feeding insects	Haas et al. (2018)
Maize (*Zea mays* L.)	Green belly stink bugs (*Dichelops melacanthus*)	Hyperspectral imaging	Leaf phytocompound levels	Hyperspectral imaging may be considered as a powerful tool to increase our understanding on plant communication after herbivory	do Prado Ribeiro et al. (2018)

Arabidopsis (*Arabidopsis thaliana*)	Western flower thrips (*Frankliniella occidentalis*)	Video tracking/ EthoVision	Thrip behavior	The analysis using software resulted in much larger contrasts in behavior traits of thrips than previously reported	Jongsma et al. (2019)
Cassava (*Manihot esculenta*), sweet potato (*Ipomoea batatas*), tomato (*Solanum lycopersicum*), and cotton (*Gossypium hirsutum*)	Whitefly (*Bemisia tabaci*)	EPG	Feeding behavior	Preference of host species was recorded, and host-specific whiteflies' feeding was not restricted to its host	Milenovic et al. (2019)
Five capsicum species (*Capsicum annuum.*, *C. chinense*, *C. baccatum*, *C. pubescens*, and *C. eximium*)	Thrips (*Frankliniella occidentalis*)	ilastik and ImageJ	Quantification of feeding damage	In capsicum, leaf position and ontogeny are important determinants of resistance against thrips	Visschers et al. (2019)
Cotton (*Gossypium hirsutum*)	Cotton bollworm (*Helicoverpa armigera*) and cotton leafworm (*Spodoptera litura*)	ImageJ	Trichome and stomatal densities and length	Tolerant cotton variety had longer trichome and high stomatal density compared to susceptible variety	Diaz-Montano et al. (2006)
Arabidopsis (*Arabidopsis thaliana*)	Shield bug (*Eurydema oleracea*)	ImageJ	Leaf damage	*E. oleracea* suppresses plant defense by manipulating hormone signaling components in plants	Grover et al. (2020)
Sorghum (*Sorghum bicolor*)	Sugarcane aphid, (*Melanaphis sacchari*)	LI-COR and metabolomics	Photosynthesis parameters	Sorghum-tolerant line has robust photosynthetic machinery along with higher levels of OPDA	Grover et al. (2020)

(continued)

Table 12.1 (continued)

Crop(s)	Insect(s)	Tool(s)/technique (s) used	Parameter(s) recorded	Conclusions	References
Grass species (*Brachypodium distachyon*)	Cotton bollworm (*Helicoverpa armigera*) and house cricket (*Acheta domesticus*)	ImageJ	Leaf surface morphology	Silicon changes the morphology of defense structures on leaves and acts as a direct defense against leaf-chewing insects	Hall et al. (2020)
Potato (*Solanum tuberosum*)	Aphid (*Myzus persicae*)	LI-COR	Stomatal conductance	With drought stress and aphid infestation, stomatal conductance was not affected	Quandahor et al. (2020)
Kidney bean (*Phaseolus vulgaris*)	Western flower thrips (*Frankliniella occidentalis*)	LI-COR	Leaf area	Kidney bean leaf damage caused by thrips was significantly affected by CO_2 level and wounding time	Qian et al. (2021)
Apple (*Malus domestica*)	Woolly apple aphid (*Eriosoma lanigerum*)	EPG	Feeding behavior	Aphids prefer to feed on shoots of plant compared to other tree parts	Zhou et al. (2021b)
Tomato (*Solanum lycopersicum*)	Mirid bug (*Tupiocoris cucurbitaceus*)	EthoVision	Prey-searching behavior	Mirid bug preferred to prey on eggs of leaf miner (*Tuta absoluta*)	Cáceres et al. (2022)
Grapevine (*Vitis vinifera*)	Two spittlebugs [meadow spittlebug (*Philaenus spumarius*) and spittlebug (*Neophilaenus campestris*)]. Two sharpshooter leafhoppers	DC-EPG	Feeding behavior	Sharpshooters tended to perform significantly more probes than spittlebugs	Markheiser et al. (2022)

	[rhododendron leafhopper (*Graphocephala fennahi*) and green leafhopper (*Cicadella viridis*)				
Cotton genotypes	Different sucking pests	Leaf area meter	Leaf area	Leaf area was significantly declined with increasing sucking insect pest infestation levels	Singh et al. (2022)

sucrose negatively affect aphid performance (Douglas 2006; Grover et al. 2022b). Drought stress also contributes to the production of plant defensive compounds such as peroxidases, phenolics, glucosinolates, and physical structures such as callose, which deters insect pest infestation (Kansman et al. 2022). Hence, measuring physiological parameters pertaining to water stress may be recommended as a key component of HTP of insect performance on crops.

12.2.6 Development of Software Packages to Analyze Image Data

Several imaging platforms such as fluorescence imaging, multispectral imaging, hyperspectral imaging, thermal imaging, 3D imaging, and visible imaging have enabled high-throughput phenotyping in detecting and controlling crop losses to disease and pests (Singh et al. 2020). In their review, Minervini et al. (2015) describe that the speed of phenotype extraction through image analysis (the software) is not matching the speed of genotyping and remains a bottleneck in modern-day sustainable agricultural practices. However, in recent years, a cascade of publicly available software has been developed to analyze high-throughput plant phenotyping data. For chewing insects, digital imaging has been employed with the help of scanners and public domains to feed the software with the entire and fed portions of leaves, thus largely cutting the time for large-scale plant phenotyping (O'Neal et al. 2002). Maize-IAS, an easy-to-use image analysis software, uses artificial intelligence and deep learning for high-throughput maize phenotyping (Zhou et al. 2021a). Maize-IAS offers a plethora of functions for the processing of large datasets and enables automation, which greatly minimizes human error in performing quantitative analysis. Similarly, HSI-PP, a stand-alone open-source software that combines advanced image processing with the extraction of meaningful information from hyperspectral images, enables decision-making in agricultural research (ElManawy et al. 2022). HSI-PP delivers by preparing images for deep leaning analysis and by enabling the machine learning model for classification and regression. HSI-PP is expected to be revolutionary in improving the efficiency of high-throughput phenotyping in agricultural management, plant physiology, breeding programs, and functional genomics. Gehan et al. (2017) developed and rewritten Plant CV v2, an improved version of Plant CV v1.0, on the Python platform to add novel functionality in image analysis to support high-throughput phenotyping. Plant CV v2 offers new functionalities such as white balancing, size marker normalization, auto-thresholding, multiplant detection, watershed segmentation, combined image processing, improved image segmentation, and landmarking. Gehan et al. (2017) explicitly elaborate the way each of these applications substantially improves image analysis in a user-convenient way.

12.3 High-Throughput Phenotyping for Insects

In addition to studying plant traits, insect parameters (behavior, growth, and reproduction) are other important aspects that can be studied for screening plants for insect resistance. Unlike plant parameters, insect parameters are quicker to study and record. For instance, insect growth differences can be measured in a few days compared to the effects of insect attacks on plants that can last for several days. Studying insect-feeding behavior on varied plant varieties deduces the status of resistant and susceptible plants. Moreover, insect behavior studies usually range from a few minutes to hours. Robust phenotyping systems can fasten the process of selecting resistant plant types with a high level of accuracy. The basis for designing HTP to understand insect behavior depends on parameters like feeding behavior, damage caused to crops, and growth stages of the insects. In this section, we have discussed the currently available tools for identifying resistant plants by studying insect behavior.

12.3.1 Video Tracking/EthoVision

Plant images can provide a greater amount of information about plant traits, feeding damage, and physiological changes in plants in response to insect feeding. However, tracking the behavior and movement of insects on plants can help to analyze the acceptance or rejection of plant material. An automated video tracking software called EthoVision was developed, which has been developed over the years that showed high potential for screening a large number of host plants for insect resistance, tracking the movement of insects, and recording their behavior pattern (Noldus et al. 2002; Kloth et al. 2015). Additionally, video tracking can also reduce resource use, such as using leaf discs instead of the whole plant. For aphids, automated video tracking was employed to track the movement, stylet penetrations, and feeding patterns on leaf discs of *Arabidopsis* and lettuce to identify the level of resistance. The analyses were performed in EthoVisionXT software to record the duration of aphid penetration and detailed movements over the leaf disc (Kloth et al. 2015). This study also offers unique opportunities to employ video tracking system for other piercing and sucking-type and/or rasping-sucking-type insects like thrips. Recently, a choice assay setup for western flower thrips (WFT, *Frankliniella occidentalis*) was developed, and an automated video tracking system was used to track the choice made by thrips on *Arabidopsis* plant accessions for resistance (Thoen et al. 2016). Further T-maze system was developed for insect choice assay (WFT) automation and analysis of the tracking of insect movement simultaneously. The integrated approach was named EntoLabs (Jongsma et al. 2019). Recently, a Trax software was developed to track the movement of color-tagged insects in video recording (Gal et al. 2020). Collectively, EthoVision has been extensively used in plant-insect interactions in a wide range of systems such as clover seed weevils on white and red clovers (Nyabuga et al. 2015), corn root worm on maize (Bernklau and Bjostad 2009), whitefly on eggplants (Moreno-Delafuente et al. 2013), tobacco hornworm on corn pericarp (Tayal et al. 2020), aphid on lettuce (Booij et al. 2013), green rice leafhoppers on rice plants (Chang et al. 2021), and *Eucalyptus* weevils on *Eucalyptus* (Branco et al. 2019). Further automation in assay setups, data recordings, and analyses would immensely enhance the use of video tracking in plant-insect interaction studies.

12.3.2 Electrical Penetration Graph (EPG) Technique for Sap-Sucking Insects

Electrical penetration graph (EPG) technique is a powerful tool to precisely measure the feeding behavior of piercing and sucking type of insects (Tjallingii 1985). This technique is widely used to understand plant-insect interactions, characterize plant resistance to insects, and understand insecticide resistance in insects and virus transmission (Moreno et al. 2012; Garzo et al. 2016; Grover et al. 2019). Hemipteran insects use their stylets to pierce the cell wall and penetrate inside to ingest the cell sap and derive away the nutrients from plants causing minimal visible injury to plants. Therefore, EPG system allows to track the movement of hemipteran insects' stylets and to study the probing behavior of insects (Backus and Shih 2020). It can precisely tell where the insect is exactly feeding on the plant part in a nondestructive manner. On the contrary, stylectomy is a destructive technique where aphid stylet fed on phloem tissue is severed to quantify the proteins present in phloem sap, which could potentially be involved in plant defenses, hence not preferred over EPG. A wide adaptability of EPG in all crops such as sorghum, corn, soybean, tea, wheat, cotton, tomato, tobacco, beans, potato, and *Arabidopsis*, to study the feeding behavior of hemipterans, has significantly contributed to enormous improvements over the last three decades (Pegadaraju et al. 2007; Trębicki et al. 2012; Civolani et al. 2014; Pearson et al. 2014; Cervantes et al. 2016; Garzo et al. 2016; Grover et al. 2019, 2022b; Souza and Davis 2021; Jhou et al. 2021; Rech et al. 2021; Salsabillah et al. 2021).

EPG system consists of two electrodes, namely the plant electrode and the insect electrode (Van Helden and Tjallingii 2000). When an insect stylet pierces through plant tissues, it completes the circuit leading to the flow of current. Before sustained feeding starts on plants, insects insert their stylet intracellularly or intercellularly to sample the cell contents. EPG generates distinct waveforms associated with different activities performed by insects. The waveforms are mainly categorized into three phases: (1) pathway phase where an insect is looking for potential feeding sites, inserts the stylet into plant tissue, samples the cell content, and accepts or rejects the plant; (2) phloem phase where the insect salivates and aggressively ingests the feeds on the phloem tissue or sieve elements for longer periods; and (3) xylem phase where it feeds on water (Nalam et al. 2018). In non-probing phase, insects generally do not perform any feeding behavior other than walking or relaxing on the plant (Nalam et al. 2018). EPG has revolutionized the plant-insect interaction area of study due to its ability to screen insect-resistant host plants faster and understand host resistance barriers to insects precisely. This technique can collectively demonstrate host resistance based on antibiosis or antixenosis (Diaz-Montano et al. 2006). Phloem-based host resistance is usually attributed to the presence of antibiotic factors in phloem, which negatively affects insect physiology. Although EPG provides accurate data, it is a sensitive system and requires high human precision while handling insects and can generate data only from a limited number of plants at a time. Over time, advancements have been made in automating the EPG analysis of different waveforms as manually calculating each waveform is very tedious and time-consuming. Software like MacStylet, Aphid-AutoEPG, EPF-Calc, and A2EPG (Febvay et al. 1996, Giordanengo 2014, Adasme-Carreño et al. 2015, Prufer et al. 2014) have been developed to bring automation in EPG analysis.

12.4 Functional Phenomics

Functional phenomics refers to the interdisciplinary emerging approach integrating high-throughput phenotyping, plant physiology, bioinformatics, and "*omic*" approaches (York 2019; Jammer et al. 2022). The overall goal of functional phenomics is to understand and connect plant phenotypes with plant functions. The use of various phenotyping methods for understanding plant phenotypes and insect behavior has significantly advanced the knowledge in the area of plant-insect interactions (Table 1). Despite these significant outcomes, understanding and establishing the relationship between plant phenotypes and plant function have been a bottleneck for breeding superior and resilient cultivars despite the enormous progress in knowledge available in genetics. For instance, it has been shown that the root crown architectural phenes are correlated to rooting depth, yield, and senescence phenology (Slack et al. 2018). This information would be immensely helpful in breeding cultivars with desired traits. Functional phenomics in the area of plant-insect interactions has not been explored much. For example, a recent study from Australian *Vitis* sp. in abiotic stress reports the use of RGB-based digital imaging and *LI-COR* measurements to correlate the high temperature-mediated accumulation of Cl^{-1} ions in the lamina of the *Vitis* leaves (Dunlevy et al. 2022). This technique could be further extrapolated for its employment in herbivory, where leaf-chewing caterpillars lead to a significant accumulation of Ca^{2+} ions in the plant cell within seconds. Moreover, plant morphological characteristics such as the presence of trichomes, leaf thickness, lamina type, and stem thickness can also impact the plant's resistance to herbivory (Heinze 2020; War et al. 2012). Functional phenomics has the potential to identify biomarkers/biosignatures directly related to plant resistance, possibly by combining HTP and "*omic*" approaches (Fig. 12.2). Going forward, this emerging approach offers unique opportunities to combat global challenges by accelerating the process of developing pest-resistant genotypes.

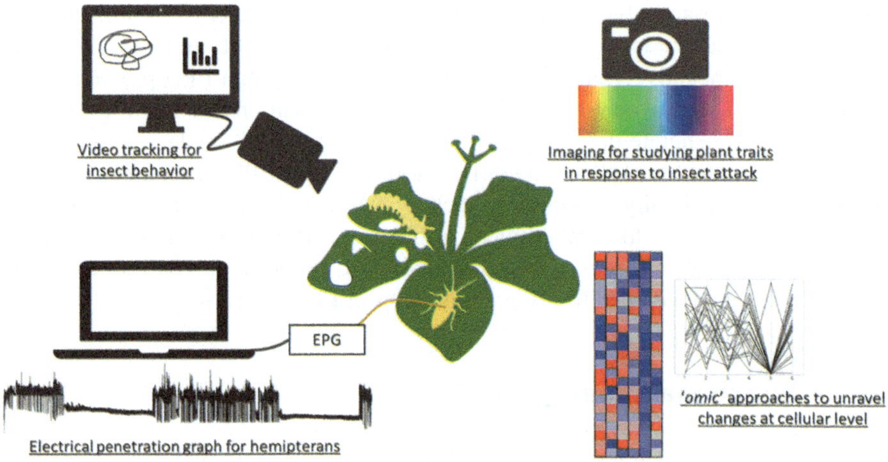

Fig. 12.2 A holistic representation of functional phenomics approach to understand plant-insect interactions using multiple methods and linking cellular and physiological phenotypes to insect phenotype

12.5 Bottlenecks in the Advancement of HTP in Plant-Insect Interactions

Despite the progress of HTP in plant sciences, it is still not being largely used for the screening of insect-resistant plants. The cost of HTP in plants is still very high, and scientists with limited funding for research cannot afford to set up this system in individual labs and departments. For such a kind of initiative, scientists need to collaborate within and beyond the area of entomology. Though working on different plant-insect systems, entomologists can still use common tools to address specific research questions. Furthermore, interdisciplinary collaborations are highly required with plant scientists, engineers, data scientists, and statisticians. Entomologists are primary participants in the area of host plant resistance to insects (Peterson et al. 2017). There are a handful of scientists with interdisciplinary expertise in plant physiology, biochemistry, metabolism, plant anatomy, and water relations. The lack of interdisciplinary training to entomologists is still an obstacle to the development of novel and innovative tools to move the discipline forward. Additionally, to streamline the phenotyping process globally, several organizations like European Plant Phenotyping Network (EPPN) and International Plant Phenotyping Network (IPPN) have been working on setting the standards for the quantification of the phenotypic data to be used worldwide (Williams et al. 2018).

12.6 Conclusions and Future Directions

Food security is a common challenge where all agricultural scientists are working towards using different paths and trajectories. To meet the global food security challenges of the future in a sustainable manner, research on plant-insect interactions is one of the top-most priorities where the natural potential of plants to fight against insects can be unlocked. Considering the ill effects of chemicals on the environment, the public perception of GMOs, and the increasing cost of seeds for transgenic plants, research on plant-insect interactions can accelerate the development of insect-resistant plants. To meet the pace of food security challenges, sustainable agricultural products have to be developed sooner than later. HTP in plants for insect resistance needs to be utilized to match the pace of molecular breeding and genomic studies. In order to achieve this goal, we need to foster interdisciplinary teams of scientists' workforce. Consequently, it is crucial that entomologists, plant scientists, statisticians, data scientists, and engineers work collaboratively to develop and utilize HTP capabilities for accelerating the development of sustainable insect control products. HTP would enormously increase the capacity to screen plant populations against insect infestations. This would ultimately assist basic and applied studies in the area of plant-insect interactions to understand the genetic basis of insect resistance.

References

Adasme-Carreño F, Muñoz-Gutiérrez C, Salinas-Cornejo J, Ramírez CC (2015) A2EPG: a new software for the analysis of electrical penetration graphs to study plant probing behaviour of hemipteran insects. Comput Electron Agric 113:128–135

Ali MM, Hashim N, Abd Aziz S, Lasekan O (2022) Quality prediction of different pineapple (*Ananas comosus*) varieties during storage using infrared thermal imaging technique. Food Control 138:108988

Araus JL, Kefauver SC, Zaman-Allah M, Olsen MS, Cairns JE (2018) Translating high-throughput phenotyping into genetic gain. Trends in Plant Sci 23:451–466

Arya S, Sandhu KS, Singh J (2022) Deep learning: as the new frontier in high-throughput plant phenotyping. Euphytica 218:1–22

Azzam O, Chancellor TC (2002) The biology, epidemiology, and management of rice tungro disease in Asia. Plant Dis 86(2):88–100

Backus EA, Shih HT (2020) Review of the EPG waveforms of sharpshooters and spittlebugs including their biological meanings in relation to transmission of *Xylella fastidiosa* (Xanthomonadales: Xanthomonadaceae). J Insect Sci 20(4):1–14. https://doi.org/10.1093/jisesa/ieaa055

Banerjee I, Madhumathy P (2022) IoT based agricultural business model for estimating crop health management to reduce farmer distress using SVM and machine learning. In: Pattnaik PK, Kumar R, Pal S, eds. Internet of things and analytics for agriculture, Volume 3. Singapore: Springer Singapore, 165–183

Bernklau EJ, L Bjostad LB, Meihls LN, Coudron TA, Lim E, Hibbard BE (2009) Localized search cues in corn roots for western corn rootworm (Coleoptera: Chrysomelidae) larvae. J Econ Entomol 102(2):558–562. https://doi.org/10.1603/029.102.0213

Bhuvaneswari K, Mani M, Suganthi A, Manivannan A (2022) Novel insecticides and their application in the management of horticultural crop pests. Trends Hort Entomol, pp 419–454

Blundell R, Schmidt JE, Igwe A, Cheung AL, Vannette RL, Gaudin A, Casteel CL (2020) Organic management promotes natural pest control through altered plant resistance to insects. Nat Plants 6(5):483–491

Booij MW, Kloth KJ, Jongsma MA, Dicke M, Hemerik L (2013) Analysing aphid behaviour with time-to-event techniques to discriminate between susceptible and resistant plants. In: *Proceedings of the Netherlands Entomological Society Meeting, 2012*, vol 24. Netherlands Entomological Society, pp 9–16. https://edepot.wur.nl/305581

Branco S, Mateus EP, da Silva MDRG, Mendes D, Rocha S, Mendel Z, Schutz S, Paiva MR (2019) Electrophysiological and behavioural responses of the eucalyptus weevil, *Gonipterus platensis*, to host plant volatiles. J Pest Sci 92:221–235. https://doi.org/10.1007/s10340-018-1055-0

Cáceres RIM, Rocca M, Luna MG (2022) Searching behavior and prey preference by the predatory mirid bug *Tupiocoris cucurbitaceus*, a potential biocontrol agent of tomato crops in Argentina. PREPRINT (version 1) available at Research Square, doi: https://doi.org/10.21203/rs.3.Rs-1547879/v1

Campbell ZC, Acosta-Gamboa LM, Nepal N, Lorence A (2018) Engineering plants for tomorrow: how high-throughput phenotyping is contributing to the development of better crops. Phytochem Rev 17:1329–1343

Carvalho MR, Wilf P, Barrios H, Windsor DM, Currano ED, Labandeira CC, Jaramillo CA (2014) Insect leaf-chewing damage tracks herbivore richness in modern and ancient forests. PLoS One 9:e94950

Cervantes FA, Backus EA, Godfrey L, Akbar W, Clark TL (2016) Characterization of an EPG waveform library for adult *Lygus lineolaris* and *Lygus hesperus* (Hemiptera: Miridae) feeding on cotton squares. Ann Entomol Soc Am 109:684–697. https://doi.org/10.1093/aesa/saw039

Chang X, Wang F, Fang Q, ChenF YH, Gatehouse AMR, Ye G (2021) Virus-induced plant volatiles mediate the olfactory behaviour of its insect vectors. Plant Cell Environ 44(8): 2700–2715

Chen Z, Wang L, Dai Y, Wan X, Liu S (2017) Phenology-dependent variation in the non-structural carbohydrates of broadleaf evergreen species plays an important role in determining tolerance to defoliation (or herbivory). Sci Rep 7(1):1–11

Chen C, Ye S, Hu H, Xue C, Yu X (2018) Use of electrical penetration graphs (EPG) and quantitative PCR to evaluate the relationship between feeding behaviour and *Pandora neoaphidis* infection levels in green peach aphid, *Myzus persicae*. J Insect Physiol 104:9–14. https://doi.org/10.1016/j.jinsphys.2017.11.003

Cheng L, Huang F, Jiang Z, Lu B, Zhong X, Qiu Y (2021) Improved phenotyping procedure for evaluating resistance in rice against gall midge (*Orseolia oryzae*, wood-Mason). Plant Methods 17:1–11. https://doi.org/10.1186/s13007-021-00823-5

Civolani S, Cassanelli S, Chicca M, Rison JL, Bassi A, Alvarez JM, Annan IB, Parrella G, Giorgini M, Fano EA (2014) An EPG study of the probing behavior of adult *Bemisia tabaci* biotype Q (Hemiptera: Aleyrodidae) following exposure to cyantraniliprole. J Econ Entomol 107:910–919. https://doi.org/10.1603/EC13511

Cornara D, Garzo E, Morente M, Moreno A, Alba-Tercedor J, Fereres A (2018) EPG combined with micro-CT and video recording reveals new insights on the feeding behavior of *Philaenus spumarius*. PLoS One 13(7):e0199154

Dake W, Chengwei M (2006) The support vector machine (SVM) based near-infrared spectrum recognition of leaves infected by the leafminers. In: First international conference on innovative computing, information and control-volume, vol I (ICICIC'06), Vol. 3. IEEE, pp 448–451

Dhondt S, Wuyts N, Inzé D (2013) Cell to whole-plant phenotyping: the best is yet to come. Trends Plant Sci 18:428–439. https://doi.org/10.1016/j.tplants.2013.04.008

Diaz-Montano J, Reese JC, Schapaugh WT, Campbell LR (2006) Characterization of antibiosis and antixenosis to the soybean aphid (Hemiptera: Aphididae) in several soybean genotypes. J Econ Entomol 99:1884–1889. https://doi.org/10.1093/jee/99.5.1884

do Prado Ribeiro L, ALS K, JAW F, Tramontin MA, Trapp MA, Mithöfer A, Nansen C (2018) Hyperspectral imaging to characterize plant-plant communication in response to insect herbivory. Plant Methods 6(14):54. https://doi.org/10.1186/s13007-018-0322-7

Douglas A (2006) Phloem-sap feeding by animals: problems and solutions. J Exp Bot 57(4): 747–754

Dunlevy J, Blackmore D, Betts A, Jewell N, Brien C, Berger B, Walker R, Edwards E, Walker A (2022) Investigating the effects of elevated temperature on salinity tolerance traits in grapevine rootstocks using high-throughput phenotyping. Aust J Grape Wine Res 28:276–291

ElManawy AI, Sun D, Abdalla A, Zhu Y, Cen H (2022) HSI-PP: a flexible open-source software for hyperspectral imaging-based plant phenotyping. Comput Electron Agric 200:107248

Fan Y, Wang T, Qiu Z, Peng J, Zhang C, He Y (2017) Fast detection of striped stem-borer (*Chilo suppressalis* Walker) infested rice seedling based on visible/near-infrared hyperspectral imaging system. Sensors (Switzerland) 17. https://doi.org/10.3390/s17112470

Febvay G, Rahbe Y, van Helden M (1996) MacStylet, software to analyse electrical penetration graph data on the Macintosh. Entomol Exp Appl 80:105–108

Filho FHI, Pazini J de B, de Medeiros AD, Rosalen DL, Yamamoto PT (2022) Assessment of injury by four major pests in soybean plants using hyperspectral proximal imaging. Agron 12. doi: https://doi.org/10.3390/agronomy12071516

Finch S, Collier RH (2000) Host-plant selection by insects – a theory based on 'appropriate/inappropriate landings' by pest insects of cruciferous plants. Entomol Exp Appl 96:91–102. https://doi.org/10.1046/j.1570-7458.2000.00684.x

Follett PA (2017) Insect-plant interactions: host selection, herbivory, and plant resistance – an introduction. Entomol Exp Appl 162:1–3. https://doi.org/10.1111/eea.12524

Fuentes-Jacques LJ, Hanson-Snortum P, Hernández-Ortiz V, Díaz-Castelazo C, Mehltreter K (2021) A global review and network analysis of phytophagous insect interactions with ferns and lycophytes. Plant Ecol 223:27–40

Gago J, Fernie AR, Nikoloski Z, Tohge T, Martorell S, Escalona JM, Ribas-Carbó M, Flexas J, Medrano H (2017) Integrative field scale phenotyping for investigating metabolic components of water stress within a vineyard. Plant Methods 13(1):1–14

Gal A, Saragosti J, Kronauer D (2020) anTraX, a software package for high-throughput video tracking of color-tagged insects. elife 9:e58145

Garzo E, Moreno A, Hernando S, Mariño V, Torne M, Santamaria E, Díaz I, Fereres A (2016) Electrical penetration graph technique as a tool to monitor the early stages of aphid resistance to insecticides. Pest Manag Sci 72:707–718. https://doi.org/10.1002/ps.4041

Gehan MA, Fahlgren N, Abbasi A, Berry JC, Callen ST, Chavez L, Doust AN, Feldman MJ, Gilbert KB, Hodge JG, Hoyer JS, Lin A, Liu S, Lizarraga C, Lorence A, Miller M, Platon E, Tessman M, Sax T (2017) PlantCV v2: image analysis software for high-throughput plant phenotyping. PeerJ 5:e4088

Giordanengo P (2014) EPG-Calc: a PHP-based script to calculate electrical penetration graph (EPG) parameters. Arthropod Plant Interact 8:163–169

Girma M, Kofoid KD, Reese JC (1998) Sorghum germplasm tolerant to greenbug (Homoptera: Aphididae) feeding damage as measured by reduced chlorophyll loss. J Kansas Entomol Soc 71(2):108–115

Goggin FL, Lorence A, Topp CN (2015) Applying high-throughput phenotyping to plant–insect interactions: picturing more resistant crops. Curr Opin Insect Sci 9:69–76. https://doi.org/10.1016/j.cois.2015.03.002

Grover S, Wojahn B, Varsani S, Sattler SE, Louis J (2019) Resistance to greenbugs in the sorghum nested association mapping population. Arthropod Plant Interact 13:261–269. https://doi.org/10.1007/s11829-019-09679-y

Grover S, Agpawa E, Sarath G, Sattler SE, Louis J (2020) Interplay of phytohormones facilitate sorghum tolerance to aphids. Plant Mol Biol 109:639–650. https://doi.org/10.1007/s11103-020-01083-y

Grover S, Shinde S, Puri H, Palmer N, Sarath G, Sattler SE, Louis J (2022a) Dynamic regulation of phenylpropanoid pathway metabolites in modulating sorghum defense against fall armyworm. Front Plant Sci 13. Available at: https://www.frontiersin.org/articles/10.3389/fpls.2022.1019266. Accessed January 7, 2023

Grover S, Puri H, Xin Z, Sattler S, Louis J (2022b) Dichotomous role of jasmonic acid in modulating sorghum defense against aphids. Mol Plant-Microbe Interact. https://doi.org/10.1094/MPMI-01-22-0005-R

Guo H, Huang L, Sun Y, Guo H, Ge F (2016) The contrasting effects of elevated CO_2 on TYLCV infection of tomato genotypes with and without the resistance gene, *Mi-1.2*. Front. Plant Sci 9(7):1680. https://doi.org/10.3389/fpls.2016.01680

Haas J, Lozano ER, Haida KS, Mazaro SM, Vismara EDS, Poppy GM (2018) Getting ready for battle: do cabbage seeds treated with jasmonic acid and chitosan affect chewing and sap-feeding insects? Entomol Exp Appl 166(5):412–419. https://doi.org/10.1111/eea.12678

Hall CR, Dagg V, Waterman JM, Johnson SN (2020) Silicon alters leaf surface morphology and suppresses insect herbivory in a model grass species. Plants (Basel) 9(5):643. https://doi.org/10.3390/plants9050643

Han P, Desneux N, Michel T, Blot JL, Seassau A. Wajnberg E, Amiens-Desneux E, Lavoir AV (2016) Does plant cultivar difference modify the bottom-up effects of resource limitation on plant-insect herbivore interactions? J Chem Ecol 42:1293–1303

Havko NE, Das MR, McClain AM, Kapali G, Sharkey TD, Howe GA (2020) Insect herbivory antagonizes leaf cooling responses to elevated temperature in tomato. Proc Natl Acad Sci 117(4):2211–2217

Heinze J (2020) Herbivory by aboveground insects impacts plant root morphological traits. Plant Ecol 221:725–732. https://doi.org/10.1007/s11258-020-01045-w

Hibino H, Roechan M, Sudarisman S (1978) Association of two types of virus particles with penyakit habang (tungro disease) of rice in Indonesia. Phytopathology 68(10):1412–1416

Hillnhütter C, Mahlein AK, Sikora R, Oerke EC (2011) Remote sensing to detect plant stress induced by *Heterodera schachtii* and *Rhizoctonia solani* in sugar beet fields. Field Crop Res 122:70–77

Horgan FG, Jauregui A, Peñalver Cruz A, Crisol Martínez E, Bernal CC (2020) Changes in reflectance of rice seedlings during planthopper feeding as detected by digital camera: potential applications for high-throughput phenotyping. PLoS One 15(8):0238173

Hosseini B, Montagne R, Hammer B (2020) Deep-aligned convolutional neural network for skeleton-based action recognition and segmentation. Data Sci Eng 5:126–139

Hull R (1996) Molecular biology of rice tungro viruses. Annu Rev Phytopathol 34(1):275–297

Hussain A, Razaq M, Zaka SM, Shahzad W, Mahmood K (2015) Effect of aphid infestation on photosynthesis, growth and yield of *Brassica carinata* A. Braun Pak J Zool 47(5):1335–1340

James KMF, Sargent DJ, Whitehouse A, Cielniak G (2022) High-throughput phenotyping for breeding targets—current status and future directions of strawberry trait automation. Plants People Planet 4:432–443

Jammer A, Akhtar SS, Amby DB, Pandey C, Mekureyaw MF, Bak F, Roth PM, Roitsch T (2022) Enzyme activity profiling for physiological phenotyping within functional phenomics: plant growth and stress responses. J Exp Bot 73(15):5170–5198

Jhou YS, Poovendhan S, Huang LH, Tsai CW (2021) Host acceptance and plant resistance: a comparative behavioral study of *Myzus persicae* and *Acyrthosiphon pisum*. Insects 12. https://doi.org/10.3390/insects12110975

Jongsma MA, Thoen MPM, Poleij LM, Wiegers GL, Goedhart PW, Dicke M, Noldus LPJJ, Kruisselbrink JW (2019) An integrated system for the automated recording and analysis of insect behavior in T-maze arrays. Front Plant Sci 10:1–16. https://doi.org/10.3389/fpls.2019.00020

Kansman J, Basu S, Casteel CL, Crowder D, Lee BW, Nihranz CT, Finke DL (2022) Plant water stress reduces aphid performance: exploring mechanisms driven by water stress intensity. Front Ecol Evol 10:846908. https://doi.org/10.3389/fevo.2022.846908

Kessler A, Baldwin IT (2002) Plant responses to insect herbivory: the emerging molecular analysis. Annu Rev Plant Biol 53:299–328. https://doi.org/10.1146/annurev.arplant.53.100301.135207

Kloth KJ, Thoen MPM, Bouwmeester HJ, Jongsma MA, Dicke M (2012) Association mapping of plant resistance to insects. Trends Plant Sci 17:311–319. https://doi.org/10.1016/j.tplants.2012.01.002

Kloth KJ, Ten Broeke CJ, Thoen MP, Hanhart-van den Brink M, Wiegers GL, Krips OE, Noldus LP, Dicke M, Jongsma MA (2015) High-throughput phenotyping of plant resistance to aphids by automated video tracking. Plant Methods 11:1–14

Kloth KJ, Wiegers GL, Busscher-Lange J, van Haarst JC, Kruijer W, Bouwmeester HJ, Dicke M, Jongsma MA (2016) *AtWRKY22* promotes susceptibility to aphids and modulates salicylic acid and jasmonic acid signalling. J Exp Bot 67(11):3383–3396. https://doi.org/10.1093/jxb/erw159

Kundu P, Sahu R (2021) *GIGANTEA* confers susceptibility to plants during spot blotch attack by regulating salicylic acid signalling pathway. Plant Physiol Biochem 167:349–357

Kundu P, Vadassery J (2021) Role of *WRKY* transcription factors in plant defense against lepidopteran insect herbivores: an overview. J Plant Biochem Biotechnol 30:698–707

Kundu A, Mishra S, Kundu P, Jogawat A, Vadassery J (2022) *Piriformospora indica* recruits host-derived putrescine for growth promotion in plants. Plant Physiol 188:2289–2307

Labandeira CC (2013) A paleobiologic perspective on plant–insect interactions. Curr Opin Plant Biol 16:414–421. https://doi.org/10.1016/j.pbi.2013.06.003

Lage J, Skovmand B, Andersen SB (2003) Characterization of greenbug (Homoptera: Aphididae) resistance in synthetic hexaploid wheats. J Econ Entomol 96:1922–1928. https://doi.org/10.1093/jee/96.6.1922

Lin PA, Paudel S, Afzal A, Shedd NL, Felton GW (2021) Changes in tolerance and resistance of a plant to insect herbivores under variable water availability. Environ Exp Bot 183:104334

Liu Z, Guo P, Liu H, Fan P, Zeng P, Liu X, Feng C, Wang W, Yang F (2021) Gradient boosting estimation of the leaf area index of apple orchards in UAV remote sensing. Remote Sens 13(16):3263

Liu L, Gao X, Ren C, Cheng X, Zhou Y, Huang H, Zhang J, Ba Y (2022) Applicability of the crop water stress index based on canopy–air temperature differences for monitoring water status in a cork oak plantation, northern China. Agric Forest Meteorol 327:109226

Mahanti NK, Pandiselvam R, Kothakota A, Chakraborty SK, Kumar M, Cozzolino D (2022) Emerging non-destructive imaging techniques for fruit damage detection: image processing and analysis. Trends Food Sci Technol 120:418–438

Mangrauthia SK, Malathi P, Agarwal S, Ramkumar G, Krishnaveni D, Neeraja CN, Viraktamath BC (2012) Genetic variation of coat protein gene among the isolates of Rice tungro spherical virus from tungro-endemic states of the India. Virus Genes 44(3):482–487

Manickavasagan A, Jayas D, White N (2008) Thermal imaging to detect infestation by *Cryptolestes ferrugineus* inside wheat kernels. J Stored Prod Res 44:186–192

Marček T, Hamow KÁ, Végh B, Janda T, Darko E (2019) Metabolic response to drought in six winter wheat genotypes. PLoS One 14(2):e0212411

Markheiser A, Santoiemma G, Fereres A, Kugler S, Maixner M, Cornara D (2022) DC-EPG assisted comparison of European spittlebugs and sharpshooters feeding behaviour on grapevine. J Appl Entomol 147(3):192–204

McDaniel T, Tosh CR, Gatehouse AMR, George D, Robson M, Brogan B (2016) Novel resistance mechanisms of a wild tomato against the glasshouse whitefly. Agron Sustain Dev 36(1):14. https://doi.org/10.1007/s13593-016-0351-4

Meena MK, Prajapati R, Krishna D, Divakaran K, Pandey Y, Reichelt M, Mathew M, Boland W, Mithöfer A, Vadassery J (2019) The Ca^{2+} channel CNGC19 regulates Arabidopsis defense against Spodoptera herbivory. Plant Cell 31:1539–1562

Milenovic M, Wosula EN, Rapisarda C, Legg JP (2019) Impact of host plant species and whitefly species on feeding behavior of *Bemisia tabaci*. Front Plant Sci 10:1. https://doi.org/10.3389/fpls.2019.00001

Minervini M, Scharr H, Tsaftaris SA (2015) Image analysis: the new bottleneck in plant phenotyping [applications corner]. IEEE Signal Process Mag 32(4):126–131

Mondal S, Wenninger EJ, Hutchinson PJS, Whitworth JL, Shrestha D, Eigenbrode SD, Bosque-perez NA (2016) Comparison of transmission efficiency of various isolates of potato virus Y among three aphid vectors. Entomol Exp Appl158:258–268. doi: https://doi.org/10.1111/eea.12404

Moreno A, Tjallingii WF, Fernandez-Mata G, Fereres A (2012) Differences in the mechanism of inoculation between a semi-persistent and a non-persistent aphid-transmitted plant virus. J Gen Virol 93:662–667. https://doi.org/10.1099/vir.0.037887-0

Moreno-Delafuente A, Garzo E, Moreno A, Fereres A (2013) A plant virus manipulates the behavior of its whitefly vector to enhance its transmission efficiency and spread. PLoS One 16;8(4):e61543. https://doi.org/10.1371/journal.pone.0061543

Moustaka J, Meyling NV, Hauser TP (2021) Induction of a compensatory photosynthetic response mechanism in tomato leaves upon short time feeding by the chewing insect *Spodoptera exigua*. Insects 12. https://doi.org/10.3390/insects12060562

Nabity PD, Zavala JA, Delucia EH (2013) Herbivore induction of jasmonic acid and chemical defences reduce photosynthesis in *Nicotiana attenuata*. J Exp Bot 64:685–694. https://doi.org/10.1093/jxb/ers364

Nagaraj N, Reese JC, Kirkham MB, Kofoid K, Campbell LR, Loughin TM (2002) Relationship between chlorophyll loss and photosynthetic rate in greenbug (Homoptera: Aphididae) damaged sorghum. J Kansas Entomol Soc 75(2):101–109

Nalam V, Louis J, Patel M, Shah J (2018) Arabidopsis-green peach aphid interaction: rearing the insect, no-choice and fecundity assays, and electrical penetration graph technique to study insect feeding behavior. Bio-Protocol 8:1–24. https://doi.org/10.21769/bioprotoc.2950

Nguyen D, Rieu I, Mariani C, van Dam NM (2016) How plants handle multiple stresses: hormonal interactions underlying responses to abiotic stress and insect herbivory. Plant Mol Biol 91(6): 727–740

Noldus LPJJ, Spink AJ, Tegelenbosch RAJ (2002) Computerised video tracking, movement analysis and behaviour recognition in insects. Comput Electron Agric 35(2–3):201–227

Nyabuga FN, Carrasco D, Ranåker L, Andersson MN, Birgersson G, Larsson MC, Lundin O, Rundlöf M, Svensson GP, Anderbrant O, Lankinen A (2015) Field abundance patterns and odor-mediated host choice by clover seed weevils, *Apion fulvipes* and *Apion trifolii* (Coleoptera: Apionidae). J Econ Entomol 108(2):492–503. https://doi.org/10.1093/jee/tou099

O'Neal ME, Landis DA, Isaacs R (2002) An inexpensive, accurate method for measuring leaf area and defoliation through digital image analysis. J Econ Entomol 95:1190–1194

Ortega MA, All JN, Boerma HR, Parrott WA (2016) Pyramids of QTLs enhance host-plant resistance and Bt-mediated resistance to leaf-chewing insects in soybean. Theor Appl Genet 129(4):703–715. https://doi.org/10.1007/s00122-015-2658-y

Panda N, Khush GA (1995) Host plant resistance to insects. Host plant resistance to insects. https://www.cabdirect.org/cabdirect/abstract/19951111984. Accessed 13 Apr 2020

Pearson CC, Backus EA, Shugart HJ, Munyaneza JE (2014) Characterization and correlation of EPG waveforms of *Bactericera cockerelli* (Hemiptera: Triozidae): variability in waveform appearance in relation to applied signal. Ann Entomol Soc Am 107:650–666. https://doi.org/10.1603/AN13178

Pegadaraju V, Louis J, Singh V, Reese JC, Bautor J, Feys BJ, Cook G, Parker JE, Shah J (2007) Phloem-based resistance to green peach aphid is controlled by Arabidopsis *PHYTOALEXIN DEFICIENT4* without its signaling partner *ENHANCED DISEASE SUSCEPTIBILITY1*. Plant J52(2):332–341

Pérez-Bueno ML, Pineda M, Barón M (2019) Phenotyping plant responses to biotic stress by chlorophyll fluorescence imaging. Front Plant Sci 10:1–15. https://doi.org/10.3389/fpls.2019.01135

Peterson RKD, Varella AC, Higley LG (2017) Tolerance: the forgotten child of plant resistance. PeerJ 5:e3934. https://doi.org/10.7717/peerj.3934

Poljaković-Pajnik L, Petrović-Obradović O, Drekić M, Orlović S, Vasić V, Kovacevic B, Kacprzyk M, Nikolić N (2016) Aphid feeding effects on physiological parameters of poplar cultivars. In: VII international scientific agriculture symposium," Agrosym 2016", 6–9 October 2016, Jahorina, Bosnia and Herzegovina. Proceedings. University of East Sarajevo, Faculty of Agriculture, pp 2848–2855

Powell G, Tosh CR, Hardie J (2006) HOST PLANT SELECTION BY APHIDS: behavioral, evolutionary, and applied perspectives. Annu Rev Entomol 51:309–330. https://doi.org/10.1146/annurev.ento.51.110104.151107

Prado Maluta NK, Garzo E, Moreno A, Navas-Castillo J, Fiallo-Olive E, Lopes JRS, Fereres A (2017) Stylet penetration activities of the whitefly *Bemisia tabaci* associated with inoculation of the crinivirus tomato chlorosis virus. J Gen Virol 98:1515–1520

Prufer T, Thieme T, Tjallingii WF (2014) Aphid-AutoEPG software for analysing electrically monitored feeding behaviour of aphids. Euro J Environ Sci 4(1):53–59

Qian L, Liu X, Huang Z, Wang L, Zhang Y, Gao Y, Gui F, Chen F (2021) Elevated CO_2 enhances the host resistance against the western flower thrips, *Frankliniella occidentalis*, through increased callose deposition. J Pest Sci 94(1):55–68. https://doi.org/10.1007/s10340-019-01123-7

Qu W, Robert CA, Erb M, Hibbard BE, Paven M, Gleede T, Riehl B, Kersting L, Cankaya AS, Kunert AT, Xu Y, Schueller MJ, Shea C, Alexoff D, Lee SJ, Fowler JS, Ferrieri RA (2016) Dynamic precision phenotyping reveals mechanism of crop tolerance to root herbivory. Plant Physiol 172(2):776–788. https://doi.org/10.1104/pp.16.00735

Quandahor P, Gou Y, Lin C, Mujitaba Dawuda M, A Coulter J, Liu C (2020) Phytohormone cross-talk synthesizes glycoalkaloids in potato (*Solanum tuberosum* L.) in response to aphid (*Myzus persicae* Sulzer) infestation under drought stress. Insects 11(11):724. doi: https://doi.org/10.3390/insects11110724

Rech C, Lucini T, Panizzi AR, Nardi C (2021) Feeding behavior of *Collaria scenica* (Stål) (Hemiptera: Miridae) on wheat plants: an EPG waveform characterization. Neotrop Entomol 50:366–373. https://doi.org/10.1007/s13744-021-00859-1

Rossato JA Jr, Madaleno LL, Mutton MJ, Higley LG, Fernandes OA (2019) Photosynthesis, yield and raw material quality of sugarcane injured by multiple pests. PeerJ 7:e6166

Sailaja B, Anjum N, Patil YK, Agarwal S, Malathi P, Krishnaveni D, Balachandran SM, Viraktamath BC, Mangrauthia SK (2013) The complete genome sequence of a south Indian isolate of Rice tungro spherical virus reveals evidence of genetic recombination between distinct isolates. Virus Genes 47(3):515–523

Salsabillah V, Putra NS, Aldawood AS, Soffan A (2021) Increased probing activities of green peach aphid (GPA), *Myzus persicae*, on chitosan-treated caisim (*Brassica juncea*) monitored by electrical penetration graph (EPG). Int J Trop Insect Sci 41:2805–2810. https://doi.org/10.1007/s42690-021-00461-3

Satpathy S, Gotyal BS, Babu VR (2020) Role of novel insecticides in crop protection and their selectivity to natural enemies: a review. J Environ Biol 41(2):149–160

Shahzad M, Ghani WH, Ayyub M, Ali Q, Ahmad HM, Faisal M, Ali A, Qasim MU (2019) Performance of some wheat cultivars against aphid and its damage on yield and photosynthesis. J Glob Innov Agric Soc Sci 7(3):105–109

Sharma HC, Ortiz R (2002) Host plant resistance to insects: an eco-friendly approach for pest management and environment conservation. J Environ Biol 23:11–135

Simon AL, Wellham PA, Aradottir GI, Gange AC (2017) Unravelling mycorrhiza-induced wheat susceptibility to the English grain aphid *Sitobion avenae*. Sci Rep 13(7):46497. https://doi.org/10.1038/srep46497

Singh CB, Jayas DS, Paliwal J, White ND (2010) Identification of insect-damaged wheat kernels using short-wave near-infrared hyperspectral and digital colour imaging. Comput Electron Agric 73:118–125

Singh A, Ganapathysubramanian B, Singh AK, Sarkar S (2016a) Machine learning for high-throughput stress phenotyping in plants. Trends Plant Sci 21:110–124

Singh AK, Ponnuswamy R, Donempudi K, Mangrauthia SK (2016b) The differential reaction of rice hybrids to tungro virus by phenotyping and PCR analysis. J Phytopathol 164(3):177–184

Singh V, Sharma N, Singh S (2020) A review of imaging techniques for plant disease detection. Artifi Intell Agric 4:229–242

Singh V, Mandhania S, Pal A, Kaur T, Banakar P, Sankaranarayanan K, Arya SS, Malik K, Datten R (2022) Morpho-physiological and biochemical responses of cotton (*Gossypium hirsutum* L.) genotypes upon sucking insect-pest infestations. Physiol Mol Biol Plants 28(11–12): 2023–2039. https://doi.org/10.1007/s12298-022-01253-w

Slack S, York LM, Roghazai Y, Lynch J, Bennett M, Foulkes J (2018) Wheat shovelomics II: revealing relationships between root crown traits and crop growth. BioRxiv 13:280917

Song P, Wang J, Guo X, Yang W, Zhao C (2021) High-throughput phenotyping: breaking through the bottleneck in future crop breeding. Crop J 9:633–645

Souza MF, Davis JA (2021) Characterizing host plant resistance to *Melanaphis sacchari* (Hemiptera: Aphididae) in selected sorghum plant introductions. J Econ Entomol 114:959–969. https://doi.org/10.1093/jee/toab003

Srilatha P, Yousuf F, Methre R, Vishnukiran T, Agarwal S, Poli Y, Raghurami Reddy M, Vidyasagar B, Shanker C, Krishnaveni D, Triveni S, Brajendra PS, Balachandran SM, Subrahmanyam D, Mangrauthia SK (2019) Physical interaction of *RTBV ORFI* with D1 protein of *Oryza sativa* and Fe/Zn homeostasis play a key role in symptoms development during rice tungro disease to facilitate the insect mediated virus transmission. Virology 526:117–124

Stout MJ (2014) Chapter 1-host-plant resistance in Pest management. In: Abrol DP (ed) Integrated Pest management. Academic Press, San Diego, pp 1–21. https://doi.org/10.1016/B978-0-12-398529-3.00002-6

Tayal M, Somavat P, Rodriguez I, Thomas T, Christoffersen B, Kariyat R (2020) Polyphenol-rich purple corn pericarp extract adversely impacts herbivore growth and development. Insects 11(2):98. https://doi.org/10.3390/insects11020098

Thoen MPM, Kloth KJ, Wiegers GL, Krips OE, Noldus LPJJ, Dicke M, Jongsma MA (2016) Automated video tracking of thrips behavior to assess host-plant resistance in multiple parallel two-choice setups. Plant Methods 12:1–12. https://doi.org/10.1186/s13007-016-0102-1

Tjallingii WF (1985) Electrical nature of recorded signals during stylet penetration by aphids. Entomol Exp Appl 38:177–186. https://doi.org/10.1111/j.1570-7458.1985.tb03516.x

Trębicki P, Tjallingii WF, Harding RM, Rodoni BC, Powell KS (2012) EPG monitoring of the probing behaviour of the common brown leafhopper *Orosius orientalis* on artificial diet and selected host plants. Arthropod Plant Interact 6:405–415. https://doi.org/10.1007/s11829-012-9192-5

Ubbens JR, Stavness I (2017) Deep plant phenomics: a deep learning platform for complex plant phenotyping tasks. Front plant Sci, vol 8, p 1190. https://doi.org/10.3389/fpls.2017.01190. Accessed 7 Jan 2023

van Emden HF, Harrington R (2017) Aphids as crop pests, 2nd edn. CABI

Van Helden M, Tjallingii WF (2000) Experimental design and analysis in EPG experiments with emphasis on plant resistance research. In: Walker GP, Backus EA (eds) Principles and applications of electronic monitoring and other techniques in the study of Homopteran feeding behaviour: proceedings of the symposium at the XIX International Congress of Entomology. Thomas Say Publications in Entomology, Beijing, China, pp 144–172

Verrelst J, Rivera JP, Veroustraete F, Muñoz-Marí J, Clevers JG, Camps-Valls G, Moreno J (2015) Experimental Sentinel-2 LAI estimation using parametric, non-parametric and physical retrieval methods–a comparison. ISPRS J Photogramm Remote Sens 108:260–272

Visschers IGS, Peters JL, van de Vondervoort JAH, Hoogveld RHM, van Dam NM (2019) Thrips resistance screening is coming of age: leaf position and ontogeny are important determinants of leaf-based resistance in pepper. Front Plant Sci 10:510. https://doi.org/10.3389/fpls.2019.00510

Walter A, Liebisch F, Hund A (2015) Plant phenotyping: from bean weighing to image analysis. Plant Methods 11:14. https://doi.org/10.1186/s13007-015-0056-8

War AR, Paulraj MG, Ahmad T, Buhroo AA, Hussain B, Ignacimuthu S, Sharma HC (2012) Mechanisms of plant defense against insect herbivores. Plant Signal Behav 7:1306–1320. https://doi.org/10.4161/psb.21663

Williams D, Aitkenhead M, Karley AJ, Graham J, Jones HG (2018) Use of imaging technologies for high throughput phenotyping. Springer, Raspberry, pp 145–158

Wu WB (2007) Ground measurements of leaf area index of fruit trees. Ph.D. Thesis,. South China Agricultural University, Guangzhou, China

Wu X, Zhang W, Qiu Z, Cen H, He Y (2016) A novel method for detection of *Pieris rapae* larvae on cabbage leaves using NIR hyperspectral imaging. Appl Eng Agric 32:311–316. https://doi.org/10.13031/aea.32.11444

Wu S, Deng W, Wu G (2017) Research on LAI detection in precision pesticide application. J Agric Mech Res 39:262–268

Wu M, Dong Y, Zhang Q, Li S, Chang L, Loiacono FV, Ruf S, Zhang J, Bock R (2022) Efficient control of western flower thrips by plastid-mediated RNA interference. Proc Natl Acad Sci U S A 119(15):e2120081119. https://doi.org/10.1073/pnas.2120081119

Xing G, Liu K, Gai J (2017) A high-throughput phenotyping procedure for evaluation of antixenosis against common cutworm at early seedling stage in soybean. Plant Methods 13:66

Xu S, Liu Z, Zhao L, Zhao H, Ren S (2018) Diurnal response of sun-induced fluorescence and PRI to water stress in maize using a near-surface remote sensing platform. Remote Sens 10(10):1510

Yang W, Duan L, Chen G, Xiong L, Liu Q (2013) Plant phenomics and high-throughput phenotyping: accelerating rice functional genomics using multidisciplinary technologies. Curr Opin Plant Biol 16:180–187

Yi XU (2020) Aphids and their transmitted potato viruses: a continuous challenges in potato crops. J Integr Agric 19(2):367–375

York LM (2019) Functional phenomics: an emerging field integrating high-throughput phenotyping, physiology, and bioinformatics. J Exp Bot 70(2):379–386

Zavala JA, Nabity PD, DeLucia EH (2013) An emerging understanding of mechanisms governing insect herbivory under elevated CO_2. Annu Rev Entomol 58(1):79–97

Zavala JA, Mazza CA, Dillon FM, Chludil HD, Ballaré CL (2015) Soybean resistance to stink bugs (*Nezara viridula* and *Piezodorus guildinii*) increases with exposure to solar UV-B radiation and correlates with isoflavonoid content in pods under field conditions. Plant Cell Environ 38(5): 920–928. https://doi.org/10.1111/pce.12368

Zhai C, Zhao C, Ning W, Long J, Wang X, Weckler P, Zhang H (2018) Research progress on precision control methods of air-assisted spraying in orchards. Trans Chin Soc Agric Eng 34(10):1–15

Zhang Y, Zhang N (2018) Imaging technologies for plant high-throughput phenotyping: a review. Front Agric Sci Eng 5:406–419. https://doi.org/10.15302/J-FASE-2018242

Zhao YR, Yu KQ, Feng C, Cen HY, He Y (2017) Early detection of aphid (*Myzus persicae*) infestation on Chinese cabbage by hyperspectral imaging and feature extraction. Trans ASABE 60:1045–1051. https://doi.org/10.13031/trans.11886

Zhou S, Chai X, Yang Z, Wang H, Yang C, Sun T (2021a) Maize-IAS: a maize image analysis software using deep learning for high-throughput plant phenotyping. Plant Methods 17(1):1–17

Zhou H, Tan X, Teng Z, Du L, Zhou H (2021b) EPG analysis of stylet penetration preference of woolly apple aphid on different parts of apple trees. PLoS One 16(8):e0256641. https://doi.org/10.1371/journal.pone.0256641

Zogli P, Pingault L, Grover S, Louis J (2020) Ento(o)mics: the intersection of 'omic' approaches to decipher plant defense against sap-sucking insect pests. Curr Opin Plant Biol 56:153–161. https://doi.org/10.1016/j.pbi.2020.06.002

The manufacturer's authorised representative in the EU is Springer Nature Customer Service Centre GmbH, Europaplatz 3, 69115 Heidelberg, Germany. If you have any concerns regarding our products, please contact ProductSafety@springernature.com

Printed and bound by CPI Group (UK) Ltd, Croydon, CR0 4YY
30/09/2026
02225318-0014